CAMBRIDGE TRACTS IN MATHEMATICS

General Editors

B. BOLLOBAS, H. HALBERSTAM & C. T. C. WALL

---

93   On L¹-approximation

T0275781

ALLAN M. PINKUS

Professor of Mathematics
Israel Institute of Technology

On L^1-approximation

CAMBRIDGE UNIVERSITY PRESS

Cambridge

London   New York   Port Chester

Melbourne   Sydney

CAMBRIDGE UNIVERSITY PRESS
Cambridge, New York, Melbourne, Madrid, Cape Town, Singapore, São Paulo

Cambridge University Press
The Edinburgh Building, Cambridge CB2 8RU, UK

Published in the United States of America by Cambridge University Press, New York

www.cambridge.org
Information on this title: www.cambridge.org/9780521366502

First published 1989
This digitally printed version 2008

A catalogue record for this publication is available from the British Library

Library of Congress Cataloguing in Publication data

Pinkus, Allan, 1946–
On L¹–approximation.
(Cambridge tracts in mathematics; 93)
On t.p. 1 is superscript.
Bibliography: p.
Includes indexes.
1. Approximation theory.   2. Least absolute
deviations (Statistics) I. Title. II. Series.
QA211.P55 1989   511'.4   88-35343

ISBN 978-0-521-36650-2 hardback
ISBN 978-0-521-05769-1 paperback

For D. E. T. and D.

Contents

Preface

Any monograph should speak for itself, and this is no exception. However a few words of explanation would certainly do no harm. The linear theory of best uniform approximation is well documented both in journals and in books. The same cannot be said for the linear theory of best L^1-approximation, which has generally received little attention.

This monograph (aside from a few digressions) is about the qualitative theory of best L^1-approximation from finite-dimensional subspaces, where the approximation is either two-sided or one-sided. The questions considered are 'classical'. What is, to me, surprising is that it is only in the last few years that many (but not all) of these questions have been answered. Thus most of the contents of Chapters 4 and 5, as well as some of the contents of Chapters 3 and 6, are the result of very recent research in this area.

This work is not all-encompassing. Various topics which could have been included are not. The most glaring of these omissions is the non-linear theory. I had originally intended to say something about this topic. But I found that I was being forced far afield in order to say relatively little. As of now, non-linear L^1-approximation theory is a poorly developed subject which deserves more attention.

Following each of the first six chapters is a series of exercises. These exercises are an integral part of this work and should be studied. They serve two main functions. Firstly, they provide an opportunity for the reader to check whether he or she has understood the contents of the chapter. Secondly, they helped me to write this monograph without being constantly forced to digress from the main body. Thus they contain much relevant material.

This monograph is primarily intended for those with an interest in approximation theory, but can and hopefully will be read by others. The main prerequisite is only a good course in advanced calculus or real analysis. However a smidgen of knowledge of measure theory, functional analysis, and numerical analysis will prove useful.

My thanks are due to various people. Professors D. Braess, H. Strauss, A. Kroó, G. A. Watson, C. de Boor and M. Sommer made a number of helpful comments and suggestions. Any errors and/or omissions are totally my responsibility. Parts of the manuscript were typed by the technical typists of the School of Mathematics at Tel-Aviv University. To them I owe my appreciation. As a new TEX user, I am also indebted to them for their patience. Thanks are also due to D. Hershkowitz, and to the Fund for the Promotion of

Research at the Technion. Finally, I wish to thank David Tranah and the staff at Cambridge University Press for their constant helpfulness and cooperation.

It is my hope that you, the reader, will enjoy this work as much as I enjoyed writing it.

Technion, Haifa, 1988   Allan Pinkus

1
Preliminaries

In this chapter we introduce certain basic general facts from approximation theory. These will be used in later chapters. The topics we touch upon are the classic problems of approximation theory, namely existence, characterization, uniqueness, and continuity of the best approximation operator. We assume that most of these results are familiar to the reader, for they are contained, in one form or another, in various introductory texts on approximation theory. For the sake of completeness, at the very least, we also include most of their proofs. Most readers should skim the contents of this chapter simply to familiarize themselves with the notation and certain definitions.

1. Existence

We first fix some notation. X will always denote a normed linear space over the reals $I\!\!R$. A subset Y of X is given. Our problem is to approximate elements $f \in X$ from elements of Y. The 'error' in this problem we denote by

$$E(f;Y) = \inf\{\|f - g\| : g \in Y\}.$$

The subset Y of X is said to be an *existence set* for X (often termed a *proximinal set*) if to each $f \in X$ there exists a $g^* \in Y$ for which

$$\|f - g^*\| \leq \|f - g\|$$

for all $g \in Y$, i.e., for each $f \in X$ the above infimum is attained. Such g^* (if they exist) are called *best approximants* to f from Y.

Much is known concerning existence sets. However we shall only review some very elementary results.

Theorem 1.1. *Let Y be a compact subset of X. Then Y is an existence set for X.*

Proof. Let $f \in X$ and

$$E = E(f;Y) = \inf\{\|f - g\| : g \in Y\}.$$

From the definition of E, there exists a sequence $\{g_n\}$, $g_n \in Y$, with the property that $\lim_{n\to\infty} \|f - g_n\| = E$.

Since Y is compact, there exists a subsequence $\{g_{n_k}\}$ of $\{g_n\}$ which converges to a $g^* \in Y$, i.e., $\lim_{k\to\infty} \|g_{n_k} - g^*\| = 0$. Now, for every k,

$$\|f - g^*\| \leq \|f - g_{n_k}\| + \|g_{n_k} - g^*\|.$$

The left-hand-side of the inequality is independent of k. Let $h \to \infty$. The first term on the right-hand-side approaches E, while the second term tends to zero. Thus

$$\|f - g^*\| \leq \lim_{k \to \infty} [\|f - g_{n_k}\| + \|g_{n_k} - g^*\|] = E.$$

However $\|f - g^*\| \geq E$ since $g^* \in Y$. Hence $\|f - g^*\| = E$, and g^* is a best approximant to f from Y.   □

It is not necessary that Y be compact in order for it to be an existence set.

Theorem 1.2. *Let C be a closed subset of a finite-dimensional subspace U of X. Then C is an existence set for X.*

Proof. Let $f \in X$ and $v \in C$. When best approximating f from C, it suffices to consider only those $u \in C$ for which

$$\|f - u\| \leq \|f - v\| = M.$$

Define

$$A = \{u : u \in C, \|f - u\| \leq M\}.$$

Let $\|f\| = N$. Then, for each $u \in A$,

$$\|u\| \leq \|f\| + \|f - u\| \leq N + M.$$

Thus A is a closed, bounded subset of U. Any closed, bounded subset of a finite-dimensional subspace is compact. From Theorem 1.1, A is an existence set for X. We therefore have a $u^* \in A$ for which

$$\|f - u^*\| \leq \|f - u\|, \quad \text{all } u \in A.$$

This in turn implies that

$$\|f - u^*\| \leq \|f - u\|, \quad \text{all } u \in C.$$ □

As a special case of the above theorem we have a classic result which, for convenience, we now formally state.

Corollary 1.3. *Let U be a finite-dimensional subspace of a normed linear space X. Then U is an existence set for X.*

2. Characterization

We present here two types of characterization theorems. The first is based on the one-sided Gateaux derivatives, while the second is a consequence of the Hahn-Banach Theorem.

Let $f, g \in X$. If

$$\lim_{t \to 0} \frac{\|f + tg\| - \|f\|}{t}$$

exists, then the limit is said to be the Gateaux derivative of f in the direction g. Such limits do not necessarily exist. However the one-sided limits always exist.

Proposition 1.4. *Let* $f, g \in X$, *and set*

$$r(t) = \frac{\|f + tg\| - \|f\|}{t}.$$

On $(0, \infty), r(t)$ *is a non-decreasing function of* t *and is bounded below.*

Proof. We first prove that $r(t)$ is bounded below on $(0, \infty)$. From the triangle inequality,

$$\|f + tg\| \ge \|f\| - \|tg\| = \|f\| - t\|g\|.$$

Thus for $t \in (0, \infty), r(t) \ge -\|g\|$.

It remains to prove that $r(t)$ is non-decreasing on $(0, \infty)$. Let $0 < s < t < \infty$. Then,

$$t\|f + sg\| = \|tf + stg\| = \|s(f + tg) + (t - s)f\| \le s\|f + tg\| + (t - s)\|f\|.$$

Thus

$$t(\|f + sg\| - \|f\|) \le s(\|f + tg\| - \|f\|),$$

whence we obtain $r(s) \le r(t)$. $\square$

For $f, g \in X$, set

$$\tau_+(f, g) = \lim_{t \to 0^+} \frac{\|f + tg\| - \|f\|}{t}.$$

On the basis of Proposition 1.4, $\tau_+(f, g)$ exists for every $f, g \in X$.

Our first characterization theorem now follows.

Theorem 1.5. *Let* M *be a linear subspace of* X, *and* $f \in X \backslash \overline{M}$. *Then* $g^* \in M$ *is a best approximant to* f *from* M *if and only if* $\tau_+(f - g^*, g) \ge 0$ *for all* $g \in M$.

Proof. ($\Rightarrow$). Assume g^* is a best approximant to f from M. Since M is a subspace,

$$\|f - g^* + tg\| \ge \|f - g^*\|$$

for every $g \in M$ and $t \in \mathbb{R}$. Thus $\tau_+(f - g^*, g) \ge 0$, essentially by definition. ($\Leftarrow$). Assume $\tau_+(f - g^*, g) \ge 0$ for all $g \in M$. From Proposition 1.4, $r(t)$ is a non-decreasing function of t on $(0, \infty)$. Setting $t = 1$ and remembering that M is a linear subspace, we obtain

$$\|f - g\| - \|f - g^*\| \ge \tau_+(f - g^*, g^* - g) \ge 0.$$

Thus $\|f - g\| \ge \|f - g^*\|$ for all $g \in M$. $\square$

A totally analogous proof allows us to obtain this next result.

Theorem 1.6. *Let K be a convex subset of X, and assume $f \in X\backslash\overline{K}$. Then g^* is a best approximant to f from K if and only if $\tau_+(f - g^*, g^* - g) \geq 0$ for all $g \in K$.*

Heuristically the above technique should be considered as a generalized perturbation technique. The next set of results, although formally equivalent, are more in the spirit of separating hyperplanes.

For a normed linear space X over $\mathbb{R}$, let X^* denote the space of continuous (bounded) real-valued linear functionals on X. For $h \in X^*$,

$$|||h||| = \sup\{|h(f)| : f \in X, \|f\| \leq 1\}$$

defines a norm on X^*. With this norm, X^* is a Banach space, i.e., is complete.

The Hahn-Banach (Extension) Theorem, in one of its simpler forms, may be stated as follows.

Theorem 1.7 (Hahn-Banach). *Let M be a linear subspace of X. Assume H is a continuous linear functional on M. There then exists an $h \in X^*$ for which*

$$1)\ h(g) = H(g),\ \text{all } g \in M$$
$$2)\ |||h|||_X = |||H|||_M.$$

The subscript on the $||| \cdot |||$ in (2) indicates where this norm is taken. In general it is well understood from the context and is deleted. One consequence of the above result is:

Proposition 1.8. *Let M be a linear subspace of X. Assume $f \in X$, and*

$$E(f; M)\ (= E) = \inf\{\|f - g\| : g \in M\} > 0.$$

There exists an $h \in X^*$ satisfying

$$1)\ h(g) = 0,\ \text{all } g \in M$$
$$2)\ |||h||| = 1$$
$$3)\ h(f) = E.$$

Proof. Let L denote the linear span of M and f. Define a continuous linear functional H on L as follows: For $\alpha \in \mathbb{R}$, $g \in M$,

$$H(\alpha f + g) = \alpha E.$$

By definition $H(g) = 0$ for all $g \in M$, and $H(f) = E$. Furthermore, it is easily seen that $|||H|||_L = 1$. Now apply Theorem 1.7. □

As a consequence of Proposition 1.8 we have this next main result paralleling Theorem 1.5.

Theorem 1.9. *Let M be a linear subspace of X, and $f \in X \backslash \overline{M}$. Then g^* is a best approximant to f from M if and only if there exists an $h \in X^*$ for which*

1) $h(g) = 0$, all $g \in M$

2) $|||h||| = 1$

3) $h(f - g^*) = \|f - g^*\|$.

Proof. Since $f \in X \backslash \overline{M}$,

$$E = \inf\{\|f - g\| : g \in M\} > 0.$$

($\Rightarrow$). Assume g^* is a best approximant to f from M. Thus $\|f - g^*\| = E$. Let h be as given by Proposition 1.8. Then (1) and (2) are valid. Furthermore, from (1) and (3) of Proposition 1.8,

$$\|f - g^*\| = E = h(f) = h(f - g^*).$$

Thus (3) of this theorem holds.

($\Leftarrow$). Assume that (1), (2) and (3) hold. Let $g \in M$. Then,

$$\|f - g^*\| = h(f - g^*) = h(f - g) \le |||h||| \cdot \|f - g\| = \|f - g\|.$$

Thus g^* is a best approximant to f from M. □

Remark. Note that $\widetilde{g}$ is any other best approximant to f from M if and only if $h(f - \widetilde{g}) = \|f - \widetilde{g}\|$ for the h satisfying (1), (2) and (3) of Theorem 1.9.

To obtain a result parallel to Theorem 1.6, we use the following generalization of the Hahn-Banach Theorem.

Theorem 1.10 (Basic Separation Theorem). *Let A and B be disjoint convex subsets of X. Assume A has interior. There exists a non-zero $h \in X^*$ and $c \in \mathbb{R}$ such that $h(f) \ge c$ for all $f \in A$, and $h(f) \le c$ for all $f \in B$.*

This next result also generalizes Theorem 1.9.

Theorem 1.11. *Let K be a convex subset of X, and assume $f \in X \backslash \overline{K}$. Then g^* is a best approximant to f from K if and only if there exists an $h \in X^*$ satisfying*

1) $h(g^*) \ge h(g)$, all $g \in K$

2) $|||h||| = 1$

3) $h(f - g^*) = \|f - g^*\|$.

Proof. ($\Rightarrow$). Assume

$$\|f - g^*\| = \inf\{\|f - g\| : g \in K\} = E > 0.$$

Let $A = \{f_0 : f_0 \in X, \|f - f_0\| < E\}$. The sets A and K satisfy the conditions of Theorem 1.10. As such there exists an $\widetilde{h} \in X^*, \widetilde{h} \ne 0$, and a $\widetilde{c} \in \mathbb{R}$ for which $\tilde{\tilde{h}}(f_0) \geq \tilde{c}$ for all $f_0 \in A$ and $\tilde{\tilde{h}}(y) \leq \tilde{c}$ for all $g \in K$. By continuity, $\tilde{h}(f_0) \geq \tilde{c}$ for all $f_0 \in \overline{A}$, and therefore $\tilde{h}(g^*) = \tilde{c}$.

Translating by $\tilde{c}$, there exists a $c \in \mathbb{R}$ ($c = \tilde{h}(f) - \tilde{c}$) with the property that $\tilde{h}(f - f_0) \leq c$ for all $f_0 \in \overline{A}$, and $\tilde{h}(f - g) \geq c$ for all $g \in K$. Since A is a ball of positive radius about f, we necessarily have $c > 0$. Set $h = (E/c)\tilde{h}$. It is now easily checked that h satisfies (1), (2) and (3).

($\Leftarrow$). Assume (1), (2) and (3) hold for some $h \in X^*$ and $g^* \in K$. Then, for any $g \in K$,

$$\|f - g^*\| = h(f - g^*) \leq h(f - g) \leq \|\|h\|\| \cdot \|f - g\| = \|f - g\|,$$

and g^* is a best approximant to f from K.   □

If K is a convex cone, i.e., $g \in K$ implies $\alpha g \in K$ for all $\alpha > 0$, then Theorem 1.11 can be somewhat sharpened.

Corollary 1.12. *Let K be a convex cone in X. Assume that $f \in X \backslash \overline{K}$. Then g^* is a best approximant to f from K if and only if there exists an $h \in X^*$ satisfying*

$$1)\ 0 = h(g^*) \geq h(g),\ \text{all}\ g \in K$$
$$2)\ \|\|h\|\| = 1$$
$$3)\ h(f - g^*) = \|f - g^*\|,$$

or, equivalently,

$$1')\ 0 \geq h(g),\ \text{all}\ g \in K$$
$$2')\ \|\|h\|\| = 1$$
$$3')\ h(f) = \|f - g^*\|.$$

3. Uniqueness and Strong Uniqueness

Let Y be a subset of X. For each $f \in X$, set

$$P_Y(f) = \{g^* : g^* \in Y, \|f - g^*\| = E(f; Y)\}.$$

$P_Y(f)$ is the subset of Y containing all the best approximants to f from Y. $P_Y(f)$ may, of course, be the empty set. $P_Y(f)$ is not the empty set for every $f \in X$ if and only if Y is an existence set for X. In general P_Y is a set-valued map from X onto Y. P_Y is referred to as the *metric projection* onto Y. We are naturally interested in the various general properties enjoyed by P_Y. These depend on both Y and X. Convexity is one geometric property which the metric projection inherits directly from Y.

Proposition 1.13. *If K is convex, then $P_K(f)$ is convex for each $f \in X$.*

Proof. Assume $f \in X$ and $P_K(f)$ contains at least two distinct elements g_1 and g_2. Then $E(f; K) = E = \|f - g_i\|$, $i = 1, 2$. For each $\lambda \in [0, 1]$, set $g_\lambda = \lambda g_1 + (1 - \lambda)g_2$. Then

$$\|f - g_\lambda\| \leq \lambda \|f - g_1\| + (1 - \lambda)\|f - g_2\| = E.$$

Since K is convex, $g_\lambda \in K$, and $\|f - g_\lambda\| \geq E$. Thus $\|f - g_\lambda\| = E$ and, by definition, $g_\lambda \in P_K(f)$. □

We continue to assume that K is a convex subset of X. For each $f \in X$, $P_K(f)$ may be the empty set, a unique element of K, or a convex subset of K containing more than one element. In certain cases, simple norm properties eliminate this third option, i.e., $P_K(f)$ will contain at most one element.

Definition 1.1. The normed linear space X is said to be *strictly convex* if, for any $f, g \in X$ satisfying $f \neq g$ and $\|f\| = \|g\| = 1$, we have $\|\lambda f + (1-\lambda)g\| < 1$ for every $\lambda \in (0, 1)$.

The above definition is equivalent to the statement that if $\|f\| = \|g\| = \|(f+g)/2\|$, then $f = g$.

Theorem 1.14. *Assume K is a convex subset of a strictly convex normed linear space X. Then, for each $f \in X$, $P_K(f)$ contains at most one element.*

Proof. Assume that $g_1, g_2 \in P_K(f)$, and $E = \|f - g_i\|$, $i = 1, 2$. From the proof of Proposition 1.13, it follows that $\|\lambda(f - g_1) + (1 - \lambda)(f - g_2)\| = \lambda\|f - g_1\| + (1 - \lambda)\|f - g_2\| = E$ for all $\lambda \in [0, 1]$. Since the norm is strictly convex, this implies that $f - g_1 = f - g_2$, i.e., $g_1 = g_2$. □

Strict convexity is a global property of the norm. A local property dependent on the one-sided Gateaux derivatives will sometimes give even more.

Assume that for a given $f \in X$ there exists a best approximant g^* from Y. Thus

$$\|f - g^*\| \leq \|f - g\|$$

for all $g \in Y$. If, in addition, there exists a $\gamma > 0$ for which

$$\gamma\|g - g^*\| \leq \|f - g\| - \|f - g^*\|$$

for all $g \in Y$, then we say that g^* is a *strongly unique* best approximant to f from Y. The reason for this terminology is simply that 'strong uniqueness' is stronger than 'uniqueness'. If g^* is a strongly unique best approximant to f from Y, then it is most certainly the unique best approximant to f from Y. The converse need not and generally does not hold. If strong uniqueness is present, then we shall denote by $\gamma(f)$ the largest constant satisfying the above inequality.

Strong uniqueness and the identification of $\gamma(f)$ is intimately connected with one-sided Gateaux derivatives. We state and prove our result for a subspace M of X.

Theorem 1.15. *Let M be a subspace of X and $f \in X \backslash \overline{M}$. Assume g^* is a best approximant to f from M. Set*

$$\gamma = \inf\{\tau_+(f - g^*, g) : g \in M, \|g\| = 1\}.$$

Then $\gamma \geq 0$ and, for all $g \in M$,

$$\gamma\|g - g^*\| \leq \|f - g\| - \|f - g^*\|.$$

Furthermore, if $\gamma' > \gamma$ there exists a $\widetilde{g} \in M$ for which

$$\gamma'\|\widetilde{g} - g^*\| > \|f - \widetilde{g}\| - \|f - g^*\|.$$

Therefore strong uniqueness holds if and only if $\gamma > 0$, and in this case $\gamma(f) = \gamma$.

Proof. Since g^* is a best approximant to f from M, we have from Theorem 1.5 that $\gamma \geq 0$. Assume $\gamma > 0$. From Exercise 2(a) and the definition of γ, we have

$$\tau_+(f - g^*, -g) \geq \gamma\|g\|$$

for all $g \in M$. From Proposition 1.4,

$$\frac{\|f - g^* - tg\| - \|f - g^*\|}{t} \geq \gamma\|g\|$$

for all $t > 0$ and $g \in M$. Therefore

$$\|f - g^* - g\| - \|f - g^*\| \geq \gamma\|g\|$$

for all $g \in M$. Since M is a subspace, we immediately obtain

$$\gamma\|g - g^*\| \leq \|f - g\| - \|f - g^*\|$$

for all $g \in M$.

Assume $\gamma' > \gamma$. By definition there exists a $\overline{g} \in M$, $\|\overline{g}\| = 1$, for which

$$\tau_+(f - g^*, -\overline{g}) < \gamma'\|\overline{g}\|.$$

Thus for $t_0 > 0$, sufficiently small,

$$\|f - g^* - t_0\overline{g}\| - \|f - g^*\| < \gamma'\|t_0\overline{g}\|.$$

Set $\widetilde{g} = g^* + t_0\overline{g}$. Then

$$\gamma'\|\widetilde{g} - g^*\| > \|f - \widetilde{g}\| - \|f - g^*\|,$$

which proves the theorem.   □

4. Continuity

Let Y be a subset of X, and recall that

$$E(f; Y) = \inf\{\|f - g\| : g \in Y\}.$$

The first simple fact to be shown is the following.

Proposition 1.16. *$E(f; Y)$ is a continuous function of f. In fact, for any $f_1, f_2 \in X$,*

$$|E(f_1; Y) - E(f_2; Y)| \leq \|f_1 - f_2\|.$$

Proof. Assume without loss of generality that $E(f_1; Y) \geq E(f_2; Y)$. Given $\varepsilon > 0$, let $g_0 \in Y$ satisfy

$$\|f_2 - g_0\| \leq E(f_2; Y) + \varepsilon.$$

Such a g_0 necessarily exists. Then

$$E(f_1; Y) \leq \|f_1 - g_0\| \leq \|f_1 - f_2\| + \|f_2 - g_0\| \leq \|f_1 - f_2\| + E(f_2; Y) + \varepsilon.$$

Thus, for every $\varepsilon > 0$, we have

$$E(f_1; Y) - E(f_2; Y) \leq \|f_1 - f_2\| + \varepsilon$$

which implies the desired result. □

An important application of Proposition 1.16 is in this next result.

Theorem 1.17. *Assume Y is a subset of X. Let $f, f_n \in X$, $n \in \mathbb{N}$, satisfy $\lim_{n \to \infty} \|f - f_n\| = 0$. Assume $g_n \in P_Y(f_n)$ all n, and there exists a $g \in Y$ for which $\lim_{n \to \infty} \|g - g_n\| = 0$. Then $g \in P_Y(f)$.*

Proof. From the triangle inequality,

$$\|f - g\| \leq \|f - f_n\| + \|f_n - g_n\| + \|g_n - g\|$$

for all n. By assumption, $\lim_{n \to \infty} \|f - f_n\| = \lim_{n \to \infty} \|g - g_n\| = 0$. From Proposition 1.16,

$$\lim_{n \to \infty} \|f_n - g_n\| = \lim_{n \to \infty} E(f_n; Y) = E(f; Y).$$

Thus $\|f - g\| \leq E(f; Y)$. Since $g \in Y$, this implies that $g \in P_Y(f)$. □

From Theorem 1.17 we deduce the following.

Proposition 1.18. *Let U be a finite-dimensional subspace of X. Assume $f, f_n \in X$, $n \in \mathbb{N}$, and $\lim_{n \to \infty} \|f - f_n\| = 0$. Further assume that $P_U(f) = \{u^*\}$. Then, for any choice of $u_n \in P_U(f_n)$, we have $\lim_{n \to \infty} \|u_n - u^*\| = 0$.*

Proof. Since $\lim_{n \to \infty} \|f - f_n\| = 0$, there exists a $c \in \mathbb{R}$ such that $\|f_n\| \leq c$ for all n. Thus $\|u_n\| \leq 2c$ for all n. Each element of the sequence $\{u_n\}$ is in the compact set

$$U \cap \{g : g \in X, \|g\| \leq 2c\}.$$

Thus there exists a subsequence of $\{u_n\}$ which converges to some $u \in U$. From Theorem 1.17, $u = u^*$. Since this is valid for any convergent subsequence, it follows that $\lim_{n \to \infty} \|u_n - u^*\| = 0$. □

Before ending this chapter we make the following formal definition.

Definition 1.2. Let Y be an existence set for X. Then Y is said to be a *unicity set* if $P_Y(f)$ is a singleton for all $f \in X$. That is, to each $f \in X$ there exists a unique best approximant from Y. If Y is a subspace of X and a unicity set, then we shall say that Y is a *unicity space*.

Some authors use the term unicity set, without the assumption of Y being an existence set, to mean that $P_Y(f)$ contains at most one element for each $f \in X$. Other authors use the terms Chebyshev and semi-Chebyshev, respectively. In this work the term Chebyshev will be used in a different context.

As an immediate application of Proposition 1.18, we have:

Corollary 1.19. *Let U be a finite-dimensional unicity space of X. Then the single-valued operator $P_U(\cdot)$ is continuous on X.*

That is, if $f, f_n \in X$, and $\lim_{n\to\infty} \|f - f_n\| = 0$, then necessarily $\lim_{n\to\infty} \|P_U(f) - P_U(f_n)\| = 0$.

Exercises

1. Prove that, if Y is an existence set for X, then Y is closed.

2. Prove that, for every $f, g, h \in X$,

$$a)\, \tau_+(f, \alpha g) = \alpha\, \tau_+(f, g) \text{ for all } \alpha \geq 0;$$
$$b)\, \tau_+(f, g + h) \leq \tau_+(f, g) + \tau_+(f, h).$$

3. For $f, g \in X$, set

$$\tau_-(f, g) = \lim_{t\to 0^-} \frac{\|f + tg\| - \|f\|}{t}.$$

Prove that $\tau_-(\cdot, \cdot)$ always exists and $\tau_-(f, g) = -\tau_+(f, -g)$.

4. Let $f \in X$, $f \neq 0$. Assume that

$$\tau(f, g) = \lim_{t\to 0} \frac{\|f + tg\| - \|f\|}{t}$$

exists for all $g \in X$. Prove that $\tau(f, \cdot) \in X^*$, i.e., $\tau(f, \cdot)$ is a continuous linear functional on X.

5. Assume that $\tau(f, g)$ exists for all $f, g \in X$, with $f \neq 0$. Let M be a linear subspace of X and $f \in X\backslash\overline{M}$. Prove that $g^* \in P_M(f)$ if and only if $\tau(f - g^*, g) = 0$ for all $g \in M$.

6. Let X be an inner product space and $f \neq 0$. Prove that

$$\tau(f,g) = (f,g)/\|f\|.$$

7. Prove Theorem 1.6.

8. Let M be a linear subspace of X. Assume $f \in X$, and

$$E = \inf\{\|f - g\| : g \in M\} > 0.$$

Prove that

$$E = \max\{h(f) : h \in X^*, \|\|h\|\| \leq 1, h(g) = 0, \text{ all } g \in M\}.$$

9. Prove Corollary 1.12.

10. Let K be a convex subset of X and $f \in X\backslash\overline{K}$. Prove that g^* is a strongly unique best approximant to f from K if and only if

$$\inf\{\tau_+(f - g^*, g^* - g)/\|g^* - g\| : g \in K, g \neq g^*\} > 0.$$

11. Let U be a finite-dimensional subspace of X. Prove that, for each $f \in X$, $P_U(f)$ is bounded and closed (hence compact).

12. Let $u_1, \ldots, u_n$ be a basis for the n-dimensional subspace U of X. Set

$$H(a_1, \ldots, a_n) = \left\|f - \sum_{i=1}^{n} a_i u_i\right\|.$$

Prove that H is continuous, convex, and $\lim_{\|\|\mathbf{a}\|\| \to \infty} H(\mathbf{a}) = \infty$, where $\|\| \cdot \|\|$ is any norm on $I\!\!R^n$ and $\mathbf{a} = (a_1, \ldots, a_n)$.

13. For $H(\mathbf{a})$ as in Exercise 12, set

$$A = \{\mathbf{a}^* : H(\mathbf{a}^*) = \min_{\mathbf{a}} H(\mathbf{a})\}.$$

Prove that A is a convex, closed, bounded subset of $I\!\!R^n$.

Notes and References

Most of the material of this chapter may be found in either Sections 1 of Chapter I and Appendix I of Singer [1970], Chapter 1 of Cheney [1966], or Chapter 1 of Watson [1980]. Additional material on Gateaux derivatives is contained in Dunford, Schwartz [1958, pp.445–451] and Chapter 26 of Köthe [1969]. There is a direct interconnection between linear functionals and what we have called one-sided Gateaux derivatives. It is given by the fact that the range of $h(g)$ for $h \in X^*$ satisfying $\|\|h\|\| = 1$ and $h(f) = \|f\|$, is exactly the interval $[-\tau_+(f, -g), \tau_+(f, g)]$ (see Dunford, Schwartz [1958, p.447] and Köthe [1969, p.349]). The Basic Separation Theorem was lifted from Dunford, Schwartz [1958, p.417]. Singer [1970] is the best reference for a historical development of this material. The concept of strong uniqueness was introduced by Newman, Shapiro [1963]. The approach taken here may be found in Papini [1978], see also Wulbert [1971].

If the functional $\tau(f,g)$ of Exercise 4 exists for all $f, g \in X$, $f \neq 0$, then the space X is said to be *smooth*. This corresponds to the existence of a unique $h \in X^*$ satisfying $|||h||| = 1$ and $h(f) = \|f\|$ for each $f \in X$, $f \neq 0$. From Exercise 5 we have that in a smooth space strong uniqueness never holds with respect to any subspace. Smoothness and strict convexity are essentially dual concepts (Köthe [1969, p.346]). If X^* is strictly convex (smooth), then X is smooth (strictly convex). If X is a reflexive Banach space, then the converse holds. This is one explanation for the fact that in any L^p space, $1 < p < \infty$, strong uniqueness from a subspace never holds.

2
Approximation from Finite-Dimensional
Subspaces of L^1

---

1. Introduction and Notation

We first fix some notation. B is a set, Σ a σ-field of subsets of B, and ν a positive measure defined on Σ, i.e., $\nu(E) \geq 0$ for all $E \in \Sigma$. By $L^p(B,\nu)$ (we suppress the Σ for brevity), $1 \leq p < \infty$, we denote the set of all real-valued ν-measurable functions f defined on B for which $|f|^p$ is ν-integrable over B. We consider two functions of $L^p(B,\nu)$ as equivalent if they are equal ν a.e. (almost everywhere). Under this convention $L^p(B,\nu)$ with norm

$$\|f\|_p = \left(\int_B |f(x)|^p d\nu(x) \right)^{1/p}$$
$$\left(= \left(\int_B |f|^p d\nu \right)^{1/p} \right)$$

is a normed linear space and is in fact a Banach space, i.e., is also complete. $L^\infty(B,\nu)$ is defined analogously with norm

$$\|f\|_\infty = \operatorname*{ess\ sup}_{x \in B} |f(x)|$$

where the ess sup (essential supremum) is the infimum of all real constants c for which $|f(x)| \leq c$, ν a.e.. $L^\infty(B,\nu)$ is also a Banach space.

We may identify the dual of $L^p(B,\nu)$ with $L^q(B,\nu)$ for all $1 < p < \infty$, where $1/p + 1/q = 1$. Using standard notation we write

$$\left(L^p(B,\nu) \right)^* = L^q(B,\nu) \ .$$

We are interested in the case $p = 1$. If $p = 1$, the dual space is not always given by this equality. However we shall assume that ν is σ-finite, in which case we necessarily have

$$\left(L^1(B,\nu) \right)^* = L^\infty(B,\nu) \ .$$

In this chapter we consider the problem of approximating functions in the $L^1(B,\nu)$ norm. Only a very few results are presented, but some of their proofs are rather lengthy. The central results are as follows. In Theorem 2.1 we prove the main criterion for characterizing best approximants from linear subspaces of $L^1(B,\nu)$. A variant of this result is given in Theorem 2.3 for finite-dimensional subspaces U of $L^1(B,\nu)$, where ν is a non-atomic positive measure. This latter result is applied in Theorem 2.7 to prove that, if ν is a non-atomic positive measure, then no U is a unicity space for $L^1(B,\nu)$. This result is further strengthened in Theorem 2.9 where it is shown that the set of $f \in L^1(B,\nu)$, for which the best approximant from U is not unique, is dense in $L^1(B,\nu)$. In Theorem 2.13 we prove that this set, while dense, is only of first category. Finally in Theorem 2.14 we prove that in this setting one cannot even choose a best approximant to each $f \in L^1(B,\nu)$ such that the resulting map is continuous. While most of these results are negative in character, the reader should not be unduly disturbed. In Chapter 3 we shall see that the situation is radically altered if we restrict ourselves to continuous functions.

2. Characterization

For each $f \in L^1(B,\nu)$, we define its zero set

$$Z(f) = \{x : f(x) = 0\}$$

and $N(f) = B \backslash Z(f)$. Note that $Z(f)$ is ν-measurable. In addition, for $f \in L^1(B,\nu)$, set

$$\operatorname{sgn}(f(x)) = \left\{ \begin{array}{ll} 1 , & f(x) > 0 \\ 0 , & f(x) = 0 \\ -1 , & f(x) < 0 . \end{array} \right.$$

Let M be a subspace of $L^1(B,\nu)$. We first present a condition for characterizing best $L^1(B,\nu)$ approximants from M.

Theorem 2.1. *Let M be a subspace of $L^1(B,\nu)$ and $f \in L^1(B,\nu)\backslash\overline{M}$. Then g^* is a best $L^1(B,\nu)$ approximant to f from M if and only if*

$$(2.1) \qquad \left| \int_B \operatorname{sgn}(f - g^*)g \, d\nu \right| \le \int_{Z(f-g^*)} |g| d\nu$$

for all $g \in M$.

Two techniques for obtaining characterization theorems were presented in Chapter 1. As such we present two proofs of the above result. Each is interesting in and of itself. The first is an application of Theorem 1.5, while the second is a consequence of Theorem 1.9.

Proof 1. From Theorem 1.5, g^* is a best approximant to f from M if and only if $\tau_+(f - g^*, g) \ge 0$ for all $g \in M$. We now explicitly calculate $\tau_+(f - g^*, g)$.

By definition,

$$\tau_+(f - g^*, g) = \lim_{t \to 0^+} \frac{\|f - g^* + tg\|_1 - \|f - g^*\|_1}{t} .$$

For $t > 0$,

$$\frac{\|f - g^* + tg\|_1 - \|f - g^*\|_1}{t} = \frac{1}{t}\left[\int_B (|f - g^* + tg| - |f - g^*|)d\nu\right]$$

$$= \frac{1}{t}\left[\int_{Z(f-g^*)} t|g|d\nu\right.$$

$$\left. + \int_{N(f-g^*)} (|f - g^* + tg| - |f - g^*|)d\nu\right]$$

$$= \int_{Z(f-g^*)} |g|d\nu$$

$$+ \frac{1}{t}\int_{N(f-g^*)} (|f - g^* + tg| - |f - g^*|)d\nu .$$

On $N(f - g^*)$,

$$\left|\frac{|f - g^* + tg| - |f - g^*|}{t}\right| \leq |g| ,$$

and

$$\frac{|f - g^* + tg| - |f - g^*|}{t} = \frac{|f - g^* + tg|^2 - |f - g^*|^2}{t[|f - g^* + tg| + |f - g^*|]}$$

$$= \frac{2(f - g^*)g + t|g|^2}{|f - g^* + tg| + |f - g^*|} .$$

Thus on $N(f - g^*)$,

$$\lim_{t\to 0^+} \frac{|f - g^* + tg| - |f - g^*|}{t} = \frac{2(f - g^*)g}{2|f - g^*|} = \text{sgn}(f - g^*)g .$$

Applying Lebesgue's Dominated Convergence Theorem, we obtain

$$\tau_+(f - g^*, g) = \int_{Z(f-g^*)} |g|d\nu + \int_B \text{sgn}(f - g^*)g \, d\nu .$$

Thus g^* is a best approximant to f from M if and only if

$$-\int_B \text{sgn}(f - g^*)g \, d\nu \leq \int_{Z(f-g^*)} |g|d\nu$$

for all $g \in M$. Since M is a subspace ($g \in M$ implies $-g \in M$) this is in turn equivalent to (2.1). □

Proof 2. ($\Leftarrow$). Assume (2.1) holds and $g \in M$. Then

$$\|f - g^*\|_1 = \int_B \text{sgn}(f - g^*)(f - g^*)d\nu$$

$$= \int_B \text{sgn}(f - g^*)(f - g)d\nu + \int_B \text{sgn}(f - g^*)(g - g^*)d\nu$$

$$\leq \int_{N(f-g^*)} |f - g|d\nu + \int_{Z(f-g^*)} |g - g^*|d\nu$$

$$= \int_{N(f-g^*)} |f - g|d\nu + \int_{Z(f-g^*)} |g - f|d\nu$$

$$= \|f - g\|_1 .$$

Thus g^* is a best approximant to f from M.

($\Rightarrow$). To prove the converse, assume that g^* is a best approximant to f from M. From Theorem 1.9 there exists an $h \in L^\infty(B, \nu)$ ($= \left(L^1(B,\nu)\right)^*$) satisfying

$$1)\ \int_B hg\, d\nu = 0, \quad \text{all } g \in M$$

$$2)\ \|h\|_\infty = 1$$

$$3)\ \int_B h(f - g^*)d\nu = \int_B |f - g^*|d\nu.$$

From (2) and (3), it follows that $h = \text{sgn}(f - g^*)$ ν a.e. on $N(f - g^*)$. Using (1) and (2),

$$\left| \int_B \text{sgn}(f - g^*)g\, d\nu \right| = \left| -\int_{Z(f-g^*)} hg\, d\nu \right| \le \int_{Z(f-g^*)} |g| d\nu$$

for all $g \in M$.   □

Every characterization result is in some sense a tautology. Its importance is measured in terms of its usefulness and the insight it provides. While Theorem 2.1 is the main characterization result in the problem of best $L^1(B, \nu)$-approximation from subspaces, it is often an insufficient theoretical tool for proving additional facts. If $M = U$ is a finite-dimensional subspace of $L^1(B, \nu)$ and ν is a non-atomic positive measure, then a sharpened form of Theorem 1.9 will prove more useful. The key to this sharpened form of Theorem 1.9 is Liapounoff's Theorem. Before stating this latter theorem, we note that a positive measure ν is *non-atomic* if to each $E \in \Sigma$ with $\nu(E) > 0$, and each $\lambda \in (0, 1)$, there exists an $E_\lambda \subseteq E$ ($E_\lambda \in \Sigma$) for which $\nu(E_\lambda) = \lambda\nu(E)$.

Theorem 2.2 (Liapounoff). *Assume that ν is a non-atomic positive measure, and let $u_1, \ldots, u_n$ be any given n functions in $L^1(B, \nu)$. Set*

$$A = \left\{ \left(\int_B hu_1 d\nu, \ldots, \int_B hu_n d\nu \right) \ : \ h \in L^\infty(B, \nu), \|h\|_\infty \le 1 \right\},$$

and

$$\widetilde{A} = \left\{ \left(\int_B hu_1 d\nu, \ldots, \int_B hu_n d\nu \right) \ : \ h \in L^\infty(B, \nu), |h(x)| = 1 \text{ all } x \in B \right\}.$$

Then $A = \widetilde{A}$.

Liapounoff's Theorem is not true if ν has atoms. As an immediate application of the above form of Liapounoff's Theorem and Theorem 1.9, we obtain this next result, which is well worth highlighting as it shall be repeatedly used.

Theorem 2.3. *Let ν be a non-atomic positive measure and U a finite-dimensional subspace of $L^1(B, \nu)$. Let $f \in L^1(B, \nu)$. Then $u^* \in P_U(f)$,* i.e., u^* is a best approximant to f from U, if and only if there exists an $h \in L^\infty(B, \nu)$ satisfying

1) $|h(x)| = 1$, all $x \in B$

2) $\int_B hu \, d\nu = 0$, all $u \in U$

3) $\int_B h(f - u^*)d\nu = \|f - u^*\|_1$.

Furthermore, $\widetilde{u} \in P_U(f)$ if and only if $h(f - \widetilde{u}) \geq 0$ ν a.e., where h satisfies (1) and (2).

3. Uniqueness

A natural question to ask is the following. Given f and a best approximant $g^* \in M$ to f, when is g^* the unique best approximant to f from M? We first present, as a simple consequence of the second proof of Theorem 2.1, a criterion for determining when g^* is the unique best approximant to f from M.

Proposition 2.4. Let $f \in L^1(B, \nu)$. Assume g^* is a best approximant to f from M. Then $\widetilde{g} \in M$, $\widetilde{g} \neq g^*$, is also a best approximant to f from M if and only if

a) $(f - g^*)(f - \widetilde{g}) \geq 0$, ν a.e. on B

b) $\int_B \mathrm{sgn}(f - g^*)(\widetilde{g} - g^*)d\nu = \int_{Z(f-g^*)} |\widetilde{g} - g^*|d\nu$.

Proof. ($\Rightarrow$). Assume $\widetilde{g}$ is also a best approximant to f from M. From the sufficiency part of Proof 2 of Theorem 2.1, we obtain

$$\|f - g^*\|_1 = \int_B \mathrm{sgn}(f - g^*)(f - g^*)d\nu$$
$$= \int_B \mathrm{sgn}(f - g^*)(f - \widetilde{g})d\nu + \int_B \mathrm{sgn}(f - g^*)(\widetilde{g} - g^*)d\nu$$
$$= \int_{N(f-g^*)} |f - \widetilde{g}|d\nu + \int_{Z(f-g^*)} |\widetilde{g} - g^*|d\nu$$
$$= \|f - \widetilde{g}\|_1 .$$

Since (2.1) holds, it necessarily follows that

1) $\mathrm{sgn}(f - g^*)(f - \widetilde{g}) = |f - \widetilde{g}|$, ν a.e. on $N(f - g^*)$

2) $\int_B \mathrm{sgn}(f - g^*)(\widetilde{g} - g^*)d\nu = \int_{Z(f-g^*)} |\widetilde{g} - g^*|d\nu$.

(a) follows from (1), while (b) is (2).

($\Leftarrow$). If $\widetilde{g} \in M$, $\widetilde{g} \neq g^*$, satisfies (a) and (b), then $\|f - g^*\|_1 = \|f - \widetilde{g}\|_1$ as a consequence of the above sequence of equalities.   □

Obviously the conditions of Proposition 2.4 are difficult to verify, and generally useless in practice. However, the following simple consequence will (perhaps surprisingly) prove useful.

Corollary 2.5. *Let g^* be a best approximant to f from M. Assume that*

$$(2.2) \qquad \left| \int_B \mathrm{sgn}(f - g^*) g \, d\nu \right| < \int_{Z(f-g^*)} |g| d\nu$$

for all $g \in M$, $g \neq 0$. Then the best approximant to f from M is unique.

Inequality (2.2) may be restated in the form $\tau_+(f - g^*, g) > 0$ for all $g \in M$, $g \neq 0$. As an immediate consequence of Theorem 1.15 and the calculation of $\tau_+(f - g^*, g)$, we obtain in this same vein a characterization of when we have strong uniqueness.

Corollary 2.6. *Let M be a subspace of $L^1(B, \nu)$ and $f \in L^1(B, \nu) \backslash \overline{M}$. Then g^* is the strongly unique best approximant to f from M if and only if*

$$\gamma = \inf \left\{ \int_{Z(f-g^*)} |g| d\nu - \int_B \mathrm{sgn}(f - g^*) g \, d\nu \ : \ g \in M, \ \|g\|_1 = 1 \right\}$$

satisfies $\gamma > 0$. Furthermore if $\gamma > 0$, then $\gamma(f) = \gamma$, i.e., γ is the largest possible strong uniqueness constant.

We now address a different and more central question. Namely, do there exist subspaces which are unicity spaces for $L^1(B, \nu)$ and, if so, can we characterize them? The answer to this first question is no if $M = U$ is a finite-dimensional subspace of $L^1(B, \nu)$ and ν is a non-atomic positive measure. Here we make use of Theorem 2.3.

Theorem 2.7. *Let ν be a non-atomic positive measure. No finite-dimensional subspace U of $L^1(B, \nu)$ is a unicity space for $L^1(B, \nu)$.*

Proof. From Theorem 2.2 or 2.3, there exists an $h \in L^\infty(B, \nu)$ satisfying

$$1) \ |h(x)| = 1, \ \text{all } x \in B$$

$$2) \int_B hu \, d\nu = 0, \ \text{all } u \in U.$$

Let $u^* \in U$, $u^* \neq 0$, and set $f = h|u^*|$. Since $hf \geq 0$ on B, we have from Theorem 2.3 that $0 \in P_U(f)$. For any $\alpha \in \mathbb{R}$, $|\alpha| \leq 1$,

$$h(f - \alpha u^*) = h(h|u^*| - \alpha u^*) = |u^*| - \alpha hu^* \geq 0$$

since $|\alpha h| \leq 1$. Therefore $\alpha u^* \in P_U(f)$ for any $\alpha \in [-1, 1]$ and U is not a unicity space for $L^1(B, \nu)$. □

Is it really necessary that ν be non-atomic for the result of Theorem 2.7 to hold? Not quite (see Exercise 6) but let us now to go the other extreme and assume that ν is a purely atomic measure. For convenience we shall consider the space

$$\ell_1 = \left\{ \mathbf{x} = (x_1, x_2, \dots) : \|\mathbf{x}\|_1 = \sum_{i=1}^{\infty} |x_i| < \infty \right\}.$$

That is, ν is the purely atomic measure with mass one at each positive integer. The theorem to be proved is the following.

Theorem 2.8. *An n-dimensional subspace U of ℓ_1 is a unicity space for ℓ_1 if and only if for each $\mathbf{y} \in \ell_\infty$, $\mathbf{y} \neq \mathbf{0}$, satisfying $(\mathbf{y}, \mathbf{u}) = 0$ for all $\mathbf{u} \in U$, there exist at least n indices $1 \leq i_1 < \cdots < i_n$ for which $|y_{i_r}| < \|\mathbf{y}\|_\infty$, $r = 1, \dots, n$.*

Remark. By $\|\mathbf{y}\|_\infty$ we mean $\sup_i |y_i|$, and by $(\mathbf{y}, \mathbf{u})$ we mean the usual inner product $\sum_{i=1}^{\infty} y_i u_i$.

Proof. ($\Rightarrow$). Assume there exists a $\mathbf{y} \in \ell_\infty$, $\|\mathbf{y}\|_\infty = 1$, satisfying $(\mathbf{y}, \mathbf{u}) = 0$, all $\mathbf{u} \in U$, and such that $|y_i| = \|\mathbf{y}\|_\infty$ except for $i \in \{i_1, \dots, i_k\}$ where $0 \leq k \leq n - 1$. Since U is n-dimensional and $k < n$, there exists a $\mathbf{u}^* \in U$, $\mathbf{u}^* \neq \mathbf{0}$, satisfying $|u_{i_r}^*| = 0$, $r = 1, \dots, k$.

Set $x_j = y_j |u_j^*|$, all j, and $\mathbf{x} = (x_1, x_2, \dots)$. By our choice of $\mathbf{u}^*$, $|x_j| = |u_j^*|$ for all j, and

$$\operatorname{sgn} x_j = \begin{cases} y_j, & u_j^* \neq 0 \\ 0, & u_j^* = 0, \end{cases}$$

i.e., $|y_j| = 1$ if $u_j^* \neq 0$. For each $\alpha \in \mathbb{R}$, $|\alpha| < 1$, we have $\operatorname{sgn}(x_j) = \operatorname{sgn}(x_j - \alpha u_j^*)$ for all j, and $Z(\mathbf{u}^*) = Z(\mathbf{x}) = Z(\mathbf{x} - \alpha \mathbf{u}^*)$. Thus for any $\mathbf{u} = (u_1, u_2, \dots) \in U$,

$$\left| \sum_{j=1}^{\infty} \operatorname{sgn}(x_j - \alpha u_j^*) u_j \right| = \left| \sum_{j=1}^{\infty} y_j u_j - \sum_{j \in Z(\mathbf{x})} y_j u_j \right| = \left| -\sum_{j \in Z(\mathbf{x})} y_j u_j \right|$$

$$\leq \sum_{j \in Z(\mathbf{x})} |u_j| = \sum_{j \in Z(\mathbf{x} - \alpha \mathbf{u}^*)} |u_j|.$$

From Theorem 2.1, $\alpha \mathbf{u}^*$ is a best approximant to $\mathbf{x}$ from U for every $\alpha \in (-1, 1)$. Thus U is not a unicity space.

($\Leftarrow$). Assume U is not a unicity space. By translation and normalization, there exists an $\mathbf{x} \in \ell_1$ and a $\mathbf{u}^1 \in U$, $\mathbf{u}^1 \neq \mathbf{0}$, such that

$$1 = \|\mathbf{x}\|_1 = \|\mathbf{x} - \mathbf{u}^1\|_1 = E(\mathbf{x}; U).$$

From Theorem 1.9 there exists a $\mathbf{y} \in \ell_\infty$ satisfying

$$1) \; (\mathbf{y}, \mathbf{u}) = 0, \text{ all } \mathbf{u} \in U$$

$$2) \; \|\mathbf{y}\|_\infty = 1$$

$$3) \; (\mathbf{y}, \mathbf{x}) = (\mathbf{y}, \mathbf{x} - \mathbf{u}^1) = 1.$$

Set $C = \{i : |y_i| < \|\mathbf{y}\|_\infty = 1\}$. From (2) and (3), we obtain

$$C \subseteq \{i : x_i = 0\} \cap \{i : x_i - u_i^1 = 0\} \subseteq \{i : u_i^1 = 0\}.$$

We claim that there exists a **y** satisfying both (1) and (2), and with associated C for which $|C| < n$ (where $|C|$ denotes the number of indices in C).

Assume $|C| \geq n$. (Otherwise we are finished.) Let $S_1 = \{i_1, \ldots, i_n\}$ consist of the first (smallest) n of these integers. Thus $u_{i_r}^1 = 0$, $r = 1, \ldots, n$. Let $\mathbf{u}^2, \ldots, \mathbf{u}^n \in U$, where $\mathbf{u}^1, \mathbf{u}^2, \ldots, \mathbf{u}^n$ is a basis for U. On the indices $\{i_1, \ldots, i_n\}$, U is of dimension at most $n - 1$ (since $u_{i_r}^1 = 0$, $r = 1, \ldots, n$). Thus there exists a $\mathbf{z} \in \ell_\infty$, $\mathbf{z} \neq \mathbf{0}$, for which $z_i = 0$, $i \notin S_1$, and $(\mathbf{z}, \mathbf{u}) = 0$ for all $\mathbf{u} \in U$. Therefore $(\mathbf{y} + a\mathbf{z}, \mathbf{u}) = 0$, for all $\mathbf{u} \in U$ and all $a \in \mathbb{R}$. Choose a^* so that $|y_i - a^* z_i| \leq 1$ for all $i \in S_1$ with equality for at least one $i \in S_1$. Set $\mathbf{y}^2 = \mathbf{y} - a^*\mathbf{z}$, and $C_2 = \{i : |y_i^2| < 1\}$. Then $C_2 \subset C, C_2 \neq C$. If $|C_2| < n$ we are finished. Otherwise let $S_2 = \{i_1', \ldots, i_n'\}$ be the smallest n integers in C_2. Continue as above. In this way we either prove our result after a finite number of steps, or obtain a sequence $\{\mathbf{y}^j\}_2^\infty$ of vectors satisfying (1) and (2), and such that for each index k there exists a j_k such that $y_k^j = y_k^{j_k}$ and $|y_k^{j_k}| = 1$ for all $j \geq j_k$, with at most $n-1$ exceptions. It is now easily seen (via weak compactness) that the sequence $\{\mathbf{y}^j\}_{j=2}^\infty$ has a limit vector $\mathbf{y}^*$ satisfying (1) and (2), and $|y_i^*| < 1$ for at most $n - 1$ indices.   □

This characterization of unicity spaces in ℓ_1 still does not imply their existence. However if $U = \operatorname{span}\{\mathbf{e}^1, \ldots, \mathbf{e}^n\}$, where $\mathbf{e}^i$ is the ith unit vector, then the conditions of the theorem obviously hold (and it is easy to determine the unique best approximant). In addition, small perturbations of U will not alter this property.

Let us now return to the problem of best approximating from U where ν is a non-atomic positive measure. We proved (Theorem 2.7) that no U (finite-dimensional) is a unicity space for $L^1(B, \nu)$. How large is the set of functions $f \in L^1(B, \nu)$ for which $P_U(f)$ is a singleton (i.e., for which the best approximant from U is unique)? Or alternatively, how large is the set of functions $f \in L^1(B, \nu)$ for which $P_U(f)$ is *not* a singleton? We shall prove that both these sets are dense in $L^1(B, \nu)$. In fact we shall show more in each case. We start with the latter problem.

Set

$$Q = \{f : f \in L^1(B, \nu), \ P_U(f) \text{ is not a singleton}\}$$

and let

$$Q_0 = \{f : f \in Q, \ P_U(f) \text{ contains an open neighborhood in } U\}.$$

Our first result concerns the size of Q_0, and is a direct generalization of Theorem 2.7. In fact the proof is also the same.

Theorem 2.9. *Let ν be a non-atomic positive measure, and U a finite-dimensional subspace of $L^1(B, \nu)$. Then Q_0 is dense in $L^1(B, \nu)$.*

Proof. Let $f \in L^1(B, \nu)$. For convenience we shall assume that $0 \in P_U(f)$ (translate f by any element of $P_U(f)$). From Theorem 2.3, there exists an $h \in L^\infty(B, \nu)$ for which

$$1) \, |h(x)| = 1, \text{ all } x \in B$$

$$2) \int_B hu \, d\nu = 0, \text{ all } u \in U$$

$$3) \int_B hf \, d\nu = \|f\|_1 .$$

Let $u_1, \ldots, u_n$ be any basis for U. Given $\varepsilon > 0$, let $\eta > 0$ be chosen such that $\eta \sum_{i=1}^n \|u_i\|_1 < \varepsilon$. Set

$$f_\varepsilon = f + \eta h \sum_{i=1}^n |u_i| .$$

Therefore $\|f - f_\varepsilon\|_1 < \varepsilon$.

We claim that $\eta \sum_{i=1}^n \alpha_i u_i \in P_U(f_\varepsilon)$ for every $\{\alpha_i\}_{i=1}^n$ satisfying $|\alpha_i| \le 1$, $i = 1, \ldots, n$. Thus $f_\varepsilon \in Q_0$. To see this note that

$$h(f_\varepsilon - \eta \sum_{i=1}^n \alpha_i u_i) = h(f + \eta h \sum_{i=1}^n |u_i| - \eta \sum_{i=1}^n \alpha_i u_i)$$

$$= hf + \eta \Big[\sum_{i=1}^n \big(|u_i| - h \alpha_i u_i \big) \Big] .$$

From (1) and (3), $hf \ge 0$ ν a.e. Since $|\alpha_i h| = |\alpha_i| \le 1$, for each i, we also have that $\sum_{i=1}^n (|u_i| - h\alpha_i u_i) \ge 0$ on B. Thus

$$h(f_\varepsilon - \eta \sum_{i=1}^n \alpha_i u_i) \ge 0 \quad \nu \text{ a.e.}$$

which implies that $\eta \sum_{i=1}^n \alpha_i u_i \in P_U(f_\varepsilon)$. □

While Q_0, and therefore Q, is dense in $L^1(B, \nu)$, we shall now prove that it is in fact only a set of first category in $L^1(B, \nu)$.

We first remind the reader that a set E of X is said to be of *first category* if it is a countable union of *nowhere dense sets*, i.e., sets for which the complement of its closure is dense.

Definition 2.1. Let X be a normed linear space. A subspace M of X is said to be *almost Chebyshev* if the set of points of X which do not possess a unique best approximant from M is of first category.

This notation is due to Garkavi. The phrase 'almost Chebyshev' is used since it parallels the use of the term 'Chebyshev'. We, on the other hand, are using the term 'unicity' rather than 'Chebyshev', and consistency demands that we should therefore use the terminology 'almost unicity'. We shall, however, be inconsistent.

The following is a variant of a result of Garkavi. We shall not reproduce the proof here.

Proposition 2.10. *Let X be a Banach space, and U a finite-dimensional subspace of X. If the set of points of X which possess a unique best approximant from U is dense in X, then U is almost Chebyshev.*

Before applying this result we prove two lemmas, the first of which will also be used in the next section.

Lemma 2.11. *Let ν be a non-atomic positive measure, and U a finite-dimensional subspace of $L^1(B, \nu)$. Given $\varepsilon > 0$, there exists a ν-measurable set A satisfying the following.*

1) $0 < \nu(A) < \varepsilon$.
2) *If $u \in U$ and $u = 0$ ν a.e. on A, then $u = 0$.*

Proof. Our proof is via induction on n, the dimension of U. For $n = 1$, set $C = \{x : u_1(x) \neq 0\}$, where $U = \text{span}\{u_1\}$. Then $\nu(C) > 0$. Let $0 < \lambda < \min\{\varepsilon, \nu(C)\}$. Since ν is non-atomic, there exists a set $A \subseteq C$, ν-measurable, for which $\nu(A) = \lambda$. If $u \in U$ satisfies $u = 0$ ν a.e. on A, then as is easily seen, $u = 0$.

We now assume that the lemma is valid for every subspace of dimension $\leq n - 1$. Assume $\dim U = n$, and $U = \text{span}\{u_1, \ldots, u_n\}$. From the induction hypothesis there exists a ν-measurable set C with $0 < \nu(C) < \varepsilon$ and such that if $\sum_{i=1}^{n-1} a_i u_i = 0$ ν a.e. on C, then $a_i = 0$, $i = 1, \ldots, n - 1$. If the $u_1, \ldots, u_n$ are linearly independent on C, set $A = C$ and we are finished.

Assume therefore that the $u_1, \ldots, u_n$ are linearly dependent on C. Thus the dimension of U restricted to C is $n - 1$. As such there exists a $u^* \in U$, $u^* \neq 0$, which is unique up to multiplication by a constant, satisfying $u^* = 0$ ν a.e. on C. Let C' be any ν-measurable subset of $\{x : u^*(x) \neq 0\}$ for which $\nu(C') > 0$ and $\nu(C \cup C') < \varepsilon$. Set $A = C \cup C'$. If $u \in U$, $u = 0$ ν a.e. on A, then $u = 0$ ν a.e. on C and thus $u = \alpha u^*$ for some $\alpha \in \mathbb{R}$. But $u = \alpha u^* = 0$ ν a.e. on C', and therefore $\alpha = 0$, i.e., $u = 0$.   □

The above lemma simply says that, given $\varepsilon > 0$, there exists a set A of measure less than ε for which the dimension of U restricted to A equals the full dimension of U. This is not necessarily true if ν has atoms.

We use Lemma 2.11 in this next lemma.

Lemma 2.12. *Let ν be a non-atomic positive measure, and W a finite-dimensional subspace of $L^1(B, \nu)$ with the property that $\int_B w \, d\nu = 0$ for all $w \in W$. Given $\varepsilon > 0$, there exists a ν-measurable set C satisfying the following.*

1) $0 < \nu(C) < \varepsilon$.
2) *If $w \in W$ and $w \leq 0$ ν a.e. on C, then $w = 0$.*

Proof. From Lemma 2.11, there exists a set A for which $\nu(A) < \varepsilon/2$, and if $w = 0$ ν a.e. on A, then $w = 0$. If $w \le 0$ ν a.e. on A implies $w = 0$, then we are finished. Assume not.

Let $W = \text{span}\{w_1, \dots, w_n\}$, and set

$$W_1 = \left\{ w = \sum_{i=1}^{n} a_i w_i : \|\mathbf{a}\| = 1 \right\}$$

where $\mathbf{a} = (a_1, \dots, a_n)$ and $\|\cdot\|$ is any norm on $\mathbb{R}^n$. A compactness argument implies that there exists a $c > 0$ such that if $w \in W_1$ and $w \le 0$ ν a.e. on A, then $\int_A w\, d\nu \le -c$.

For each $x \in B$, let B_x be any open set in B such that $x \in B_x$ and $0 < \nu(B_x) < \varepsilon/2(n+1)$. For each $x \in B$, set

$$D_x = \left\{ \mathbf{a} : \|\mathbf{a}\| \le 1, \ w = \sum_{i=1}^{n} a_i w_i \le 0 \ \ \nu \text{ a.e. on } A \cup B_x, \ \int_A w\, d\nu \le -c \right\}.$$

If there exist $x_1, \dots, x_{n+1}$ for which $\bigcap_{i=1}^{n+1} D_{x_i} = \emptyset$, then setting $C = A \cup (\bigcup_{i=1}^{n+1} B_{x_i})$, we are finished. This follows since $\nu(C) < \varepsilon$, and if $w \in W$ satisfies $w \le 0$, ν a.e. on C, and $w \ne 0$, then there exists an $\alpha > 0$ such that $\alpha w \in W_1$. It then easily follows that for $\alpha w = \sum_{i=1}^{n} a_i w_i$, we have $\mathbf{a} \in \bigcap_{i=1}^{n+1} D_{x_i}$, and thus $\bigcap_{i=1}^{n+1} D_{x_i} \ne \emptyset$, a contradiction.

We therefore assume that $\bigcap_{i=1}^{n+1} D_{x_i} \ne \emptyset$ for every choice of $x_1, \dots, x_{n+1}$, as above. As is easily checked D_x is a compact, convex subset of $\mathbb{R}^n$. From Helly's Theorem, it therefore follows that

$$\bigcap_{x \in B} D_x \ne \emptyset.$$

However, if $\mathbf{a}^* \in \bigcap_{x \in B} D_x$, then $w^* = \sum_{i=1}^{n} a_i^* w_i$ satisfies $w^* \le 0$ ν a.e. on B, and $\int_A w^*\, d\nu \le -c$. Since $\int_B w\, d\nu = 0$ for all $w \in W$, this is a contradiction.

Thus there exist points $x_1, \dots, x_{n+1}$ for which $\bigcap_{i=1}^{n+1} D_{x_i} = \emptyset$, proving the lemma. $\quad\square$

We can now prove:

Theorem 2.13. *Let ν be a non-atomic positive measure. Then each finite-dimensional subspace U of $L^1(B, \nu)$ is almost Chebyshev.*

Proof. From Proposition 2.10, it suffices to prove that the set $\{f : P_U(f)$ a singleton$\}$ is dense in $L^1(B, \nu)$. (Recall that U is an existence set.)

Let $f \in L^1(B, \nu)$ and assume that $P_U(f)$ is not a singleton. Translating f by a carefully chosen element of $P_U(f)$, we may assume that $0 \in P_U(f)$, and if $u \in P_U(f)$, then $-\alpha u \in P_U(f)$ for some $\alpha > 0$. (This is easily done since $P_U(f)$ is a convex bounded subset, with interior, of $u_0 + N$, where $u_0 \in P_U(f)$ and N is a subspace of U.)

Let $u \in P_U(f)$, and assume without loss of generality that $-u \in P_U(f)$. (Otherwise take αu with α small.) For any $x \in B$

$$2|f(x)| \le |(f - u)(x)| + |(f + u)(x)|$$

and

$$2\|f\|_1 \le \|f - u\|_1 + \|f + u\|_1 .$$

But $0, \pm u \in P_U(f)$, implying $\|f\|_1 = \|f \pm u\|_1$. Thus

$$2|f(x)| = |(f - u)(x)| + |(f + u)(x)| \qquad \nu \text{ a.e.},$$

which in turn implies that $Z(f) \subseteq Z(u)$ ν a.e. for all $u \in P_U(f)$.

Since $0 \in P_U(f)$, we have from Theorem 2.3 the existence of a function $h \in L^\infty(B, \nu)$ satisfying:

$$1)\ |h| = 1$$

$$2)\ \int_B hu\, d\nu = 0, \text{ all } u \in U$$

$$3)\ \int_B hf\, d\nu = \|f\|_1 .$$

Set

$$W = \{hu : u \in U\} .$$

W is a finite-dimensional subspace of $L^1(B, \nu)$ and, from (2), $\int_B w\, d\nu = 0$ for all $w \in W$.

Given $\varepsilon > 0$, it follows from the absolute continuity of integrals that there exists a $\delta > 0$ such that if $C \subseteq B$ and $\nu(C) < \delta$, then $\int_C |f| d\nu < \varepsilon$. Let $C \subseteq B$, $\nu(C) < \delta$, satisfy the conditions of Lemma 2.12 with respect to W. Set

$$f_\varepsilon = \begin{cases} f, & \text{off } C \\ 0, & \text{on } C . \end{cases}$$

Then $\|f - f_\varepsilon\|_1 = \int_C |f| d\nu < \varepsilon$. We shall prove that $P_U(f_\varepsilon) = \{0\}$, establishing the theorem. (To be precise, we then reverse the translation in the second paragraph of the proof to obtain our result.)

Since $0 \in P_U(f)$, we have for every $u \in U$

$$\|f_\varepsilon\|_1 = \int_B |f| d\nu - \int_C |f| d\nu \le \int_B |f - u| d\nu - \int_C |f| d\nu$$

$$\le \int_B |f - u| d\nu + \left[\int_C |u| d\nu - \int_C |f - u| d\nu \right]$$

$$= \int_{B \backslash C} |f - u| d\nu + \int_C |u| d\nu = \|f_\varepsilon - u\|_1 .$$

Thus $0 \in P_U(f_\varepsilon)$. Assume $v \in P_U(f_\varepsilon)$. Then equality holds in the above, and it is necessary that

$$a)\ \int_B |f| d\nu = \int_B |f - v| d\nu$$

$$b)\ \int_C |f - v| d\nu = \int_C |f| d\nu + \int_C |v| d\nu .$$

From (a) we obtain $v \in P_U(f)$, while (b) implies that $fv \le 0$ ν a.e. on C.

Now on $N(f)$, $h = \mathrm{sgn}\, f$ ν a.e. from (1) and (3), and thus $hv \le 0$ ν a.e. on $C \cap N(f)$. On $C \cap Z(f)$ we have $hv = 0$ ν a.e. since $v \in P_U(f)$ and as was shown $Z(f) \subseteq Z(v)$ ν a.e. for all $v \in P_U(f)$. Thus $hv \le 0$ ν a.e. on C. Since $hv \in W$, our choice of C implies that $hv = 0$. From (1) we obtain $v = 0$. Thus $P_U(f_e) = \{0\}$. □

4. Continuous Selections

Let Y be an existence set for the normed linear space X. Then the metric projection (as defined in Chapter 1, Section 3) $P_Y(f)$ is non-empty for every $f \in X$. Any single-valued map s from X onto Y for which $s(f) \in P_Y(f)$ is called a *metric selection*. We delete the subscript Y on $s(\cdot)$ for brevity. A metric selection s is said to be a *continuous selection* if s is continuous, i.e., for every $f, f_n \in X$ satisfying $\lim_{n \to \infty} \|f - f_n\| = 0$, we necessarily have $\lim_{n \to \infty} \|s(f) - s(f_n)\| = 0$.

If U is a finite-dimensional unicity space of X then, from Corollary 1.19, P_U is itself a continuous selection. Moreover there are known examples of X, U, and metric projections on U which possess continuous selections, where U is *not* a unicity space. One such example is the following: $U_1 = \mathrm{span}\{(1,1)\}$ is a one-dimensional subspace of ℓ_1^2, i.e., $I\!\!R^2$ with norm $\|(x,y)\|_1 = |x| + |y|$. It is easily checked, using Theorem 2.1, that given $(x,y) \in I\!\!R^2$ the set of best approximants from U_1 is exactly (α, α) for all $\alpha \in [x,y]$ if $x \le y$, or all $\alpha \in [y,x]$ if $y \le x$. In particular U_1 is not a unicity space. Nonetheless there are many continuous selections. For example, for any fixed $\lambda \in [0,1]$, the vector $(\lambda x + (1-\lambda)y, \lambda x + (1-\lambda)y)$ is a best approximant to (x,y) from U_1, and is a continuous selection.

While the lack of uniqueness of the best approximant often makes it difficult, in practice, to choose one, the existence of a continuous selection is a good recipe for such a choice. The existence of a continuous selection also generally tells us a good deal about the set of best approximants. Having said all this, we prove in this section that no finite-dimensional subspace of $L^1(B, \nu)$ possesses a continuous selection if ν is a non-atomic positive measure.

Theorem 2.14. *Let ν be a non-atomic positive measure and U a finite-dimensional subspace of $L^1(B, \nu)$. There does not exist a continuous selection on U.*

Remark. As a consequence of Corollary 1.19, we note that Theorem 2.14 is another generalization of Theorem 2.7.

Proof. The idea of the proof is as follows. We construct an $f \in L^1(B, \nu)$ and two sequences of functions $\{g_n\}, \{k_n\}$ in $L^1(B, \nu)$ satisfying:

1) $\lim_{n\to\infty} \|f - g_n\|_1 = \lim_{n\to\infty} \|f - k_n\|_1 = 0$

2) $P_U(g_n) = \{u^1\}$, for all n

3) $P_U(k_n) = \{u^2\}$, for all n

4) $u^1 \neq u^2$.

There then cannot exist a continuous selection on U. For if such an s did exist, then necessarily both $s(f) = u^1$ and $s(f) = u^2$ would hold, which is impossible.

Let $\{B_n\}_1^\infty$ be a non-increasing sequence of measurable subsets of B (set $B = B_0$) with the property that $0 < \nu(B_n) < 1/n$, $n = 1, 2, \ldots$, and if $u \in U$, $u = 0$ ν a.e. on B_n, then $u = 0$. To obtain such a sequence replace B in Lemma 2.11 by B_{n-1} and set $\varepsilon = 1/n$. From Theorem 2.3 there exists, for each ν-measurable set A, an $h_A \in L^\infty(A, \nu)$ for which $|h_A| = 1$ on A and $\int_A h_A u \, d\nu = 0$ for all $u \in U$.

Set $h = h_{B_n \backslash B_{n-1}}$ on $B_n \backslash B_{n-1}$, $n = 0, 1, 2, \ldots$, and $h = 1$ on $\bigcap_{n=0}^\infty B_n$. Then $\int_{B \backslash B_n} h u \, d\nu = 0$ for every $u \in U$, and $n = 1, 2, \ldots$.

Let $u^* \in U$, $u^* \neq 0$. Set $f = h|u^*|$, and

$$g_n = \begin{cases} h|u^*|, & \text{on } B \backslash B_n, \\ 0, & \text{on } B_n, \end{cases}$$

$$k_n = \begin{cases} h|u^*|, & \text{on } B \backslash B_n, \\ u^*, & \text{on } B_n. \end{cases}$$

Then, as is easily seen,

$$\lim_{n\to\infty} \|f - g_n\|_1 = \lim_{n\to\infty} \|f - k_n\|_1 = 0 .$$

To complete the proof we show that $P_U(g_n) = \{0\}$ and $P_U(k_n) = \{u^*\}$ for all n.

For each $n = 1, 2, \ldots$, and $u \in U$, $u \neq 0$, $\int_{B_n} |u| d\nu > 0$ by our construction of B_n. Thus

$$\left| \int_B (\text{sgn } g_n) u \, d\nu \right| = \left| \int_{B \backslash B_n} (\text{sgn } g_n) u \, d\nu \right| = \left| \int_{B \backslash B_n} h u \, d\nu - \int_{Z(g_n) \backslash B_n} h u \, d\nu \right|$$

$$= \left| - \int_{Z(g_n) \backslash B_n} h u \, d\nu \right| \leq \int_{Z(g_n) \backslash B_n} |u| d\nu$$

$$< \int_{Z(g_n) \backslash B_n} |u| d\nu + \int_{B_n} |u| d\nu = \int_{Z(g_n)} |u| d\nu .$$

That is, for each $u \in U$, $u \neq 0$,

$$\left| \int_B (\text{sgn } g_n) u \, d\nu \right| < \int_{Z(g_n)} |u| d\nu.$$

Applying Corollary 2.5, we obtain $P_U(g_n) = \{0\}$ for all n.

Set
$$k_n^* = k_n - u^* = \begin{cases} h|u^*| - u^*, & \text{on } B \backslash B_n \\ 0, & \text{on } B_n. \end{cases}$$

Thus on $N(k_n^*)$, $\operatorname{sgn} k_n^* = h$. By the above method of proof it follows that

$$\left| \int_B (\operatorname{sgn} k_n^*) u \, d\nu \right| < \int_{Z(k_n^*)} |u| d\nu$$

for all $u \in U$, $u \neq 0$. Again applying Corollary 2.5, we obtain $P_U(k_n^*) = \{0\}$ for all n. Thus $P_U(k_n) = \{u^*\}$ for all n. This proves the theorem. $\qquad \square$

Exercises

1. Show that Theorem 2.1 is valid without the assumption that $(L^1(B,\nu))^* = L^\infty(B,\nu)$, by using the first method of proof thereof.

2. Let $L^1_{\mathbb{C}}(B,\nu)$ denote the complex-valued ν-measurable functions on B for which $|f|$ is ν-integrable over B, with associated norm. Prove that Theorem 2.1 is valid with (2.1) replaced by

$$\left| \int_{N(f-g^*)} \frac{\operatorname{Re}(g(\overline{f - g^*}))}{|f - g^*|} d\nu \right| \le \int_{Z(f-g^*)} |g| d\nu.$$

3. Let ν be a positive measure on B. Set

$$M = \left\{ g : g \in L^1(B,\nu), \int_B g \, d\nu = 0 \right\}.$$

For $f \in L^1(B,\nu)$, prove that $E(f; M) = |\int_B f \, d\nu|$ and find a best approximant to f from M.

4. Let ν be Lebesgue measure on $[0,1]$. Set

$$M = \left\{ g : g \in L^1([0,1],\nu), \int_0^1 gw \, d\nu = 0 \right\},$$

where $w(x) = (1 + x)/2$. For $f \in L^1([0,1],\nu) \backslash \overline{M}$, prove that $E(f; M) = |\int_0^1 fw \, d\nu|$, but f has no best approximant from M.

5. Let ν be Lebesgue measure on $[0,1]$. Construct a finite-dimensional subspace U of $L^1([0,1],\nu)$ and an $f \in L^1([0,1],\nu)$ for which

$$\int_0^1 (\operatorname{sgn} f) u \, d\nu = 0$$

for all $u \in U$, and yet $P_U(f)$ is not a singleton.

6. Put together the ideas of Theorems 2.7 and 2.8 to prove the following:

Let U be an n-dimensional subspace of $L^1(B,\nu)$, where ν is a σ-finite positive measure. Then U is not a unicity space for $L^1(B,\nu)$ if and only if there exists an $h \in L^\infty(B,\nu)$, $h \neq 0$, satisfying $\int_B hu \, d\nu = 0$ for all $u \in U$, and such that the set

$$\{x : |h(x)| < \|h\|_\infty\}$$

(to within a set of measure zero) is purely atomic and contains at most $n-1$ atoms.

7. Let $L^{1,2}(B,\nu)$ denote the space of $\underline{f} = (f_1, \ldots, f_m)$ where m is finite, each f_i is a real-valued ν-measurable function defined on B, and

$$\|\underline{f}\| = \int_B \Big[\sum_{i=1}^m (f_i(x))^2 \Big]^{1/2} d\nu(x) < \infty.$$

For $\underline{f} \in L^{1,2}(B,\nu)$, set

$$Z(\underline{f}) = \{x : \underline{f}(x) = 0\},$$

and

$$\operatorname{sgn} \underline{f}(x) = \begin{cases} \underline{f}(x) / \Big[\sum_{i=1}^m (f_i(x))^2 \Big]^{1/2}, & x \in B \backslash Z(\underline{f}) \\ 0, & x \in Z(\underline{f}). \end{cases}$$

For $\underline{f}, \underline{g} \in L^{1,2}(B,\nu)$, set $(\underline{f}, \underline{g}) = \sum_{i=1}^m f_i(x) g_i(x)$.

Prove that if M is a subspace of $L^{1,2}(B,\nu)$, then $\underline{g}^*$ is a best $L^{1,2}(B,\nu)$ approximant to $\underline{f}$ from M if and only if

$$\left| \int_B (\operatorname{sgn}(\underline{f} - \underline{g}^*), \underline{g}) d\nu \right| \le \int_{Z(\underline{f} - \underline{g}^*)} |\underline{g}| d\nu$$

for all $\underline{g} \in M$.

8. Let C be a convex subset of $L^1(B,\nu)$ and $f \in L^1(B,\nu) \backslash \overline{C}$. Prove that g^* is a best $L^1(B,\nu)$ approximant to f from C if and only if

$$\int_B \operatorname{sgn}(f - g^*)(g - g^*) d\nu \le \int_{Z(f-g^*)} |g^* - g| d\nu$$

for all $g \in C$.

9. Let U be a finite-dimensional subspace of $L^1(B,\nu)$, where ν is a non-atomic positive measure. Let C be a convex subset of U. Prove that g^* is a best $L^1(B,\nu)$ approximant to f from C if and only if there exists an $h \in L^\infty(B,\nu)$ satisfying

1) $|h| = 1$ on B

2) $\int_B h g^* d\nu \ge \int_B h g \, d\nu$ for all $g \in C$

3) $\int_B h(f - g^*) d\nu = \|f - g^*\|_1$.

10. Prove Theorem 2.14 using the ideas and method of proof of Theorem 2.13.

Notes and References

See Dunford, Schwartz [1958, p.289] for a proof of the fact that if ν is a σ-finite positive measure, then $(L^1(B,\nu))^* = L^\infty(B,\nu)$. Theorem 2.1 may be found in James [1947, p.291], see also Kripke, Rivlin [1965], and Singer [1956]. In Kripke, Rivlin [1965] is also to be found a history of the theorem, as well as an extension to complex-valued functions (Exercise 2). Rozema [1974, Theorem 4.1] contains generalizations of Theorem 2.1 to Banach space valued functions. Exercise 7 is a special case thereof and may also be found in Kroó [1984]. For a proof of Liapounoff's Theorem (Theorem 2.2), see Liapounoff [1940]. Theorem 2.3 may be found in Phelps [1966]. An equivalent version of Proposition 2.4 appears in Berdyshev [1975]. Theorem 2.7 was first proved for $B = [0,1]$ and ν Lebesgue measure in 1938 by Krein [1962]. The extension to this more general setting is given in Phelps [1960] (and attributed to H. Dye, see Phelps [1966]) and was proved independently by Moroney [1961]. Exercise 6, of which Theorem 2.8 is a special case, is due to Phelps [1966]. Theorem 2.9 is due to Havinson, Romanova [1972]. Our proof is somewhat simpler. Proposition 2.10 is a variant of a result of Garkavi [1964], and is a special case of Lemma A in Rozema [1974]. Lemma 2.11 may be found in Lazar, Wulbert, Morris [1969]. Helly's Theorem used in the proof of Lemma 2.12 is in Helly [1923], see also Danzer, Grünbaum, Klee [1963]. Theorem 2.13 was independently proved by Havinson, Romanova [1972] and Rozema [1974]. The proof given here is a variant thereof. This proof can also be modified to prove a result of Angelos, Schmidt [1983]. Namely, if ν is a non-atomic positive measure on B, and U is a finite-dimensional subspace of $L^1(B,\nu)$, then the set of $f \in L^1(B,\nu)$ which have a strongly unique best approximant from U is dense in $L^1(B,\nu)$. Theorem 2.14 is proved in Lazar, Wulbert, Morris [1969].

3
Approximation from Finite-Dimensional Subspaces in $C_1(K,\mu)$

---

1. Introduction and Notation

On the basis of the rather negative results of the last chapter, the reader may be excused from thinking that any further study in the directions considered therein is fruitless. Nothing is further from the truth. We have however been dealing with too general a situation.

In this chapter we restrict our attention to continuous functions and also impose additional constraints. What we shall have to say holds in a somewhat more general setting, but for ease of exposition we make rather stringent assumptions on our space of functions. Nevertheless we are restricting ourselves to the cases of interest. To emphasize this altered situation, we shall change our notation.

Firstly we replace B by the set K where

    1) $K \subset \mathbb{R}^d$ is compact, and

    2) $K = \overline{\operatorname{int} K}$.

Secondly, we assume that we are dealing with non-atomic positive finite measures μ defined on K, such that every real-valued continuous function is μ-measurable. Furthermore μ is also assumed to have the important property that if

$$\|f\|_1 = \int_K |f| d\mu = 0$$

for $f \in C(K)$ (i.e., f continuous on K), then $f = 0$. Thus $\|\cdot\|_1$ is truly a norm on $C(K)$, and it is not necessary, as when considering $L^1(B,\nu)$, to deal with equivalence classes of functions which are ν a.e. equal. Equivalently, we could demand that $\mu(A) > 0$ for every open set $A \subsetneq K$. Since we shall frequently have to refer to such measures, we shall say that such a measure is 'admissible'.

We let $C_1(K,\mu)$ denote the linear space $C(K)$ with norm $\|\cdot\|_1$. $C_1(K,\mu)$ is *not* a Banach space, i.e., it is not complete. It is however a dense linear subspace of $L^1(K,\mu)$, and as such

$$(C_1(K,\mu))^* = L^\infty(K,\mu).$$

This property permits us to apply many of the techniques developed in Chapter 2. It is especially worth noting that the characterization results of Theorems 2.1 and 2.3 remain valid. We shall consider the problem of best approximating functions in $C_1(K,\mu)$ from finite-dimensional subspaces U. It is always to be assumed, even when not explicitly stated, that $U \subset C(K)$.

In this chapter we deal with four main topics. In Section 2 we delineate two criteria (Theorems 3.1 and 3.3) for exactly determining when U is a unicity space for $C_1(K, \mu)$. Various examples are also given. In Theorem 3.6 we consider the 'size' of the set $\{f : P_U(f)$ a singleton$\}$, partially paralleling Theorem 2.13. In Section 4 we consider conditions on U which characterize the largest possible 'dimension' of $P_U(f)$, $f \in C(K)$. Finally in Section 5 we study the problem of the existence of continuous selections from $C_1(K, \mu)$ onto U.

2. Uniqueness

We first consider the problem of when a finite-dimensional subspace U of $C(K)$ is a *unicity* space for $C_1(K, \mu)$, i.e., when to each $f \in C(K)$ there exists a unique best approximant from U in the norm $\| \cdot \|_1$.

For the space $L^1(B, \nu)$ we proved the negative result contained in Theorem 2.7. However a rereading of the proof of that theorem indicates that it is not always valid in this setting. Using the same basic ideas we shall prove the following.

Theorem 3.1. *A finite-dimensional subspace U of $C(K)$ is a unicity space for $C_1(K, \mu)$ if and only if there does not exist an $h \in L^\infty(K, \mu)$ and a $u^* \in U$, $u^* \neq 0$, for which*

$$1) \, |h| = 1 \quad on \, K$$

$$2) \int_K h u \, d\mu = 0, \quad all \, u \in U$$

$$3) \, h|u^*| \quad is \, continuous.$$

Proof. ($\Rightarrow$). Assume there exist h and u^*, as above, satisfying (1), (2) and (3). Set $f = h|u^*|$. Then $f \in C(K)$ and it follows, word for word as in the proof of Theorem 2.7, that U is not a unicity space for $C_1(K, \mu)$.

($\Leftarrow$). Assume that U is not a unicity space. Let $u_1, u_2 \in U$, $u_1 \neq u_2$, be two best approximants to some $f \in C(K)$. Set $f^* = f - (u_1 + u_2)/2$ and $u^* = (u_1 - u_2)/2$. It is easily seen that $0, u^*$ and $-u^*$ are all best approximants to f^*. We claim that

$$2|f^*(x)| = |f^*(x) - u^*(x)| + |f^*(x) + u^*(x)|$$

for all $x \in K$. From the triangle inequality,

$$2|f^*(x)| \leq |f^*(x) - u^*(x)| + |f^*(x) + u^*(x)|.$$

If strict inequality holds for some $x \in K$, then from the continuity of the functions and the property of μ, we obtain $2\|f^*\|_1 < \|f^* - u^*\|_1 + \|f^* + u^*\|_1$. But $\|f^*\|_1 = \|f^* - u^*\|_1 = \|f^* + u^*\|_1$, a contradiction. Thus for all $x \in K$

$$2|f^*(x)| = |f^*(x) - u^*(x)| + |f^*(x) + u^*(x)|.$$

If $x \in Z(f^*)$, i.e., $f^*(x) = 0$, then $(f^* \pm u^*)(x) = 0$, implying that $x \in Z(u^*)$. Thus $Z(f^*) \subseteq Z(u^*)$.

Now, since the zero function is a best approximant to f^* from U, we have from Theorem 2.3 the existence of an $h \in L^\infty(K, \mu)$ for which

$$1) \ |h| = 1 \ \text{ on } K$$

$$2) \ \int_K hu \, d\mu = 0, \text{ all } u \in U$$

$$3) \ \int_K hf^* d\mu = \int_K |f^*| d\mu.$$

From (1) and (3), $h = \operatorname{sgn} f^*$ μ a.e. on $N(f^*)$. Alter h on a set of measure zero so that $h = \operatorname{sgn} f^*$ on all of $N(f^*)$. Now consider $h|u^*|$. If $h|u^*|$ is discontinuous at a point $x \in K$, then x is necessarily a point of discontinuity of h. Every point of discontinuity of h occurs, by construction, at a zero of f^*. But $Z(f^*) \subseteq Z(u^*) = Z(|u^*|)$ and $|u^*|$ is continuous. Thus $h|u^*|$ is continuous at each point of discontinuity of h. Consequently $h|u^*|$ is continuous on K. $\square$

Remark. For the h satisfying (1) of Theorem 3.1, the above argument implies that $h|u^*| \in C(K)$ if and only if $hu^* \in C(K)$.

In Chapter 4 are to be found many examples of unicity spaces for $C_1(K, \mu)$ for all μ in some large subset of 'admissible' measures. We shall here consider four different examples, applying the conditions of Theorem 3.1 in the analysis of each.

Example 3.1. $\dim U = 1$.

Let $U = \operatorname{span}\{u\}$. From our assumptions on K, the set $K \backslash Z(u)$ is composed of a possibly infinite, but most definitely countable, number m of disjoint, connected and open (relative to K) sets. Let us denote them by $K \backslash Z(u) = \bigcup_{i=1}^m A_i$. Set

$$\int_{A_i} u \, d\mu = \alpha_i, \qquad i = 1, \dots, m.$$

Because $u \in C_1(K, \mu)$,

$$\|u\|_1 = \sum_{i=1}^m |\alpha_i| < \infty.$$

Since the h of (3) of Theorem 3.1 must be constant on each A_i, a simple consequence of Theorem 3.1 is that U is a unicity space for $C_1(K, \mu)$ if and only if there does not exist a choice of $\varepsilon_i \in \{-1, 1\}$, $i = 1, \dots, m$, for which

$$\sum_{i=1}^m \varepsilon_i \alpha_i = 0.$$

Except in the case where $m = 1$, this condition depends on the choice of the measure μ.

Example 3.2. $U = \pi_m = \text{span}\{1, x, \ldots, x^m\}$, $K = [a, b] \subset \mathbb{R}$.

U is a unicity space for $C_1(K, \mu)$ independent of the choice of 'admissible' measure μ. For assume not. We then have a non-trivial algebraic polynomial u^* of degree at most m, and an $h \in L^\infty([a, b], \mu)$ satisfying (1), (2) and (3) of Theorem 3.1. Since u^* has at most m zeros in $[a, b]$, it follows from (1) and (3) that h has at most m jumps (points of discontinuity) in (a, b). It is well known that no such h can be orthogonal to all polynomials in π_m, i.e., satisfy (2). (This is a consequence of the fact that we can find a $u \in \pi_m$ with the same sign pattern as h.)

Example 3.3. $U = \pi_m$, $K = [a, b] \cup [c, d] \subset \mathbb{R}$, and $a < b < c < d$.

The geometry of the domain has an effect on the unicity space property. For each m there exist 'admissible' measures μ for which U is a unicity space for $C_1(K, \mu)$, and 'admissible' measures for which it is not. Note that as a consequence of Example 3.1, $\pi_0 = \text{span}\{1\}$ is a unicity space if and only if

$$\int_a^b d\mu \neq \int_c^d d\mu.$$

See also Exercise 9.

Example 3.4. *A 'non-trivial' example.*

Let $K = [0, 1] \times [0, 1]$ and μ denote the usual Lebesgue measure on K. We write $d\mu = dx\, dy$. Set $\pi_n = \text{span}\{1, x, \ldots, x^n\}$, and let φ be a strictly increasing continuous function on $[0, 1]$. Our subspace is

$$U = \{\varphi(y)q_k(x) + p_m(x) : q_k \in \pi_k, p_m \in \pi_m, k \leq m\}.$$

For this particular μ, we shall use Theorem 3.1 to prove:

Proposition 3.2. *U is a unicity space for $C_1(K, \mu)$.*

Proof. Assume U is not a unicity space for $C_1(K, \mu)$. There then exists a function $h(x, y) \in L^\infty(K)$ and a non-trivial $u^*(x, y) = \varphi(y)q_k^*(x) + p_m^*(x)$ in U, for which

$\quad$ 1) $|h(x, y)| = 1$ $\;$ on K

$\quad$ 2) $\int_0^1 \int_0^1 h(x, y)u(x, y)\, dx\, dy = 0$, $\;$ all $u \in U$

$\quad$ 3) $h(x, y)|u^*(x, y)|$ $\;$ is continuous.

Set

$$h^*(x) = \int_0^1 h(x, y)\, dy$$

and let $x_1, \ldots, x_r$ denote the distinct common zeros of q_k^* and p_m^* in $[0, 1]$. Thus $r \leq m$. From (2) we have that

$$\int_0^1 h^*(x)p_m(x)\, dx = 0$$

for all $p_m \in \pi_m$. It is a well-known fact, see Example 3.2, that h^* must exhibit at least $m + 1$ sign changes in $(0, 1)$, or identically vanish thereon. It therefore follows that there exist $m + 1 - r$ distinct points $\overline{x}_1, \ldots, \overline{x}_{m+1-r}$ in $(0, 1) \backslash \{x_1, \ldots, x_r\}$ such that either $\lim_{x \to \overline{x}_i^+} h^*(x) \cdot \lim_{x \to \overline{x}_i^-} h^*(x) \leq 0$ or $h^*(\overline{x}_i) = 0$ for each i.

Since $\overline{x}_i$ is not a common zero of q_k^* and p_m^*, and φ is a strictly increasing continuous function, it is easily seen that $u^*(\overline{x}_i, y)$, as a function of y, is not identically zero and in fact has at most one zero. Since $h(x, y)$ can only change values on $Z(u^*)$ (from (3)), it follows that h^* is continuous at $\overline{x}_i$, i.e., $h^*(\overline{x}_i) = 0$, and $u^*(\overline{x}_i, \overline{y}_i) = 0$ for some $\overline{y}_i \in (0, 1)$, $i = 1, \ldots, m + 1 - r$. Thus

$$h(\overline{x}_i, y) = \begin{cases} \varepsilon_i, & 0 \leq y < \overline{y}_i \\ \eta_i, & \overline{y}_i < y \leq 1 \end{cases}$$

for some $\varepsilon_i, \eta_i \in \{-1, 1\}$. Because

$$0 = h^*(\overline{x}_i) = \int_0^{\overline{y}_i} \varepsilon_i \, dy + \int_{\overline{y}_i}^1 \eta_i \, dy = \varepsilon_i \overline{y}_i + \eta_i (1 - \overline{y}_i),$$

it follows that $\eta_i = -\varepsilon_i$ and $\overline{y}_i = 1/2$, $i = 1, \ldots, m + 1 - r$.

Now,

$$0 = u^*(\overline{x}_i, 1/2) = \varphi(1/2) q_k^*(\overline{x}_i) + p_m^*(\overline{x}_i),$$

$i = 1, \ldots, m + 1 - r$, and

$$0 = u^*(x_i, 1/2) = \varphi(1/2) q_k^*(x_i) + p_m^*(x_i),$$

$i = 1, \ldots, r$, since $q_k^*(x_i) = p_m^*(x_i) = 0$. Thus $u^*(x, 1/2)$ has at least $m + 1$ distinct zeros in $[0, 1]$. Since $u^*(\cdot, 1/2) \in \pi_m$, this implies that $u^*(\cdot, 1/2) = 0$. Therefore

$$u^*(x, y) = \big(\varphi(y) - \varphi(1/2) \big) q_k^*(x).$$

The sign pattern of u^* is now easily determined. Let $x_1 < \cdots < x_n$ denote the distinct zeros of q_k^* in $(0, 1)$. For convenience, set $x_0 = 0$, $x_{n+1} = 1$. Let

$$A_i = \{(x, y) : x_{i-1} < x < x_i, \ 0 \leq y < 1/2\},$$

for $i = 1, \ldots, n + 1$, and

$$B_i = \{(x, y) : x_{i-1} < x < x_i, \ 1/2 < y \leq 1\},$$

for $i = 1, \ldots, n + 1$. Because u^* does not vanish in $\bigcup_{i=1}^{n+1} (A_i \cup B_i)$, it follows from (1) and (3) that h must be a constant, i.e., 1 or -1, on each A_i and B_i. Thus

$$h(x, y) = \begin{cases} \varepsilon_i, & (x, y) \in A_i, \ i = 1, \ldots, n + 1 \\ \eta_i, & (x, y) \in B_i, \ i = 1, \ldots, n + 1, \end{cases}$$

where $\varepsilon_i, \eta_i \in \{-1, 1\}$, $i = 1, \ldots, n + 1$. If $\varepsilon_i \eta_i = -1$, then $h^*(x) = 0$ for all $x \in (x_{i-1}, x_i)$. If $\varepsilon_i \eta_i = 1$, then $h^*(x) = \varepsilon_i$ for all $x \in (x_{i-1}, x_i)$.

Assume $\varepsilon_i \eta_i = 1$ for some i. Then h^* is not identically zero and has at most $n \le k \le m$ sign changes in $[0, 1]$. But this is a contradiction to the fact that

$$\int_0^1 h^*(x) p_m(x) dx = 0$$

for all $p_m \in \pi_m$. Thus $\varepsilon_i \eta_i = -1$ for all i. Since $n \le k$, there exists a $q_k \in \pi_k$ for which $\eta_i q_k(x) > 0$ for all $x \in (x_{i-1}, x_i)$, $i = 1, \ldots, n+1$. Set

$$u(x, y) = \big(\varphi(y) - \varphi(1/2)\big) q_k(x).$$

Because φ is strictly increasing and $\varepsilon_i \eta_i = -1$, all i, we have

$$\int_0^1 \int_0^1 h(x, y) u(x, y) dx\, dy = \int_0^1 \int_0^1 |u(x, y)| dx\, dy > 0.$$

This is a contradiction to (2), proving the proposition. □

There is yet another method of characterizing unicity spaces which does not explicitly depend upon constructing the function h of Theorem 3.1. Instead, we characterize unicity spaces by considering the best approximants on a set of 'test' functions. This alternative characterization will have far-reaching consequences. The set of test functions is the following.

Definition 3.1. Let U be a finite-dimensional subspace of $C(K)$. Then

$$U^* = \{g : g \in C(K), |g| = |u| \text{ for some } u \in U\}.$$

Note that every $g \in U^*$ must be continuous. U^* contains U, but it is not in general a subspace of $C_1(K, \mu)$. For example, if $U = \operatorname{span}\{1, x\}$ and $K = [a, b]$, then U^* consists of U plus all functions of the form $c|x - d|$ where $c \in \mathbb{R}$, $a < d < b$.

Theorem 3.3. *A finite-dimensional subspace U of $C(K)$ is a unicity space for $C_1(K, \mu)$ if and only if the zero function is not a best approximant to any $g \in U^*$, $g \ne 0$.*

Proof. ($\Rightarrow$). Assume that the zero function is a best approximant to some $g \in U^*$, and $|g| = |u^*|$ for $u^* \in U$, $u^* \ne 0$. From Theorem 2.1,

$$\left| \int_K (\operatorname{sgn} g) u \, d\mu \right| \le \int_{Z(g)} |u| d\mu,$$

for all $u \in U$. For any $\alpha \in \mathbb{R}$, $|\alpha| < 1$, we have $\operatorname{sgn}(g - \alpha u^*) = \operatorname{sgn} g$ and $Z(g - \alpha u^*) = Z(g)$ since $|g| = |u^*|$. Thus

$$\left| \int_K \operatorname{sgn}(g - \alpha u^*) u \, d\mu \right| \le \int_{Z(g - \alpha u^*)} |u| d\mu$$

for all $u \in U$. From Theorem 2.1, αu^* is a best approximant to g for all $\alpha \in (-1, 1)$. U is not a unicity space.

($\Leftarrow$). Assume U is not a unicity space. Let h and u^* be as given in Theorem 3.1. In the proof of Theorem 3.1 (see Theorem 2.7) we observed that $g = h|u^*|$ has the zero function as a best approximant. Furthermore, $|g| = |u^*|$ and $g \in C(K)$, i.e., $g \in U^*$, $g \neq 0$.   □

We can rephrase Theorem 3.3 in the following manner, based solely on Theorem 2.1.

Corollary 3.4. *A finite-dimensional subspace U of $C(K)$ is a unicity space for $C_1(K,\mu)$ if and only if for every $g \in U^*$, $g \neq 0$, there exists a $u_g \in U$ satisfying*

$$\left| \int_K (\text{sgn } g) u_g \, d\mu \right| > \int_{Z(g)} |u_g| \, d\mu.$$

Paralleling Theorem 2.13, we now investigate the size of the set $\{f : P_U(f)$ a singleton$\}$. If U is a unicity space, then this set is, by definition, all of $C(K)$. We therefore assume that U is not a unicity space. Is $\{f : P_U(f)$ a singleton$\}$ a complement of a set of first category, as in Theorem 2.13? Since $C_1(K,\mu)$ is not complete, we can not apply Proposition 2.10. (Owing to the lack of completeness, this question is not really the right question anyway.) We shall however show that this set is always dense in $C_1(K,\mu)$. This is *not* a consequence of Theorem 2.13, but the proof is along very much the same lines.

Before starting, a few words of explanation (warning) are in order. Consider the following example. Let $K = [-1,1]$, μ be Lebesgue measure, and $U = \text{span}\{x\}$. Observe (Theorem 3.1 with $h = 1$) that U is not a unicity space for $C_1(K,\mu)$. Furthermore, if $f \in C[-1,1]$ does not vanish on $[-1,1]$, then $P_U(f)$ is not a singleton. (These are not all the f for which $P_U(f)$ is not a singleton.) The set of strictly positive and strictly negative functions is exceedingly 'large'. However we are dealing with density with respect to the induced L^1-topology. In this topology this set is 'small'. Since we are investigating continuous functions, we might consider the uniform topology as being in some way appropriate (see Section 5 of this chapter). As the above example shows, with respect to the uniform topology $\{f : P_U(f)$ a singleton$\}$ need not be dense. However it is dense in the L^1-topology.

We first prove a lemma which in a certain sense both parallels and generalizes Lemma 2.12. The generalization is not used here, but will be used in Chapter 5.

Lemma 3.5. *Let W be a finite-dimensional subspace of $C(K)$. There exist points $\{x_i\}_{i=1}^m$, m finite, such that if $w \in W$ satisfies $w(x_i) \leq 0$, $i = 1, \ldots, m$, and $\int_K w \, d\mu \geq 0$, then $w = 0$.*

Proof. Set

$$W_1 = \{w : w \in W, \|w\|_1 = 1\}.$$

W_1 is compact and does not contain the zero function. Also, if $w \in W$, $w \neq 0$, then $\alpha w \in W_1$ for some $\alpha > 0$. It therefore suffices to prove that there exists no $w \in W_1$ satisfying $w(x_i) \leq 0$, $i = 1, \ldots, m$, and $\int_K w \, d\mu \geq 0$.

For each $x \in K$, let

$$D(x) = \{ w : w \in W_1, \, w(x) > 0 \},$$

and set

$$G = \{ w : w \in W_1, \int_K w \, d\mu < 0 \}.$$

The sets G and $D(x)$ are open in W_1. Furthermore

$$W_1 = \Big(\bigcup_{x \in K} D(x) \Big) \cup G.$$

Otherwise there exists a $w \in W_1$ for which $\int_K w \, d\mu \geq 0$, and $w(x) \leq 0$ for all $x \in K$. But the zero function is not in W_1, and this is impossible.

Since W_1 is compact, there exist $\{x_i\}_{i=1}^m$, m finite, for which

$$W_1 = \Big(\bigcup_{i=1}^m D(x_i) \Big) \cup G.$$

This implies that no $w \in W_1$ satisfies both $w(x_i) \leq 0$, $i = 1, \ldots, m$, and $\int_K w \, d\mu \geq 0$. □

Theorem 3.6. *Let U be a finite-dimensional subspace of $C(K)$. Then $\{f : P_U(f)$ a singleton$\}$ is dense in $C_1(K, \mu)$.*

Proof. Assume that U is not a unicity space and $f \in C(K)$ is such that $P_U(f)$ is not a singleton. Translating f by an element of U, we may assume that there exist $u_1, \ldots, u_r \in U$, linearly independent, $1 \leq r \leq n$, for which $\{\pm u_1, \ldots, \pm u_r\} \subseteq P_U(f) \subset \text{span}\{u_1, \ldots, u_r\} = N$. Since $P_U(f)$ is convex, we also have $0 \in P_U(f)$.

From Theorem 2.3 there exists an $h \in L^\infty(K, \mu)$ satisfying

1) $|h| = 1$ on K

2) $\int_K hu \, d\mu = 0$, all $u \in U$

3) $\int_K hf \, d\mu = \|f\|_1$.

We may assume that $h = \text{sgn } f$ on $N(f)$. As is readily seen, $Z(f) \subseteq Z(u_i)$, $i = 1, \ldots, r$, and therefore $h|u_i| \in C(K)$ and thus $hu_i \in C(K)$, $i = 1, \ldots, r$, (see the proof of Theorem 3.1 and the remark thereafter). Set

$$W = \{hu : u \in N\}.$$

W is a finite-dimensional subspace of $C(K)$, and from (2), $\int_K w \, d\mu = 0$ for all $w \in W$. Let $x_1, \ldots, x_m$ be as in Lemma 3.5 with respect to W.

For given $\varepsilon > 0$, set

$$A_\varepsilon(x_i) = \{x : |x - x_i| < \varepsilon\} \cap K, \qquad i = 1, \ldots, m.$$

Assume ε is sufficiently small so that $A_\varepsilon(x_i) \cap A_\varepsilon(x_j) = \emptyset$ for all $i \neq j$. Let

$$f_\varepsilon(x) = \begin{cases} f(x), & x \notin \bigcup_{i=1}^m A_\varepsilon(x_i) \\ g(x), & x \in \bigcup_{i=1}^m A_\varepsilon(x_i), \end{cases}$$

where $g(x) = f(x)(|x - x_i|/\varepsilon)$ for all $x \in A_\varepsilon(x_i)$, $i = 1, \ldots, m$. Thus $f_\varepsilon \in C(K)$, and

$$\|f - f_\varepsilon\|_1 \leq \int_{\bigcup_{i=1}^m A_\varepsilon(x_i)} |f| d\mu.$$

Therefore

$$\lim_{\varepsilon \to 0^+} \|f - f_\varepsilon\|_1 = 0.$$

Note that sgn $f_\varepsilon =$ sgn f on $K \backslash \{x_1, \ldots, x_m\}$, and $|g| \leq |f|$ on $\bigcup_{i=1}^m A_\varepsilon(x_i)$.

We shall prove that $P_U(f_\varepsilon) = \{0\}$ for all $\varepsilon > 0$, which then proves our theorem. We divide the proof of this fact into three steps.

Claim 1. $0 \in P_U(f_\varepsilon)$.

Proof. Since $0 \in P_U(f)$, we have for every $u \in U$,

$$\left| \int_{N(f_\varepsilon)} (\text{sgn } f_\varepsilon) u \, d\mu \right| = \left| \int_{N(f)} (\text{sgn } f) u \, d\mu \right| \leq \int_{Z(f)} |u| d\mu = \int_{Z(f_\varepsilon)} |u| d\mu.$$

From Theorem 2.1, $0 \in P_U(f_\varepsilon)$.

Claim 2. $P_U(f_\varepsilon) \subseteq P_U(f)$.

Proof. Assume $v \in P_U(f_\varepsilon)$, $v \neq 0$. From Proposition 2.4,

$$a) \; f_\varepsilon(f_\varepsilon - v) \geq 0$$

$$b) \int_K v \, \text{sgn } f_\varepsilon d\mu = \int_{Z(f_\varepsilon)} |v| d\mu.$$

From the definition of f_ε, it follows that

$$a') \; f(f - v) \geq 0$$

$$b') \int_K v \, \text{sgn } f d\mu = \int_{Z(f)} |v| d\mu.$$

Applying Proposition 2.4 to (a') and (b'), we obtain $v \in P_U(f)$.

Claim 3. $P_U(f_\varepsilon) = \{0\}$.

Proof. Assume $v \in P_U(f_\varepsilon)$. From Claim 2, $v \in P_U(f)$ and thus $hv \in W$. We shall prove that $(hv)(x_i) \leq 0$, $i = 1, \ldots, m$. If $x_i \in Z(f)$, then $u(x_i) = 0$ for all $u \in P_U(f)$ and there is nothing to prove. Assume that $x_i \in N(f)$. From (a) of Claim 2,

$$(\text{sgn } f)(g - v) \geq 0$$

on $N(f) \cap A_\varepsilon(x_i)$, implying that $([\text{sgn } f]v)(x_i) \leq 0$. But $h = \text{sgn } f$ on $N(f)$ and therefore $(hv)(x_i) \leq 0$. Applying Lemma 3.5, we obtain that $hv = 0$. Since $|h| = 1$ on all of K, we have $v = 0$. This proves the theorem. □

3. Best Approximation with Constraints: An Example

In the previous section we were able to determine criteria for when a finite-dimensional subspace is a unicity space for $C_1(K, \mu)$. For a closed convex set, this problem is much more difficult. In this section we consider an example of best approximating from a closed convex subset of a finite-dimensional subspace U of $C(K)$. The example is sufficiently elementary in that we are able to determine almost exactly when it is a unicity set for $C_1(K, \mu)$.

Let $U = \text{span}\{u_1, \ldots, u_n\}$ be an n-dimensional subspace of $C(K)$. For given fixed $\boldsymbol{\alpha} = (\alpha_1, \ldots, \alpha_n)$ and $\boldsymbol{\beta} = (\beta_1, \ldots, \beta_n)$ satisfying $-\infty \leq \alpha_i < \beta_i \leq \infty$, $i = 1, \ldots, n$, set

$$U(\boldsymbol{\alpha}; \boldsymbol{\beta}) = \Big\{ \sum_{i=1}^n a_i u_i : \alpha_i \leq a_i \leq \beta_i, \ i = 1, \ldots, n \Big\}.$$

It is an easy matter to characterize the best approximants from $U(\boldsymbol{\alpha}; \boldsymbol{\beta})$ to $f \in C(K)$ (and this characterization is valid without the continuity restrictions). From Exercise 8 of Chapter 2, we have:

Proposition 3.7. *Let* $f \in C(K) \backslash U(\boldsymbol{\alpha}; \boldsymbol{\beta})$. *Then* $u^* \in U(\boldsymbol{\alpha}; \boldsymbol{\beta})$ *is a best approximant to* f *from* $U(\boldsymbol{\alpha}; \boldsymbol{\beta})$ *if and only if*

$$(3.1) \qquad \int_K \text{sgn}(f - u^*)(u - u^*) d\mu \leq \int_{Z(f-u^*)} |u - u^*| d\mu$$

for all $u \in U(\boldsymbol{\alpha}; \boldsymbol{\beta})$.

A slightly different form of (3.1) is somewhat more elegant. For $u^* = \sum_{i=1}^n a_i^* u_i \in U(\boldsymbol{\alpha}; \boldsymbol{\beta})$, set

$$b_i^* = \begin{cases} 1, & a_i^* = \beta_i \\ 0, & \alpha_i < a_i^* < \beta_i \\ -1, & a_i^* = \alpha_i, \end{cases}$$

and

$$U(\mathbf{b}^*) = \Big\{ \sum_{i=1}^n c_i u_i : c_i b_i^* \leq 0, \ i = 1, \ldots, n \Big\}.$$

Corollary 3.8. *Let* $f \in C(K) \backslash U(\boldsymbol{\alpha}; \boldsymbol{\beta})$. *Then* $u^* = \sum_{i=1}^n a_i^* u_i \in U(\boldsymbol{\alpha}; \boldsymbol{\beta})$ *is a best approximant to* f *from* $U(\boldsymbol{\alpha}; \boldsymbol{\beta})$ *if and only if*

$$\int_K \text{sgn}(f - u^*) u \, d\mu \leq \int_{Z(f-u^*)} |u| d\mu$$

for all $u \in U(\mathbf{b}^*)$.

While the above characterization results are easily stated, there is yet the other characterization via linear functionals. This latter form provides us with the more precise information needed for the solution of the unicity problem.

Proposition 3.9. *Let* $f \in C(K) \backslash U(\alpha ; \beta)$. *Then* $u^* = \sum_{i=1}^{n} a_i^* u_i \in U(\alpha ; \beta)$ *is a best approximant to* f *from* $U(\alpha ; \beta)$ *if and only if there exists an* $h \in L^\infty(K, \mu)$ *satisfying*

$$1) \; |h(x)| = 1 \text{ on } K$$

$$2) \int_K h u_i d\mu \begin{cases} \geq 0, & \text{if } b_i^* = 1 \\ = 0, & \text{if } b_i^* = 0 \\ \leq 0, & \text{if } b_i^* = -1 \end{cases}$$

$$3) \int_K h(f - u^*) d\mu = \|f - u^*\|_1 .$$

Furthermore, if $\tilde{u} = \sum_{i=1}^{n} \tilde{a}_i u_i$ *is any other best approximant to* f *from* $U(\alpha ; \beta)$, *then*

a) $\int_K h(f - \tilde{u}) d\mu = \|f - \tilde{u}\|_1$

b) *if* $\int_K h u_j d\mu \neq 0$ *for* h *as above and some* $j \in \{1, \ldots, n\}$, *then*

$$\tilde{a}_j = a_j^* = \begin{cases} \beta_j, & \text{if } b_j^* = 1 \\ \alpha_j, & \text{if } b_j^* = -1 . \end{cases}$$

Proof. The characterization of the best approximant in terms of (1), (2) and (3) of the proposition is a direct consequence of Exercise 9 of Chapter 2.

Assume that $\tilde{u} = \sum_{i=1}^{n} \tilde{a}_i u_i$ is also a best approximant to f from $U(\alpha ; \beta)$. From (1), (2) and (3),

$$\|f - u^*\|_1 = \int_K h(f - u^*) d\mu \leq \int_K h(f - \tilde{u}) d\mu \leq \|f - \tilde{u}\|_1 .$$

Since $\tilde{u}$ is also a best approximant, equality holds at each step. Therefore (a) follows and also

$$\int_K h(u^* - \tilde{u}) d\mu = 0 .$$

That is

$$\sum_{i=1}^{n} (a_i^* - \tilde{a}_i) \int_K h u_i d\mu = 0 .$$

From (2), we have that

$$(a_i^* - \tilde{a}_i) \int_K h u_i d\mu \geq 0$$

for each $\alpha_i \leq a_i \leq \beta_i$ and $i \in \{1, \ldots, n\}$. Thus

$$(a_i^* - \tilde{a}_i) \int_K h u_i d\mu = 0, \qquad i = 1, \ldots, n .$$

If $\int_K h u_j d\mu \neq 0$, it follows that $\tilde{a}_j = a_j^*$. From (2), result (b) now follows.   □

This result has an important consequence. Set

$$I = \left\{ i \; : \; \int_K h u_i d\mu = 0 \right\}$$

and $J = \{1, \ldots, n\} \backslash I$. From Proposition 3.9, for every best approximant to f from $U(\boldsymbol{\alpha}\,;\boldsymbol{\beta})$ of the form $\sum_{i=1}^n \tilde{a}_i u_i$, we have $\tilde{a}_j = a_j^*$ for all $j \in J$. Let

$$\overline{f} = f - \sum_{j \in J} a_j^* u_j \,.$$

Then $\tilde{u} = \sum_{i=1}^n \tilde{a}_i u_i$ is a best approximant to f from $U(\boldsymbol{\alpha}\,;\boldsymbol{\beta})$ if and only if $\sum_{i \in I} \tilde{a}_i u_i$ is a best approximant to $\overline{f}$ from $U_I(\boldsymbol{\alpha};\boldsymbol{\beta})$, where

$$U_I(\boldsymbol{\alpha};\boldsymbol{\beta}) = \left\{ \sum_{i \in I} a_i u_i \; : \; \alpha_i \le a_i \le \beta_i, \; i \in I \right\}.$$

Set

$$U_I = \operatorname{span}\{u_i \; : \; i \in I\} \,.$$

For the h of Proposition 3.9, we have

1) $|h(x)| = 1$ on K

2) $\displaystyle\int_K h u_i d\mu = 0$, all $i \in I$

3) $\displaystyle\int_K h(\overline{f} - \sum_{i \in I} a_i^* u_i) d\mu = \|\overline{f} - \sum_{i \in I} a_i^* u_i\|_1$

4) $\displaystyle\int_K h(\overline{f} - \sum_{i \in I} \tilde{a}_i u_i) d\mu = \|\overline{f} - \sum_{i \in I} \tilde{a}_i u_i\|_1 \,.$

This implies (see Theorem 2.3) that both $\sum_{i \in I} a_i^* u_i$ and $\sum_{i \in I} \tilde{a}_i u_i$ are best approximants to $\overline{f}$ from U_I without any constraints on the coefficients of u_i, $i \in I$. We have therefore proven:

Theorem 3.10. *Let* $N = \{i \; : \; \alpha_i = -\infty, \; \beta_i = \infty\}$. *If* $\operatorname{span}\{u_{i_1}, \ldots, u_{i_k}\}$ *is a unicity space for* $C_1(K, \mu)$ *for every choice of distinct* $\{i_1, \ldots, i_k\}$ *satisfying* $N \subseteq \{i_1, \ldots, i_k\} \subseteq \{1, \ldots, n\}$, *then* $U(\boldsymbol{\alpha}\,;\boldsymbol{\beta})$ *is a unicity set for* $C_1(K, \mu)$.

It is not clear that the converse result is valid in this generality. It is however true with a minor additional assumption.

Theorem 3.11. *Let* $N = \{i \; : \; \alpha_i = -\infty, \; \beta_i = \infty\}$. *Assume that, for all* $i \notin N$, *we have* $-\infty < \alpha_i < \beta_i < \infty$. *If* $U(\boldsymbol{\alpha}\,;\boldsymbol{\beta})$ *is a unicity set for* $C_1(K, \mu)$, *then* $\operatorname{span}\{u_{i_1}, \ldots, u_{i_k}\}$ *is a unicity space for* $C_1(K, \mu)$ *for every choice of distinct indices* $\{i_1, \ldots, i_k\}$ *satisfying* $N \subseteq \{i_1, \ldots, i_k\} \subseteq \{1, \ldots, n\}$.

Proof. Let $N \subseteq \{i_1, \ldots, i_k\} \subseteq \{1, \ldots, n\}$. Assume that the linear space $U_k = \operatorname{span}\{u_{i_1}, \ldots, u_{i_k}\}$ is not a unicity space for $C_1(K, \mu)$. By definition there exists an $f \in C(K)$ and $u^1, u^2 \in U_k$, $u^1 \ne u^2$, such that both u^1 and

u^2 are best approximants to f from U_k. Since the set of best approximants is convex and $u^r + u$, $r = 1, 2$, are best approximants to $f + u$ from U_k for every $u \in U_k$, we may assume that

$$u^1 = \sum_{j=1}^{k} a_{i_j}^1 u_{i_j}, \qquad u^2 = \sum_{j=1}^{k} a_{i_j}^2 u_{i_j}$$

where

$$\alpha_{i_j} < a_{i_j}^r < \beta_{i_j}, \qquad r = 1, 2; \ j = 1, \ldots, k.$$

From Theorem 2.3 it follows that there exists an $h \in L^\infty(K, \mu)$ for which

1) $|h(x)| = 1$ on K

2) $\displaystyle\int_K h u_{i_j} d\mu = 0, \quad j = 1, \ldots, k$

3) $\displaystyle\int_K h(f - u^r) d\mu = \|f - u^r\|_1, \quad r = 1, 2.$

Set $Q = \{1, \ldots, n\} \backslash \{i_1, \ldots, i_k\}$. For each $i \in Q$ we have $-\infty < \alpha_i < \beta_i < \infty$ by assumption. Let $i \in Q$. If $\int_K h u_i d\mu = 0$, let $\gamma_i \in [\alpha_i, \beta_i]$. If $\int_K h u_i d\mu > 0$, set $\gamma_i = \beta_i$, while if $\int_K h u_i d\mu < 0$, set $\gamma_i = \alpha_i$. Define

$$\widehat{u} = \sum_{i \in Q} \gamma_i u_i$$

and $\widehat{f} = f + \widehat{u}$, $\widehat{u}^r = u^r + \widehat{u}$, $r = 1, 2$. It now follows from Proposition 3.9 that $\widehat{u}^1$ and $\widehat{u}^2$ are both best approximants to $\widehat{f}$ from $U(\boldsymbol{\alpha}; \boldsymbol{\beta})$. This proves the theorem.   □

4. k-Convexity

The method of proof of Theorem 3.1 also provides us with a characterization of the largest possible 'dimension' of $P_U(f)$ as f varies over $C(K)$, i.e., the size of the set of best approximants. To explain what is meant, let X be a normed linear space and U a finite-dimensional subspace of X. Recall that, for each $f \in X$, $P_U(f)$ is a convex, bounded, closed (and hence compact) non-empty subset of U (see Proposition 1.13, and Exercise 11 of Chapter 1).

Definition 3.2. $P_U(f)$ is said to be of *dimension* k, denoted $\dim P_U(f) = k$, if there exist $k + 1$ elements $u_0, u_1, \ldots, u_k$ in $P_U(f)$ for which $\{u_i - u_0\}_{i=1}^{k}$ are linearly independent, and k is maximal. If $P_U(f)$ is a singleton, then we set $k = 0$.

Definition 3.3. We say that U is *k-convex*, $0 \le k \le \dim U$, if $\dim P_U(f) \le k$ for all $f \in X$, and there exists an $f \in X$ for which $\dim P_U(f) = k$.

Note that in the above terminology U is a unicity space for X if and only if U is 0-convex. From Theorem 2.9 we explicitly have that for $X =$

$L^1(B, \nu)$, ν a non-atomic positive measure, every n-dimensional subspace U is n-convex. This is the worst possible case. We now state and prove the theorem characterizing k-convex subspaces.

Theorem 3.12. *Let U be a finite-dimensional subspace of $C(K)$. Then U is k-convex if and only if for each $h \in L^\infty(K, \mu)$ satisfying*

$$1) \ |h| = 1 \text{ on } K$$

$$2) \ \int_K hu \, d\mu = 0, \text{ all } u \in U,$$

the subspace

$$W_h = \{u : u \in U, hu \in C(K)\}$$

satisfies dim $W_h \leq k$, *and there exists an h satisfying* (1) *and* (2) *for which* dim $W_h = k$.

Proof. From Theorem 3.1, it suffices to consider only the case $k \geq 1$.

($\Rightarrow$). Assume U is k-convex. Let $f \in C(K)$ satisfy dim $P_U(f) = k$, and let $u_0, u_1, \ldots, u_k \in P_U(f)$ be such that $\{u_i - u_0\}_{i=1}^k$ are linearly independent. Set

$$f^* = f - \left(\sum_{i=0}^k u_i\right)/(k+1),$$

and $v_i = (u_i - u_0)/(k+1)$, $i = 1, \ldots, k$. Since $P_U(f)$ is convex, it is easily verified that the functions $0, \pm v_1, \ldots, \pm v_k$ are all in $P_U(f^*)$. Now, for each $x \in K$, observe that

$$2k|f^*(x)| = \sum_{i=1}^k \left[|(f^* - v_i)(x)| + |(f^* + v_i)(x)|\right].$$

Otherwise we contradict the fact that $0, \pm v_1, \ldots, \pm v_k \in P_U(f^*)$. Thus $Z(f^*) \subseteq Z(v_i)$, $i = 1, \ldots, k$. From the proof of Theorem 3.1, it follows that there exists an $h \in L^\infty(K, \mu)$ for which

$$1) \ |h| = 1 \text{ on } K$$

$$2) \ \int_K hu \, d\mu = 0, \text{ all } u \in U,$$

and $h|v_i| \in C(K)$, $i = 1, \ldots, k$. Thus $hv_i \in C(K)$, $i = 1, \ldots, k$, and dim $W_h \geq k$. (The $v_1, \ldots, v_k$ are linearly independent by assumption.)

($\Leftarrow$). Assume there exists an $h \in L^\infty(K, \mu)$ satisfying (1) and (2), and such that dim $W_h = k$. Let $W_h = $ span $\{u_1, \ldots, u_k\}$. Set $f = h(\sum_{i=1}^k |u_i|)$. Then $f \in C(K)$. From the analysis of Theorem 3.1 (see also the proof of Theorem 2.7) it follows that $0, \pm u_1, \ldots, \pm u_k \in P_U(f)$. Thus dim $P_U(f) \geq k$, and U is m-convex for some $m \geq k$. This proves the theorem. □

5. Continuous Selections

As defined in Section 4 of Chapter 2, a continuous selection from $C_1(K,\mu)$ onto U is any *single-valued* and *continuous* map s for which $s(f) \in P_U(f)$ for every $f \in C(K)$. Furthermore, as a consequence of Corollary 1.19 we have:

Proposition 3.13. *If U is a finite-dimensional unicity space for $C_1(K,\mu)$, then the best approximation operator is a continuous selection onto U.*

Unlike the situation in $L^1(K,\mu)$ (Theorem 2.14), there do exist finite-dimensional subspaces of $C_1(K,\mu)$ which possess continuous selections, simply because there exist unicity spaces for $C_1(K,\mu)$. We shall prove that a finite-dimensional subspace U of $C_1(K,\mu)$ has a continuous selection if and only if U is a unicity space for $C_1(K,\mu)$. A reading of the proof of Theorem 2.14 on the non-existence of continuous selections for finite-dimensional subspaces of $L^1(K,\mu)$ will show that the proof given therein cannot be applied here. The demands of continuity are too stringent.

We shall actually prove even more than is stated above. We shall provide an example of a rather curious phenomenon which can occur. To explain, we first present the following definition.

Definition 3.4. The metric selection s from $C_1(K,\mu)$ onto U is said to be an L^1-*continuous selection* if s is continuous with respect to L^1-convergence. That is, for any $f, f_n \in C(K)$ satisfying $\lim_{n\to\infty} \|f - f_n\|_1 = 0$, we have $\lim_{n\to\infty} \|s(f) - s(f_n)\|_1 = 0$. The metric selection s from $C_1(K,\mu)$ onto U is said to be an L^∞-*continuous selection* if s is continuous with respect to L^∞-convergence. That is, for $f, f_n \in C(K)$ satisfying $\lim_{n\to\infty} \|f - f_n\|_\infty = 0$, we have $\lim_{n\to\infty} \|s(f) - s(f_n)\|_1 = 0$.

An L^1-continuous selection is what we previously referred to simply as a continuous selection. In the above definition, $\|\cdot\|_\infty$ represents the usual uniform norm on K. Since K is, by assumption, of finite measure, i.e., $\mu(K) < \infty$, if s is an L^1-continuous selection, then it is also an L^∞-continuous selection. The converse, however, is not valid. We shall shortly provide an example of a U with an L^∞-continuous selection, but no L^1-continuous selection. Finally, we shall prove that if U is not a unicity space for $C_1(K,\mu)$ then there is no L^1-continuous selection (see Exercise 11), and if K is connected, then there exists no L^∞-continuous selection.

In a certain sense, the existence or non-existence of an L^∞-continuous selection on $C_1(K,\mu)$ is at least as interesting (and important) as the existence or non-existence of an L^1-continuous selection. After all, we are dealing with continuous functions.

One more remark is in order before starting the analysis. Since U is finite-dimensional, all norms on U are equivalent. Thus for notational ease, and to emphasize this point, rather than writing $\lim_{n\to\infty} \|s(f) - s(f_n)\|_1 = 0$ we shall simply write $\lim_{n\to\infty} s(f_n) = s(f)$.

Example. *A space with an L^∞-continuous selection, but no L^1-continuous selection.*

Let $K = [-2, -1] \cup [1, 2]$, and μ denote the usual Lebesgue measure on K. Let U be the one-dimensional space spanned by the constant functions. Set

$$h(x) = \begin{cases} 1, & x \in [-2, -1] \\ -1, & x \in [1, 2]. \end{cases}$$

Since $|h| = 1$ on K, $h \in C(K)$, and $\int_K h\, dx = 0$, it follows from Theorem 3.1 that U is not a unicity space for $C_1(K, \mu)$.

Proposition 3.14. *There is no L^1-continuous selection on U.*

Proof. Let h be as above, and set

$$f_n(x) = \begin{cases} 1, & x \in [-2, -1 - 1/n] \\ -2nx - (1 + 2n), & x \in [-1 - 1/n, -1] \\ -1, & x \in [1, 2], \end{cases}$$

and

$$g_n(x) = \begin{cases} 1, & x \in [-2, -1] \\ -2nx + (1 + 2n), & x \in [1, 1 + 1/n] \\ -1, & x \in [1 + 1/n, 2], \end{cases}$$

for all $n \geq 1$. Obviously $h, f_n, g_n \in C(K)$ and

$$\lim_{n\to\infty} \|h - f_n\|_1 = \lim_{n\to\infty} \|h - g_n\|_1 = 0.$$

It is easily seen (or obtained as a consequence of Theorem 2.1 and Proposition 2.4) that $P_U(f_n) = \{-1\}$ and $P_U(g_n) = \{1\}$ for all $n \geq 1$. This proves the proposition. □

To complete the example, we prove:

Proposition 3.15. *There are L^∞-continuous selections on U.*

Proof. For any measurable subset A of K, let $|A|$ denote its Lebesgue measure. Let $f \in C(K)$. From Theorem 2.1, the constant function $c \in P_U(f)$ if and only if

$$\left| \int_K \operatorname{sgn}(f - c) d\mu \right| \leq \int_{Z(f-c)} d\mu.$$

Set $|\{x : f(x) > c\}| = \alpha$ and $|\{x : f(x) < c\}| = \beta$. Then from the above, $c \in P_U(f)$ if and only if $|\alpha - \beta| \leq 2 - \alpha - \beta$, or equivalently, if and only if $\alpha, \beta \leq 1$. (Note that $\alpha, \beta \geq 0$, $\alpha + \beta \leq 2$.)

For each $f \in C(K)$, set

$$C(f) = \min_{-2 \leq x \leq -1} f(x), \quad c(f) = \max_{1 \leq x \leq 2} f(x),$$

and

$$d(f) = \max_{-2 \le x \le -1} f(x), \quad D(f) = \min_{1 \le x \le 2} f(x).$$

Then $P_U(f)$ is not a singleton if and only if either

a) $C(f) > c(f)$, or

b) $D(f) > d(f)$.

Also, if (a) holds, then $P_U(f) = [c(f), C(f)]$ while, if (b) holds, $P_U(f) = [d(f), D(f)]$.

For any $\lambda, \mu \in [0, 1]$, fixed, define s as follows:

1) If $P_U(f)$ is a singleton, set $s(f) = P_U(f)$.

2) If $C(f) > c(f)$, set $s(f) = \lambda C(f) + (1 - \lambda)c(f)$.

3) If $D(f) > d(f)$, set $s(f) = \mu D(f) + (1 - \mu)d(f)$.

We claim that s is an L^∞-continuous selection. Let $f, f_n \in C(K)$ and assume $\lim_{n \to \infty} \|f - f_n\|_\infty = 0$. If f satisfies (1) then $P_U(f)$ is a singleton, and from Proposition 1.18, $\lim_{n \to \infty} s(f_n) = s(f)$. If f satisfies (2), then for n sufficiently large the f_n satisfy (2), and $\lim_{n \to \infty} C(f_n) = C(f)$, $\lim_{n \to \infty} c(f_n) = c(f)$. Thus $\lim_{n \to \infty} s(f_n) = s(f)$. If (3) holds we apply the analogous reasoning.   □

We now formally state the theorem to be proved in the next pages.

Theorem 3.16. *Let U be a finite-dimensional subspace of $C(K)$. There exists an L^1-continuous selection onto U if and only if U is a unicity space for $C_1(K, \mu)$. Furthermore, if K is connected and U is not a unicity space, then there exists no L^∞-continuous selection.*

We have already noted (Proposition 3.13) that if U is a unicity space then there is a (unique) L^1-continuous selection. It remains to prove the remaining claims of the theorem. As such we shall now assume that U is *not* a unicity space.

The proof of Theorem 3.16 is rather lengthy and arduous. As such we shall first prove the theorem for the simpler case where dim $U = 1$. In the proof of this case is contained some of the main ideas of the general proof. The more secure reader may wish to skip it entirely, as the proof is not necessary for the general case.

Proof. (dim $U = 1$). Let $U = \text{span}\{u\}$, and assume, for the present, that K is connected. Since U is not a unicity space for $C_1(K, \mu)$, we have from Theorem 3.1 an $h \in L^\infty(K, \mu)$ satisfying $|h| = 1$ on K, $\int_K hu \, d\mu = 0$, and $h|u| \in C(K)$.

Set $f = h|u|$. The idea of the proof is to construct two sequences $\{g_\varepsilon\}$ and $\{k_\varepsilon\}$ in $C_1(K, \mu)$ with the following properties:

1) $\lim_{\varepsilon \to 0^+} \|f - g_\varepsilon\|_\infty = \lim_{\varepsilon \to 0^+} \|f - k_\varepsilon\|_\infty = 0$

2) $P_U(g_\varepsilon) = \{-u\}$, all $\varepsilon > 0$ sufficiently small

3) $P_U(k_\varepsilon) = \{u\}$, all $\varepsilon > 0$ sufficiently small.

Properties (1)-(3) imply that there is no L^∞-continuous selection on $C_1(K, \mu)$.

We shall now construct the $\{g_\varepsilon\}$. The $\{k_\varepsilon\}$ are constructed in a totally similar way. We begin by setting

$$A = \{x : (hu)(x) > 0\}$$

(or equivalently $A = \{x : (fu)(x) > 0\}$). Since $\int_K hu \, d\mu = 0$, A is an open, non-empty, proper subset of K. Let $\alpha \in \partial A \cap Z(u)$. Such an α exists since K is connected. For small $\varepsilon > 0$, set

$$A_\varepsilon = A \cap \{x : |x - \alpha| < \varepsilon\}.$$

A_ε is open and non-empty, and thus $\mu(A_\varepsilon) > 0$ for all $\varepsilon > 0$. We define

$$g_\varepsilon(x) = \begin{cases} f(x), & x \notin A_{2\varepsilon} \\ \sigma_\varepsilon(x), & x \in A_{2\varepsilon} \backslash \overline{A}_\varepsilon \\ -u(x), & x \in \overline{A}_\varepsilon, \end{cases}$$

where $|\sigma_\varepsilon| < |u|$ on $A_{2\varepsilon} \backslash \overline{A}_\varepsilon$, and $g_\varepsilon \in C(K)$. It is easily seen that such a construction is possible.

Now, for $\varepsilon > 0$,

$$\|f - g_\varepsilon\|_\infty \leq 2 \max_{x \in \overline{A}_{2\varepsilon}} |u(x)|$$

by construction. Since $|x - \alpha| \leq 2\varepsilon$ for all $x \in \overline{A}_{2\varepsilon}$, and $u(\alpha) = 0$, it follows that $\lim_{\varepsilon \to 0^+} \|f - g_\varepsilon\|_\infty = 0$. It remains to prove that $P_U(g_\varepsilon) = \{-u\}$.

We first consider $\mathrm{sgn}(g_\varepsilon + u)$.

(i) On $\overline{A}_\varepsilon$

$$\mathrm{sgn}(g_\varepsilon + u) = \mathrm{sgn}(-u + u) = 0.$$

(ii) On $A_{2\varepsilon} \backslash \overline{A}_\varepsilon$,

$$\mathrm{sgn}(g_\varepsilon + u) = \mathrm{sgn}(\sigma_\varepsilon + u) = \mathrm{sgn}\, u = \mathrm{sgn}\, f.$$

(iii) Off $A_{2\varepsilon}$,

$$\mathrm{sgn}(g_\varepsilon + u) = \mathrm{sgn}(f + u) = \begin{cases} 0, & f = -u \\ \mathrm{sgn}\, f, & f = u \neq 0. \end{cases}$$

Thus

$$\int_{Z(g_\varepsilon + u)} |u| d\mu = \int_{f = -u} |u| d\mu + \int_{\overline{A}_\varepsilon} |u| d\mu$$

and

$$\int_K \mathrm{sgn}(g_\varepsilon + u)u\,d\mu = \int_{\substack{f=u\neq 0 \\ x\notin A_{2\varepsilon}}} (\mathrm{sgn}\,f)u\,d\mu + \int_{A_{2\varepsilon}\setminus\overline{A}_\varepsilon} (\mathrm{sgn}\,f)u\,d\mu$$

$$= \int_A (\mathrm{sgn}\,f)u\,d\mu - \int_{\overline{A}_\varepsilon} (\mathrm{sgn}\,f)u\,d\mu = \int_A hu\,d\mu - \int_{\overline{A}_\varepsilon} hu\,d\mu.$$

Since $\int_K hu\,d\mu = 0$, we have

$$\int_K \mathrm{sgn}(g_\varepsilon + u)u\,d\mu = -\int_{f=-u} hu\,d\mu - \int_{\overline{A}_\varepsilon} hu\,d\mu = \int_{f=-u} |u|d\mu - \int_{\overline{A}_\varepsilon} |u|d\mu.$$

Therefore

$$\left| \int_K \mathrm{sgn}(g_\varepsilon + u)u\,d\mu \right| < \int_{Z(g_\varepsilon+u)} |u|d\mu.$$

From Corollary 2.5, we deduce that $P_U(g_\varepsilon) = \{-u\}$.

With a parallel analysis on the set $B = \{x : (fu)(x) < 0\}$, we construct the requisite $\{k_\varepsilon\}$ satisfying (2). This proves the non-existence of an L^∞-continuous selection under the hypothesis that K is connected.

If K is not connected then, as illustrated by our example, we can make no such claim. This is a consequence of the fact that in the above proof $\partial A \cap Z(u)$ may be empty. None the less we claim that there cannot exist an L^1-continuous selection on U. To show this we simply take α to be any point in A and apply the above proof. The only property lost in the above analysis is the convergence in the uniform norm of g_ε to f. However it is readily verified that $\lim_{\varepsilon\to 0^+} \|f - g_\varepsilon\|_1 = 0$. This proves Theorem 3.16 for dim $U = 1$. □

Because of its length, the proof of Theorem 3.16 will be divided into a series of lemmas.

We shall assume, for the present, that K is connected, and shall prove that there is no L^∞-continuous selection. A simple modification (as previously) will complete the proof in the case where K is not connected.

Since U is not a unicity space there exists, from Theorem 3.1, an $h \in L^\infty(K,\mu)$ and $u^* \in U$, $u^* \neq 0$, for which $|h| = 1$ on K, $\int_K hu\,d\mu = 0$ for all $u \in U$, and $h|u^*| \in C(K)$, i.e., $hu^* \in C(K)$. Many such h may exist. We choose one and fix it throughout.

Let

$$W = \{u : u \in U, hu \in C(K)\}.$$

W is a subspace of U. Set

$$W_1 = \{u : u \in W, \|u\|_1 = 1\}.$$

For each $u \in W$, let

$$J(u) = \{x : (hu)(x) \leq 0\}.$$

For ease of exposition, we write $|J(u)| = \mu(J(u))$. Note that $J(u) = J(cu)$ for all $c > 0$. Now, W_1 is a compact set. Furthermore it is easily seen that $|J(u)|$ is upper semi-continuous. Thus there exists a $\tilde{u} \in W_1$ satisfying $|J(\tilde{u})| \geq |J(u)|$ for all $u \in W_1$, and therefore $|J(\tilde{u})| \geq |J(u)|$ for all $u \in W$, $u \neq 0$. Since $\int_K hu \, d\mu = 0$ for all $u \in W$, we also have $0 < |J(\tilde{u})| < |K| (= \mu(K))$. We have proved:

Lemma 3.17. *There exists a $\tilde{u} \in W_1$ satisfying $|J(\tilde{u})| \geq |J(u)|$ for all $u \in W$, $u \neq 0$.*

We shall now choose a $w^* \in W_1$ satisfying $|J(w^*)| = |J(\tilde{u})|$ with a minimal zero set (in a certain sense). To this end, for each $w \in W_1$ set

$$V_w = \{v : v \in W, J(w) \subseteq J(v)\}.$$

Lemma 3.18. *There exists a $w^* \in W_1$ satisfying the following.*
 1) $|J(w^*)| \geq |J(u)|$ *for all $u \in W$, $u \neq 0$.*
 2) $Z(w^*) \subseteq Z(v)$ *for all $v \in V_{w^*}$.*
 3) *For all $v \in V_{w^*}, v \neq 0$, we have $hv > 0$, μ a.e. on $K \backslash J(w^*)$.*

Proof. Note that (3) is an immediate consequence of (1). If $w \in W_1$ satisfies (1), and $v \in V_w$, $v \neq 0$, has the property that $hv \leq 0$ on some subset of $K \backslash J(w)$ of positive μ-measure, then since $J(w) \subseteq J(v)$ it follows that $|J(v)| > |J(w)|$, contradicting (1). Thus the crux of the lemma is in statement (2).

To prove (2), let $w \in W_1$ satisfy (1). For each $v \in V_w, v \neq 0$, set

$$B_v = \{x : (hv)(x) < 0\}.$$

Since (3) holds for all $v \in V_w, v \neq 0$, we have $B_v \subseteq J(w)$, $J(w) \backslash B_v \subseteq Z(v)$, and $B_v \neq \emptyset$. For each $v \in V_w, v \neq 0$, it follows by definition that $V_v \subseteq V_w$. It thus suffices to prove the existence of a $w^* \in V_w, w^* \neq 0$, satisfying $B_v \subseteq B_{w^*}$ for all $v \in V_w$. Indeed we then have

$$Z(w^*) = J(w^*) \backslash B_{w^*} \subseteq J(w^*) \backslash B_v \subseteq Z(v)$$

for all $v \in V_{w^*}, v \neq 0$.

A proof of the existence of the desired w^* runs as follows. Assume $v_1, \ldots, v_k \in V_w$, $v_i \neq 0$, $i = 1, \ldots, k$ and $B_{v_i} \subset B_{v_{i+1}}$, $B_{v_i} \neq B_{v_{i+1}}$, $i = 1, \ldots, k-1$. We claim that the $v_1, \ldots, v_k$ are linearly independent. To see this, choose $x_i \in B_{v_i} \backslash B_{v_{i-1}}$, $i = 1, \ldots, k$, (where $B_{v_0} = \emptyset$). Now, $v_i(x_i) \neq 0$ by definition, and $v_j(x_i) = 0$ for all $j < i$ since $x_i \in B_{v_i} \backslash B_{v_{i-1}} \subseteq B_{v_i} \backslash B_{v_j} \subseteq J(w) \backslash B_{v_j} \subseteq Z(v_j)$. The matrix $(v_j(x_i))_{i,j=1}^k$ is a lower triangular matrix with non-zero diagonal entries. As such it is of rank k. Thus the $v_1, \ldots, v_k$ are linearly independent.

Now, choose $v_1 \in V_w$, $v_1 \neq 0$. If there exists a $v_2 \in V_w, v_2 \neq 0$, for which $B_{v_1} \subset B_{v_2}$, then replace v_1 by v_2. Continue this process. Since dim $V_w \leq$ dim $U < \infty$, we must stop after a finite number of steps. Thus there exists a $w^* \in V_w$, $w^* \neq 0$, with the property that if $v \in V_w$, $v \neq 0$, and $B_{w^*} \subseteq B_v$, then $B_{w^*} = B_v$.

We claim that $B_v \subseteq B_{w^*}$ for all $v \in V_w$. If this is not the case for some $v \in V_w$, then there exists a $y \in K$ such that $(hv)(y) < 0 = (hw^*)(y)$. Set $\overline{v} = v + w^*$. As is easily seen, $\overline{v} \in V_w$, and $B_{w^*} \subset B_{\overline{v}}$, $B_{w^*} \neq B_{\overline{v}}$. This is a contradiction to our choice of w^*. Lemma 3.18 is proved.   □

The following result will be used repeatedly.

Lemma 3.19. *Let $w \in W$, $w \neq 0$. If $u \in U$ and $|w| \geq hu$ on K, then $J(w) \subseteq J(w + u)$ and $w + u \in V_w$.*

Proof. We first prove that $u \in W$. Recall that $u \in W$ if and only if u vanishes at each point of discontinuity of h. Let x be such a point. Since $w \in W$, $0 = |w(x)| \geq (hu)(x)$. Assume $u(x) = c \neq 0$. Since h is discontinuous at x, in any neighborhood of x there exists a point y for which $(hu)(y) \geq |c|/2$. But $w(x) = 0$, w is continuous at x, and $|w| \geq hu$ on K. This is a contradiction. Thus $u \in W$.

Now let $x \in J(w)$. Then $(hw)(x) \leq 0$, and therefore $-(hw)(x) = |w(x)| \geq (hu)(x)$. Thus $(h(w + u))(x) \leq 0$ and so $J(w) \subseteq J(w + u)$. W is a subspace and $w, u \in W$. Thus $w + u \in W$. By definition, $w + u \in V_w$.   □

With these preliminaries, we can now begin to set up the main ingredients used in the proof of Theorem 3.16. For ease of notation, set $V = V_{w^*}$, and $V_1 = V \cap W_1$. Let

$$A = \{x : (hw^*)(x) > 0\}$$

and

$$B = \{x : (hw^*)(x) < 0\},$$

i.e., $B = B_{w^*}$.

By assumption K is connected. Thus there exists an $\alpha \in \partial A \cap Z(w^*)$. For each $\varepsilon > 0$, ε sufficiently small, let

$$A_\varepsilon = A \cap \{x : |x - \alpha| < \varepsilon\}.$$

Then $|A_\varepsilon| (= \mu(A_\varepsilon)) > 0$ for all $\varepsilon > 0$. Similarly there exists a $\beta \in \partial B \cap Z(w^*)$. Set

$$B_\varepsilon = B \cap \{x : |x - \beta| < \varepsilon\}.$$

Then $|B_\varepsilon| > 0$ for all $\varepsilon > 0$.

From Lemma 3.18, $hv > 0$ μ a.e. on A and $Z(w^*) \subseteq Z(v)$, for all $v \in V$. Since $\alpha, \beta \in Z(w^*)$, we have $v(\alpha) = v(\beta) = 0$ for all $v \in V$.

To each $\varepsilon > 0$, ε small, choose $v_\varepsilon \in V_1$ satisfying

$$\int_{B_\varepsilon} |v_\varepsilon| d\mu \geq \int_{B_\varepsilon} |v| d\mu$$

for all $v \in V_1$. V_1 is compact so that such a v_ϵ exists.

We have set up two extremal problems. The first led us to our choice of w^*. The second is the maximization, for $v \in V_1$, of the integral over B_ϵ. The first extremal problem is used in the following result.

Lemma 3.20. *Assume* $u \in U$ *satisfies* $|v_\epsilon| \geq hu$ *on* K, *and* $-|v_\epsilon| \geq hu$ *on* A_ϵ. *Then* $u = -v_\epsilon$.

Proof. Assume $u \neq -v_\epsilon$. From Lemma 3.19, $J(v_\epsilon) \subseteq J(v_\epsilon + u)$ and $v_\epsilon + u \in V_{v_\epsilon}$, $v_\epsilon + u \neq 0$. Since $v_\epsilon \in V$, $hv_\epsilon > 0$ μ a.e. on A. Thus, on A_ϵ, $-hv_\epsilon = -|v_\epsilon| \geq hu$. Therefore $h(v_\epsilon + u) \leq 0$ on A_ϵ. But $v_\epsilon + u \in V_{v_\epsilon} \subseteq V$, $v_\epsilon + u \neq 0$, and thus $h(v_\epsilon + u) > 0$ μ a.e. on A (and on A_ϵ). This is a contradiction. $\quad\square$

Using the second extremal problem we obtain this next result.

Lemma 3.21. *Assume* $u \in U$ *satisfies* $|v_\epsilon| \geq hu$ *on* K, *and* $-|v_\epsilon| \geq hu$ *on* B_ϵ. *Then* $\|v_\epsilon + u\|_1 = 2$.

Proof. On B_ϵ, $hu \leq -|v_\epsilon| = hv_\epsilon \leq 0$. If $u = -v_\epsilon$, then $v_\epsilon = 0$ on B_ϵ. But $\int_{B_\epsilon} |v_\epsilon| d\mu \geq \int_{B_\epsilon} |w^*| d\mu > 0$. Thus $u \neq -v_\epsilon$.

Let $z = (v_\epsilon + u)/\|v_\epsilon + u\|_1$. It follows from Lemma 3.19, since $|v_\epsilon| \geq hu$ on K and $V_{v_\epsilon} \subseteq V$, that $z \in V_1$. Now,

$$\|v_\epsilon + u\|_1 = \int_K [\mathrm{sgn}(v_\epsilon + u)](v_\epsilon + u)d\mu = \int_K [\mathrm{sgn}\, h(v_\epsilon + u)]\, h(v_\epsilon + u)d\mu$$
$$= \int_{h(v_\epsilon+u)>0} h(v_\epsilon + u)d\mu - \int_{h(v_\epsilon+u)\leq 0} h(v_\epsilon + u)d\mu.$$

We have $h(v_\epsilon + u) > 0$ μ a.e. on $hv_\epsilon > 0$, and $h(v_\epsilon + u) \leq 0$ μ a.e. on $hv_\epsilon \leq 0$ since $v_\epsilon + u \in V_{v_\epsilon}$, $v_\epsilon + u \neq 0$, and $|J(v_\epsilon)|$ is maximal. Thus

$$\|v_\epsilon + u\|_1 = \int_{hv_\epsilon>0} h(v_\epsilon + u)d\mu - \int_{hv_\epsilon\leq 0} h(v_\epsilon + u)d\mu$$
$$= \int_{hv_\epsilon>0} hv_\epsilon d\mu - \int_{hv_\epsilon\leq 0} hv_\epsilon d\mu$$
$$+ \int_{hv_\epsilon>0} hu\, d\mu - \int_{hv_\epsilon\leq 0} hu\, d\mu.$$

Now,

$$\int_{hv_\epsilon>0} hv_\epsilon d\mu - \int_{hv_\epsilon\leq 0} hv_\epsilon d\mu = \|hv_\epsilon\|_1 = \|v_\epsilon\|_1 = 1,$$

while

$$\int_{hv_\epsilon>0} hu\, d\mu + \int_{hv_\epsilon\leq 0} hu\, d\mu = 0,$$

for all $u \in U$. On the set of points satisfying $hv_\epsilon > 0$, we have by assumption $hv_\epsilon = |v_\epsilon| \geq hu$. Thus

$$\|v_\epsilon + u\|_1 = 1 + 2\int_{hv_\epsilon>0} hu\, d\mu \leq 1 + 2\int_{hv_\epsilon>0} hv_\epsilon d\mu = 1 + \|v_\epsilon\|_1 = 2.$$

On B_ε, $0 \geq hv_\varepsilon = -|v_\varepsilon| \geq hu$. Thus $0 \geq 2hv_\varepsilon \geq h(v_\varepsilon + u)$ implying that $|v_\varepsilon + u| \geq 2|v_\varepsilon|$ on B_ε. Finally, if $\|v_\varepsilon + u\|_1 < 2$, then

$$\int_{B_\varepsilon} |z| d\mu > \int_{B_\varepsilon} |v_\varepsilon| d\mu,$$

contradicting our choice of v_ε.   □

For v_ε as above, let $f_\varepsilon = h|v_\varepsilon|$. Construct

$$g_\varepsilon(x) = \begin{cases} f_\varepsilon(x), & x \notin A_{2\varepsilon} \\ \sigma_\varepsilon(x), & x \in A_{2\varepsilon} \backslash \overline{A}_\varepsilon \\ -v_\varepsilon(x), & x \in \overline{A}_\varepsilon, \end{cases}$$

where $|\sigma_\varepsilon| < |v_\varepsilon|$ on $(A_{2\varepsilon} \backslash \overline{A}_\varepsilon) \backslash Z(v_\varepsilon)$, $\sigma_\varepsilon = 0$ on $(A_{2\varepsilon} \backslash \overline{A}_\varepsilon) \cap Z(v_\varepsilon)$, and $g_\varepsilon \in C(K)$. Such a construction is possible. Similarly, set

$$k_\varepsilon(x) = \begin{cases} f_\varepsilon(x), & x \notin B_{2\varepsilon} \\ \gamma_\varepsilon(x), & x \in B_{2\varepsilon} \backslash \overline{B}_\varepsilon \\ v_\varepsilon(x), & x \in \overline{B}_\varepsilon, \end{cases}$$

where $|\gamma_\varepsilon| < |v_\varepsilon|$ on $(B_{2\varepsilon} \backslash \overline{B}_\varepsilon) \backslash Z(v_\varepsilon)$, $\gamma_\varepsilon = 0$ on $(B_{2\varepsilon} \backslash \overline{B}_\varepsilon) \cap Z(v_\varepsilon)$, and $k_\varepsilon \in C(K)$.

We first consider $P_U(g_\varepsilon)$ and $P_U(k_\varepsilon)$. In the proof of this theorem for the case dim $U = 1$, we showed that $P_U(g_\varepsilon) = \{-v_\varepsilon\}$ and $P_U(k_\varepsilon) = \{v_\varepsilon\}$ for all $\varepsilon > 0$. (Therein $v_\varepsilon = u$ for all ε.) Here we maintain the first equality, but not the second. However using Lemma 3.21 we shall prove that every $u \in P_U(k_\varepsilon)$ is of distance 2 from $-v_\varepsilon$, i.e., from $P_U(g_\varepsilon)$.

Lemma 3.22. $P_U(g_\varepsilon) = \{-v_\varepsilon\}$, and if $u \in P_U(k_\varepsilon)$, then $\|v_\varepsilon + u\|_1 = 2$.

Proof. On the basis of Lemmas 3.20 and 3.21, it suffices to prove:

(I) $-v_\varepsilon \in P_U(g_\varepsilon)$ and, if $u \in P_U(g_\varepsilon)$, then $|v_\varepsilon| \geq hu$ on K and $-|v_\varepsilon| \geq hu$ on A_ε.

(II) $v_\varepsilon \in P_U(k_\varepsilon)$ and, if $u \in P_U(k_\varepsilon)$, then $|v_\varepsilon| \geq hu$ on K and $-|v_\varepsilon| \geq hu$ on B_ε.

The proofs of (I) and (II) being totally analogous, we prove only (I).

We first consider $\text{sgn}(g_\varepsilon + v_\varepsilon)$.

(i) On $\overline{A}_\varepsilon$

$$\text{sgn}(g_\varepsilon + v_\varepsilon) = \text{sgn}(-v_\varepsilon + v_\varepsilon) = 0.$$

(ii) On $A_{2\varepsilon} \backslash \overline{A}_\varepsilon$

$$\text{sgn}(g_\varepsilon + v_\varepsilon) = \text{sgn}(\sigma_\varepsilon + v_\varepsilon) = \text{sgn} \, v_\varepsilon = \text{sgn} \, f_\varepsilon.$$

(iii) Off $A_{2\varepsilon}$

$$\text{sgn}(g_\varepsilon + v_\varepsilon) = \text{sgn}(f_\varepsilon + v_\varepsilon) = \begin{cases} 0, & f_\varepsilon = -v_\varepsilon \\ \text{sgn} \, f_\varepsilon, & f_\varepsilon = v_\varepsilon \neq 0. \end{cases}$$

Thus, for all $u \in U$,

$$\int_{Z(g_\varepsilon + v_\varepsilon)} |u| d\mu = \int_{f_\varepsilon = -v_\varepsilon \neq 0} |u| d\mu + \int_{\overline{A}_\varepsilon} |u| d\mu + \int_{Z(f_\varepsilon)} |u| d\mu$$

and

$$\int_K \operatorname{sgn}(g_\varepsilon + v_\varepsilon) u \, d\mu = \int_{\substack{f_\varepsilon = v_\varepsilon \neq 0 \\ x \notin A_{2\varepsilon}}} (\operatorname{sgn} f_\varepsilon) u \, d\mu + \int_{A_{2\varepsilon} \setminus \overline{A}_\varepsilon} (\operatorname{sgn} f_\varepsilon) u \, d\mu$$

$$= \int_{f_\varepsilon = v_\varepsilon \neq 0} (\operatorname{sgn} f_\varepsilon) u \, d\mu - \int_{\overline{A}_\varepsilon} (\operatorname{sgn} f_\varepsilon) u \, d\mu.$$

For every $u \in U$,

$$0 = \int_K h u \, d\mu = \int_{f_\varepsilon = v_\varepsilon \neq 0} (\operatorname{sgn} f_\varepsilon) u \, d\mu + \int_{f_\varepsilon = -v_\varepsilon \neq 0} (\operatorname{sgn} f_\varepsilon) u \, d\mu + \int_{Z(f_\varepsilon)} h u \, d\mu.$$

Therefore,

$$\int_K \operatorname{sgn}(g_\varepsilon + v_\varepsilon) u \, d\mu = - \int_{f_\varepsilon = -v_\varepsilon \neq 0} (\operatorname{sgn} f_\varepsilon) u \, d\mu - \int_{Z(f_\varepsilon)} h u \, d\mu - \int_{\overline{A}_\varepsilon} (\operatorname{sgn} f_\varepsilon) u \, d\mu.$$

From the above, we have for every $u \in U$,

$$\left| \int_K \operatorname{sgn}(g_\varepsilon + v_\varepsilon) u \, d\mu \right| \leq \int_{Z(g_\varepsilon + v_\varepsilon)} |u| d\mu.$$

Thus, from Theorem 2.1, $-v_\varepsilon \in P_U(g_\varepsilon)$. Assume $u \in P_U(g_\varepsilon)$, $u \neq -v_\varepsilon$. From Proposition 2.4,

$$A)\ (g_\varepsilon + v_\varepsilon)(g_\varepsilon - u) \geq 0 \text{ on } K$$

$$B)\ \int_K [\operatorname{sgn}(g_\varepsilon + v_\varepsilon)](u + v_\varepsilon) d\mu = \int_{Z(g_\varepsilon + v_\varepsilon)} |u + v_\varepsilon| d\mu.$$

From (i)-(iii) and the above, we obtain:

1) On $f_\varepsilon = v_\varepsilon \neq 0$, $x \notin A_{2\varepsilon}$, $f_\varepsilon(f_\varepsilon - u) \geq 0$.

2) On $A_{2\varepsilon} \setminus \overline{A}_\varepsilon$, $f_\varepsilon(\sigma_\varepsilon - u) \geq 0$.

3) On $f_\varepsilon = -v_\varepsilon \neq 0$, $-(\operatorname{sgn} f_\varepsilon)(u + v_\varepsilon) \geq 0$.

4) On $Z(f_\varepsilon)$, $-h(u + v_\varepsilon) \geq 0$.

5) On $\overline{A}_\varepsilon$, $-(\operatorname{sgn} f_\varepsilon)(u + v_\varepsilon) \geq 0$.

Since $\operatorname{sgn} f_\varepsilon = h$ off $Z(f_\varepsilon)$, it follows from the construction, and from (1)-(5), that $|v_\varepsilon| \geq hu$ on all of K and $-|v_\varepsilon| \geq hu$ on A_ε. □

We now finish the proof of Theorem 3.16.

Proof of Theorem 3.16. V_1 is compact and equicontinuous, since V_1 is a closed, hence compact, subset of W_1. Since $v_\varepsilon \in V_1$ for each $\varepsilon > 0$, there exists a subsequence $\{\varepsilon_n\}$ satisfying $\lim_{n \to \infty} \varepsilon_n = 0$, and a $v^* \in V_1$ for which

$\lim_{n \to \infty} v_{\varepsilon_n} = v^*$. Let $f = h|v^*|$. Then $f \in C(K)$, and $\lim_{n \to \infty} \|f - f_{\varepsilon_n}\|_\infty = 0$. By definition,

$$\|f_\varepsilon - g_\varepsilon\|_\infty \leq 2 \max_{x \in \overline{A}_{2\varepsilon}} |v_\varepsilon(x)|$$

$$\|f_\varepsilon - k_\varepsilon\|_\infty \leq 2 \max_{x \in \overline{B}_{2\varepsilon}} |v_\varepsilon(x)|.$$

By construction $v(\alpha) = v(\beta) = 0$ for all $v \in V_1$, and from the equicontinuity of V_1, it therefore follows that

$$\lim_{n \to \infty} \|f_{\varepsilon_n} - g_{\varepsilon_n}\|_\infty = \lim_{n \to \infty} \|f_{\varepsilon_n} - k_{\varepsilon_n}\|_\infty = 0.$$

Thus

$$\lim_{n \to \infty} \|f - g_{\varepsilon_n}\|_\infty = \lim_{n \to \infty} \|f - k_{\varepsilon_n}\|_\infty = 0.$$

We now apply Lemma 3.22 to prove our result. If there exists an L^∞-continuous selection s on U, then since $P_U(g_{\varepsilon_n}) = \{-v_{\varepsilon_n}\}$ for all n, it follows that

$$s(f) = \lim_{n \to \infty} s(g_{\varepsilon_n}) = \lim_{n \to \infty} -v_{\varepsilon_n} = -v^*.$$

Therefore

$$-v^* = s(f) = \lim_{n \to \infty} s(k_{\varepsilon_n}).$$

But, for each $u_n \in P_U(k_{\varepsilon_n})$, we have $\|v_{\varepsilon_n} + u_n\|_1 = 2$. Thus $\|v_{\varepsilon_n} + s(k_{\varepsilon_n})\|_1 = 2$, which implies that $2 = \|v^* + s(f)\|_1 = \|v^* - v^*\|_1 = 0$. A contradiction.

Assume now that K is *not* connected. We cannot apply the above proof, as is, since we cannot necessarily find the requisite α and β. Furthermore, as illustrated by our previous example, there may in fact exist an L^∞-continuous selection. In this case, take any $\alpha \in A$ and $\beta \in B$. (One may also dispense with statement (2) of Lemma 3.18.) It is not difficult to check that

$$\lim_{n \to \infty} \|f_{\varepsilon_n} - g_{\varepsilon_n}\|_1 = \lim_{n \to \infty} \|f_{\varepsilon_n} - k_{\varepsilon_n}\|_1 = 0.$$

Using the above reasoning it follows that there cannot exist an L^1-continuous selection on U (see also Exercise 11).   □

Exercises

1. Let $U = \mathrm{span}\{1, x, x_+\}$ on $K = [-1, 1]$, where

$$x_+ = \begin{cases} x, & x \geq 0 \\ 0, & x < 0. \end{cases}$$

Prove that U is a unicity space for $C_1(K, \mu)$ for every 'admissible' measure μ.

2. Let

$$P_{m,n} = \Big\{ \sum_{i=0}^{m} \sum_{j=0}^{n} a_{ij} x^i y^j : a_{ij} \in \mathbb{R} \Big\}.$$

Set $K = [0, 1] \times [0, 1]$ and let μ be Lebesgue measure on K. Prove that $P_{m,1}$ (and $P_{1,n}$) are unicity spaces for $C_1(K, \mu)$.

3. In Example 3.4 prove that, if φ is also strictly positive, then the result is valid without the restriction $k \leq m$.

4. Characterize those finite-dimensional subspaces U for which $U = U^*$.

5. Let $U = \mathrm{span}\{|x|\}$ on $[-1, 1]$ and let μ be Lebesgue measure. Prove that, if $P_U(f)$ is not a singleton, then $f(0) = 0$.

6. Let $U = \mathrm{span}\{u\}$ on $K = [-2, -1] \cup [1, 2]$, and μ be Lebesgue measure. Assume u does not vanish on K and $\int_K u \, d\mu = 0$. Find all those $f \in C(K)$ for which $P_U(f)$ is not a singleton.

7. Construct a finite-dimensional subspace U of $C(K)$ and an 'admissible' measure μ, such that U is not a unicity space for $C_1(K, \mu)$, but the set $\{f : P_U(f)$ not a singleton $\}$ is not dense in $C_1(K, \mu)$.

8. Let $U_c = \mathrm{span}\{x - c\}$ on $K = [-1, 1]$. For a given 'admissible' measure μ, prove that there exists a unique $c_\mu \in (-1, 1)$ for which U_{c_μ} is not a unicity space for $C_1(K, \mu)$. For any given $c \in (-1, 1)$, construct an 'admissible' measure μ such that U_c is not a unicity space for $C_1(K, \mu)$.

9. Let $U = \mathrm{span}\{1, x, \ldots, x^n\}$ on $K = [-2, -1] \times [1, 2]$, and μ be Lebesgue measure. Prove that U is 1-convex for n even, and 0-convex (a unicity space) for n odd. (Hint: see Appendix B.)

10. Let U, K, and μ be as in Exercise 9. Prove that for n even, $n \geq 2$, there is no L^∞-continuous selection on U.

11. In the proof of Theorem 3.6, we showed that if $f \in C(K)$ and $P_U(f)$ is not a singleton, then for any v in the relative interior of $P_U(f)$, and $\varepsilon > 0$, there exists an $f_\varepsilon \in C(K)$ satisfying $\|f - f_\varepsilon\|_1 < \varepsilon$ for which $P_U(f_\varepsilon) = \{v\}$. Use this fact to prove that if U is not a unicity space for $C_1(K, \mu)$, then there exists no L^1-continuous selection onto U.

12. Let μ be Lebesgue measure on $[a, b]$. Construct an $f \in C[a, b]$ and a subspace $U = \mathrm{span}\{u_1, \ldots, u_n\}$ of $C[a, b]$, such that

$$\left| \int_a^b \mathrm{sgn}(f) u_i \, d\mu \right| \leq \int_a^b |u_i| \, d\mu$$

for each $i = 1, \ldots, n$, and yet $0 \notin P_U(f)$.

13. Let U be a finite-dimensional subspace of $C(K)$ and μ an 'admissible' measure. Prove that there exists an $f \in C(K)$ which has a unique, but not strongly unique, best approximant from U.

Notes and References

Theorem 3.1 is due to Cheney, Wulbert [1969, Theorem 24] (see also Pták [1958] and Singer [1960]), while Theorem 3.3 and Corollary 3.4 were proved by Strauss [1981, Theorems 8 and 9]. Example 3.2 was first proved for Lebesgue measure by Jackson [1921]. Example 3.4 is a special case of a result of Kroó [1982]. Kroó proved this result for

$$U = \{\varphi(y)q_k(x) + p_m(x)\}$$

where $q_k \in U_k$, $p_m \in U_m$, with the assumptions that $U_k \subseteq U_m$ and U_k, U_m are Chebyshev (T-) systems (see Chapter 4 and Appendix A) of finite dimension on $[0, 1]$. Sommer [1987] generalized this result to 'admissible' measures of the form $d\mu = w_1(x)w_2(x)dxdy$. The proof is basically the same in all cases. The proof of Lemma 3.5 is due to A. Atzmon. Theorem 3.6 is new. The material in Section 3 is new. However it is based on parallel work in the uniform norm done in Pinkus, Strauss [1988]. Theorem 3.12 is essentially in Havinson [1957], see also Singer [1970, p.134]. All the material of Section 5 is taken from Pinkus [1988]. For a general survey on metric projections and continuous selections, see Deutsch [1983].

Exercise 2 follows from Example 3.4. It is an open question as to whether $P_{m,n}$ is a unicity space for $C_1(K, \mu)$ for all $m, n \geq 2$, if μ and K are as given in Exercise 2. Example 9 is a particular case of a more general result in Kroó [1986b].

4
Unicity Subspaces and Property A

1. Introduction

Both Theorems 3.1 and 3.3 (see also Corollary 3.4) provide us with criteria for exactly determining when a given finite-dimensional subspace U of $C(K)$ is a unicity space for $C_1(K, \mu)$. Nevertheless there are at least two drawbacks to these criteria. In the first place the criteria are generally very difficult, if not impossible, to verify. Except for certain specific subspaces it is almost always impossible to check whether there exist h and u^* satisfying the conditions of Theorem 3.1, or whether the inequalities of Corollary 3.4 are always valid. Secondly, the above-mentioned criteria are measure dependent. That is, a specific subspace U may be a unicity subspace for $C_1(K, \mu_1)$, but not for $C_1(K, \mu_2)$, where μ_1 and μ_2 are both 'admissible' measures on K. See Examples 3.1 and 3.3, and Exercise 8 of Chapter 3 where this phenomenon occurs.

It is therefore natural to ask for a characterization of finite-dimensional subspaces U of $C(K)$ with the property that U is a unicity subspace for $C_1(K, \mu)$ for all 'nice' measures μ. Perhaps such a characterization, while it would be interesting in and of itself, might also be easier to verify. Perhaps such a characterization will be an intrinsic property of the subspace U. Perhaps it might be possible to list all such subspaces. This is the problem which we investigate in this chapter.

In Section 2 we delineate a condition called Property A. We prove that U satisfies Property A if and only if U is a unicity space for $C_1(K, \mu)$ for a large class of measures μ. In Section 3 we deduce two general consequences of Property A. If $K \subset \mathbb{R}$ then, based on these two necessary conditions, we totally characterize all subspaces satisfying Property A. This we do in Section 4. In Section 5 we consider $K \subset \mathbb{R}^d$, $d \geq 2$. Here a complete characterization of U satisfying Property A is yet unknown. However various partial results are presented. In Section 6 we consider conditions for unicity in the problem of best approximation with interpolatory constraints. Section 7 contains the analogue of Property A to the case of k-convex U. Sprinkled throughout the chapter are numerous examples.

2. Property A

Recall (see Definition 3.1) that for a given finite-dimensional subspace U of $C(K)$, we defined:

$$U^* = \{g : g \in C(K), |g| = |u| \text{ for some } u \in U\}.$$

This important subset of $C(K)$ was used in Theorem 3.3 and Corollary 3.4. In Corollary 3.4 we observed that U is unicity space for $C_1(K, \mu)$ if and only if to each $g \in U^*$, $g \neq 0$, there exists a $u_g \in U$ satisfying

$$\left| \int_K (\operatorname{sgn} g) u_g \, d\mu \right| > \int_{Z(g)} |u_g| d\mu .$$

(Look back to Chapter 3 to see the conditions imposed on μ.) This is certainly an inequality which is generally unverifiable. However let us describe a condition on U which implies the above inequality.

Definition 4.1. A finite-dimensional subspace U of $C(K)$ is said to satisfy Property A_μ if to each $g \in U^*$, $g \neq 0$, there exists a $u \in U$, $u \neq 0$, for which

  a) $u = 0$, μ a.e. on $Z(g)$

  b) $gu \geq 0$ on K.

As an immediate consequence of this definition, we obtain:

Proposition 4.1. *Each finite-dimensional subspace U of $C(K)$ which satisfies Property A_μ is a unicity space for $C_1(K, \mu)$.*

Proof. Let $g \in U^*$, $g \neq 0$, and assume u is as given in Definition 4.1 satisfying (a) and (b). From (a) and (b) we have

$$\int_K (\operatorname{sgn} g) u \, d\mu = \int_K |u| d\mu > 0 = \int_{Z(g)} |u| d\mu.$$

Apply Corollary 3.4.   □

The problem with Property A_μ is that it is still μ dependent. This dependence on μ is due to (a) of Definition 4.1. It is actually not too μ dependent, but we do not wish to enter into a measuretheoretic discussion. To simplify matters we shall start with Lebesgue measure.

We say that U satisfies *Property A* if U satisfies Property A_μ where μ is Lebesgue measure on K. If U satisfies Property A, then it also satisfies Property A_μ for many other measures μ. But recall that we only considered measures μ which satisfied very particular assumptions. Many of the results of the previous chapter depended on these assumptions. For this reason we let $\mathcal{A}$ denote the set of 'admissible' measures on K which are absolutely continuous with respect to Lebesgue measure. The absolute continuity with respect to Lebesgue measure immediately implies:

Proposition 4.2. *If U is a finite-dimensional subspace of $C(K)$ which satisfies Property A, then it satisfies Property A_μ for every $\mu \in \mathcal{A}$.*

Thus U is a unicity space for $C_1(K, \mu)$ for every $\mu \in \mathcal{A}$ if U satisfies Property A. We shall soon prove a converse result.

The definition of Property A is given in terms of U and U^*. U^* is itself defined in terms of U, so that Property A is really an intrinsic property of U. (We have conveniently forgotten that we just fixed a measure.) We wish to redefine Property A without resorting to U^*. To this end we first present the following ancillary definitions.

Definition 4.2. Let $D \subseteq K$, D (relatively) open. Then $[D]$ will denote the number (possibly infinite but necessarily countable) of open connected components of D. For given $u \in U$, we set $m(u) = [K \backslash Z(u)]$. We fix an order on the open connected components $A_i = A_i(u)$ of $K \backslash Z(u)$, and write

$$K \backslash Z(u) = \bigcup_{i=1}^{m(u)} A_i .$$

Note that the A_i are not only open and connected, but are also disjoint. For notational ease we also define:

Definition 4.3. For $u \in U$, set

$$U(u) = \{v : v \in U, v = 0 \text{ a.e. on } Z(u)\}.$$

$U(u)$ is a subspace of U and, if $u \neq 0$, then $\dim U(u) \geq 1$ since $u \in U(u)$. Using Definitions 4.2 and 4.3, we can now reformulate Definition 4.1 (where the μ of Definition 4.1 is here Lebesgue measure).

Definition 4.4. A finite-dimensional subspace U of $C(K)$ is said to satisfy *Property A* if to each $u \in U$, $u \neq 0$, with $K \backslash Z(u) = \bigcup_{i=1}^{m(u)} A_i$ as above, and every choice of $\varepsilon_i \in \{-1, 1\}$, $i = 1, \ldots, m(u)$, there exists a $v \in U(u)$, $v \neq 0$, satisfying $\varepsilon_i v \geq 0$ on A_i, $i = 1, \ldots, m(u)$.

In order to develop a feel for this rather curious Property A, let us consider some examples.

Example 4.1. $\dim U = 1$.

Let $u \in U$, $u \neq 0$. It is easily checked that U satisfies Property A if and only if $m(u) = [K \backslash Z(u)] = 1$. If $m(u) = 1$, then there is essentially nothing to show. Assume $m(u) > 1$. Let A_1 and A_2 be two distinct connected components of $K \backslash Z(u)$. Assume u has sign δ_i on A_i, $i = 1, 2$. Set $\varepsilon_1 = \delta_1$ and $\varepsilon_2 = -\delta_2$. Since every $v \in U$ is of the form $v = cu$, $c \in \mathbb{R}$, it follows that if $\varepsilon_i v \geq 0$ on A_i, $i = 1, 2$, then $v = 0$. Thus Property A does not hold. Recall that this example was previously considered in Example 3.1.

Example 4.2. $U = \operatorname{span}\{1, x^2\}$, $K = [-1, 1]$.

U does not satisfy Property A. Set $u(x) = x^2 - 1/4$. If U satisfied Property A, there would then exist a $v \in U$, $v \neq 0$, for which $v \geq 0$ on $[-1, -1/2]$ and $v \leq 0$ on $[-1/2, 1]$. No such v exists.

Example 4.3. $U = \operatorname{span}\{1, x, y\}$, $K = [a, b] \times [c, d] \subset I\!\!R^2$.

U satisfies Property A. To prove this we note that each $u \in U$, $u \neq 0$, is either strictly of one sign on the interior of K (in which case we can take $v = 1$ or $v = -1$), or $K \backslash Z(u)$ has two connected components A_1 and A_2, $Z(u)$ is a straight line, and u has opposite sign on A_1 and A_2. In this latter case, if $\varepsilon_1 = \varepsilon_2$ set $v = \varepsilon_1$, while if $\varepsilon_1 = -\varepsilon_2$ set $v = u$ or $v = -u$, as appropriate.

Example 4.4. $U = \operatorname{span}\{x, y\}$, $K = [-1, 1] \times [-1, 1] \subset I\!\!R^2$.

U does not satisfy Property A since U contains no non-negative non-trivial function.

Example 4.5. $U = \pi_m = \operatorname{span}\{1, x, \ldots, x^m\}$, $K = [a, b] \subset I\!\!R$.

U satisfies Property A. If $u \in \pi_m$, $u \neq 0$, then $|Z(u)| = 0$ (recall that $|A|$ denotes the Lebesgue measure of the set A), and u has at most m distinct zeros in (a, b), i.e., $[K \backslash Z(u)] \leq m + 1$. Let $[K \backslash Z(u)] = r$, and $A_i = (c_{i-1}, c_i)$, $i = 1, \ldots, r$, where $a = c_0 < c_1 < \cdots < c_r = b$ (except that we may have $A_1 = [c_0, c_1)$ and/or $A_r = (c_{r-1}, c_r]$) satisfy $K \backslash Z(u) = \bigcup_{i=1}^{r} A_i$. Let $\varepsilon_i \in \{-1, 1\}$, $i = 1, \ldots, r$. We shall say that c_i, $i \in \{1, \ldots, r-1\}$ is *active* if $\varepsilon_i \varepsilon_{i+1} = -1$. Let $\{d_i\}_{i=1}^{k}$ denote the active c_i, $a < d_1 < \cdots < d_k < b$. Thus $k \leq m$. Set $v(x) = \varepsilon_r \prod_{i=1}^{k} (x - d_i)$. Then v satisfies the conditions of Definition 4.4. See Example 3.2.

Example 4.6. $U = \pi_m$, $K = [a, b] \cup [c, d] \subset I\!\!R$, and $a < b < c < d$.

U does not satisfy Property A. Let $a < x_1 < \cdots < x_m < b$, and set $u(x) = \prod_{i=1}^{m}(x - x_i)$. Then $K \backslash Z(u) = \bigcup_{i=1}^{m+2} A_i$, where $A_1 = [a, x_1)$, $A_i = (x_{i-1}, x_i)$, $i = 2, \ldots, m$, $A_{m+1} = (x_m, b]$, and $A_{m+2} = [c, d]$. Set $\varepsilon_i = (-1)^i$, $i = 1, \ldots, m + 2$. If $v \in \pi_m$ satisfies $\varepsilon_i v \geq 0$ on A_i, $i = 1, \ldots, m + 2$, then necessarily $v(x_i) = 0$, $i = 1, \ldots, m$, and v must admit an additional zero in $[b, c]$. v then has at least $m + 1$ distinct zeros. But $v \in \pi_m$ and therefore $v = 0$, proving that Property A does not hold. See Example 3.3.

Additional examples of subspaces U which satisfy, or do not satisfy, Property A may be found in Sections 4 and 5, and in the exercises at the end of this chapter.

Property A is a very restrictive condition. We shall later examine it in some detail. In fact we shall totally characterize all finite-dimensional subspaces of $C(K)$ satisfying Property A if $K \subset I\!\!R$. But we are getting too far ahead of ourselves. Let us first show the importance of Property A. In Proposition 4.2 we proved that if a finite-dimensional subspace $U \subset C(K)$ satisfied Property A, then U was a unicity space for $C_1(K, \mu)$ for every $\mu \in \mathcal{A}$. We shall now prove the converse. That is, if U is a finite-dimensional unicity space for $C_1(K, \mu)$ for every $\mu \in \mathcal{A}$, then U necessarily satisfies Property A.

To prove this converse, it is not in fact necessary to consider all of $\mathcal{A}$. It suffices to consider a significantly smaller class of measures.

Definition 4.5. We shall say that the set $\mathcal{B} \subseteq \mathcal{A}$ satisfies *Condition B* if
1) $\mathcal{B}$ is a convex cone
2) $f \in L^\infty(K)$ and $\int_K f \, d\mu \geq 0$ for all $\mu \in \mathcal{B}$ implies $f \geq 0$ a.e. (Lebesgue measure) on K.

Let us first give some examples of sets satisfying Condition B.
1) $\mathcal{B} = \{w(x)dx : w \in L^\infty(K), \text{ess inf } w > 0 \text{ on } K\}$
2) $\mathcal{B} = \{w(x)dx : w \in C(K), w(x) > 0 \text{ on } K\}$
3) $\mathcal{B} = \{w(x)dx : w \text{ a strictly positive polynomial on } K\}$.

From the definition of Condition B, we obtain this next lemma.

Lemma 4.3. *Assume $\mathcal{B}$ satisfies Condition B. Let $V = \text{span}\{v_1, \ldots, v_m\}$ be an m-dimensional subspace of $L^\infty(K)$. Assume that V does not contain a non-negative non-trivial function. Set*

$$A_m = \left\{ \left(\int_K v_1 d\mu, \ldots, \int_K v_m d\mu \right) : \mu \in \mathcal{B} \right\}.$$

Then $A_m = \mathbb{R}^m$.

Proof. Since $\mathcal{B}$ is a non-empty convex cone, A_m is a non-empty convex cone in $\mathbb{R}^m$. Assume $A_m \neq \mathbb{R}^m$. Then $\mathbf{0} \in \partial A_m$, and there exists a vector $\mathbf{a} = (a_1, \ldots, a_m) \neq \mathbf{0}$ for which

$$\sum_{i=1}^m a_i \int_K v_i d\mu \geq 0$$

for all $\mu \in \mathcal{B}$. Set $v = \sum_{i=1}^m a_i v_i$. Then $v \neq 0$, $v \in L^\infty(K)$, and

$$\int_K v \, d\mu \geq 0$$

for all $\mu \in \mathcal{B}$. From Condition B we obtain a contradiction to our hypothesis on V. □

We shall in practice use the following consequence of Lemma 4.3.

Corollary 4.4. *Assume $\mathcal{B}$ satisfies Condition B. If V is a finite-dimensional subspace of $L^\infty(K)$ which does not contain a non-negative non-trivial function, then there exists a $\mu \in \mathcal{B}$ satisfying*

$$\int_K v \, d\mu = 0$$

for all $v \in V$.

We can now state and prove:

Theorem 4.5. *Assume $\mathcal{B}$ satisfies Condition B. If a finite-dimensional subspace U of $C(K)$ is a unicity space for $C_1(K, \mu)$ for all $\mu \in \mathcal{B}$, then U satisfies Property A.*

Proof. Let $\dim U = n$. Assume U does not satisfy Property A. We shall construct a $\mu \in \mathcal{B}$ with the property that U is not a unicity space for $C_1(K, \mu)$.

Since U does not satisfy Property A we obtain from Definition 4.4 the existence of a $u^* \in U, u^* \neq 0$, sets A_i, $i = 1, \ldots, m(u^*)$, the open connected components of $K \backslash Z(u^*)$, and $\varepsilon_i \in \{-1, 1\}$, $i = 1, \ldots, m(u^*)$, such that if $u \in U(u^*)$ satisfies $\varepsilon_i u \geq 0$ on A_i, $i = 1, \ldots, m(u^*)$, then $u = 0$. $U(u^*) = U_1$ is a subspace of U, and since $u^* \in U_1$, we have $\dim U_1 = k$ where $1 \leq k \leq n$. Let $u_1, \ldots, u_k$ be any basis for U_1, and choose $u_{k+1}, \ldots, u_n$ so that $U = \operatorname{span}\{u_1, \ldots, u_n\}$. Set $U_2 = \operatorname{span}\{u_{k+1}, \ldots, u_n\}$. By definition, if $u \in U_2$ satisfies $u = 0$ a.e. on $Z(u^*)$, then $u = 0$. In other words, the functions $\{\chi_{Z(u^*)} u_i\}_{i=k+1}^{n}$ are linearly independent in $L^1(K)$, where $\chi_{Z(u^*)}$ is the characteristic function of the set $Z(u^*)$.

From Theorem 2.3 there exists an $\widetilde{h} \in L^\infty(Z(u^*))$ satisfying

$$1) \, |\widetilde{h}| = 1 \text{ on } Z(u^*)$$

$$2) \int_{Z(u^*)} \widetilde{h} u \, dx = 0, \text{ all } u \in U_2.$$

Set
$$h(x) = \begin{cases} \varepsilon_i, & x \in A_i, \ i = 1, \ldots, m(u^*) \\ \widetilde{h}(x), & x \in Z(u^*), \end{cases}$$

and let $V = \{hu : u \in U\}$. V is an n-dimensional subspace of $L^\infty(K)$ since $h \in L^\infty(K)$, and $|h| = 1$ on K.

We claim that V does not contain a non-negative non-trivial function. To prove this claim, assume $v \in V$ and $v \geq 0$ a.e. on K. Now, $v = hu^1 + hu^2$, where $u^i \in U_i$, $i = 1, 2$. On $Z(u^*)$, $0 \leq v = hu^2 = \widetilde{h}u^2$, and by construction $\int_{Z(u^*)} \widetilde{h}u^2 dx = 0$. Thus $v = 0$ a.e. on $Z(u^*)$ and since $|\widetilde{h}| = 1$ on $Z(u^*)$, we obtain $u^2 = 0$. Therefore $v = hu^1$. Since $v \geq 0$ a.e. on each A_i, it follows from the definition of h and U_1 that $u^1 \in U(u^*)$ and $\varepsilon_i u^1 \geq 0$ on A_i, $i = 1, \ldots, m(u^*)$. This implies, by our assumption, that $u^1 = 0$. Thus $v = 0$, proving our claim.

Applying Corollary 4.4, we obtain the existence of a $\mu \in \mathcal{B}$ for which

$$\int_K v \, d\mu = 0$$

for all $v \in V$. Hence

$$\int_K hu \, d\mu = 0$$

for all $u \in U$. Furthermore, by construction $|h| = 1$ on K, and $h|u^*| \in C(K)$. Thus from Theorem 3.1, U is *not* a unicity space for $C_1(K, \mu)$. □

Remark. As a consequence of the second example of $\mathcal{B}$ satisfying Condition B, together with Proposition 4.2 and Theorem 4.5, we can consider Property A from a slightly different perspective. The finite-dimensional subspace U of $C(K)$ satisfies Property A if and only if

$$U_w = \{\, uw : u \in U \,\}$$

is a unicity space for $C_1(K, \mu^*)$, μ^* Lebesgue measure, for every strictly positive and continuous w on K.

3. Consequences of Property A

The result of Theorem 4.5 is both pretty and elegant. If we wish to know whether U is a unicity space for $C_1(K, \mu)$ for all $\mu \in \mathcal{A}$ then it is both necessary and sufficient that U satisfy Property A. However, the examples of the previous section notwithstanding, verifying or disproving Property A is no mean matter.

It is our aim, in this section, to simplify this task. We do so by finding various consequences of Property A. If $K \subset \mathbb{R}$ then, using these results and a bit more work, we shall in Section 4 totally characterize all subspaces satisfying Property A. For $K \subset \mathbb{R}^d$, $d > 1$, the explicit characterization is a problem yet unresolved.

In Example 4.1 we noted that if dim $U = 1$, then U satisfies Property A if and only if $m(u) = 1$ for $u \in U$, $u \neq 0$. On the basis of this rather flimsy evidence it is possible to conjecture that if U satisfies Property A, then perhaps it is both necessary and sufficient that

$$[K \backslash Z(u)] \leq \dim U$$

for every $u \in U$. (There is actually somewhat more evidence to support this wild claim.) However, from Example 4.4 we see that this conjecture cannot possibly hold. For $U = \text{span}\{x, y\}$ on $K = [-1, 1] \times [-1, 1]$ we have $[K \backslash Z(u)] = 2 = \dim U$ for every $u \in U, u \neq 0$, and yet U does *not* satisfy Property A. So much for the conjecture. But this counterexample only proves that one half of the conjecture is false. The other half is in fact valid, i.e., if U satisfies Property A then the above inequality holds. Actually an even stronger version of the inequality will be proved. (In Section 4 we shall prove that this stronger version is both necessary and sufficient for U to satisfy Property A if $K \subset \mathbb{R}$.) Our first major result can now be stated.

Theorem 4.6. *If U is a finite-dimensional subspace of $C(K)$ which satisfies Property A, then*

$$[K \backslash Z(u)] \leq \dim U(u)$$

for every $u \in U$.

Note that this is a very strong restriction on U. Very few subspaces satisfy it in general.

Our proof of Theorem 4.6 is long and difficult. As such we divide it into steps. We first recall that for $u \in U, u \neq 0$, we write $K \backslash Z(u) = \bigcup_{i=1}^{m(u)} A_i$, where the A_i are the open connected components of $K \backslash Z(u)$, and $m(u) = [K \backslash Z(u)]$ is the number of these components. The number of components may be infinite, but it is necessarily countable (see Definition 4.2). Rewording Theorem 4.6, we wish to prove that if U satisfies Property A, then $m(u) \leq \dim U(u)$ for every $u \in U$, $u \neq 0$. For $u = 0$, there is nothing to prove.

The proof will be via induction on $n = \dim U$. For $n = 1$, Example 4.1 proves the theorem. We therefore assume that Theorem 4.6 is valid for all subspaces of dimension $\leq n - 1$. Now, from the definition of Property A it may be observed that, if U satisfies Property A, then so does each subspace $U(u), u \in U, u \neq 0$. Thus, if $\dim U(u) < n$, then from the induction hypothesis we may assume the validity of Theorem 4.6. We therefore presume that U satisfies Property A, and there exists a $u \in U$ for which $U(u) = U$ and $m(u) > n$. We shall eventually arrive at a contradiction. We first set up some machinery.

Definition 4.6. Let $u \in U, u \neq 0$, and $K \backslash Z(u) = \bigcup_{i=1}^{m(u)} A_i$. A non-zero vector $\mathbf{s} = (s_1, \ldots, s_{m(u)})$ is said to be an *annihilator* for u if for any function $v \in U(u)$ satisfying $s_i v \geq 0$ on A_i, $i = 1, \ldots, m(u)$, it follows that $s_i v = 0$ on A_i, $i = 1, \ldots, m(u)$.

Let $u \in U, u \neq 0, K \backslash Z(u) = \bigcup_{i=1}^{m(u)} A_i$, and assume that $\mathbf{s}$ is an annihilator for u. Set

$$U(u; \mathbf{s}) = \{v : v \in U(u), s_i v \geq 0 \text{ on } A_i, i = 1, \ldots, m(u)\}.$$

$U(u; \mathbf{s})$ is a subspace (and not simply a subset) of $U(u)$. This may seem strange at first glance. However, since $\mathbf{s}$ is an annihilator for u, it follows that, if $s_i \neq 0$ then all $v \in U(u; \mathbf{s})$ identically vanish on A_i, while if $s_i = 0$ there is no restriction whatsoever on v in A_i. Thus we could equivalently define

$$U(u; \mathbf{s}) = \{v : v \in U(u), v = 0 \text{ on } A_i \text{ if } s_i \neq 0, i = 1, \ldots, m(u)\}.$$

It may be that there exist i for which $s_i = 0$ and yet all $v \in U(u; \mathbf{s})$ vanish identically on A_i. For this reason we let $I^{\mathbf{s}}$ denote the set of indices in $\{1, \ldots, m(u)\}$ for which some $v \in U(u; \mathbf{s})$ does *not* vanish identically on A_i. If $s_i \neq 0$, then $i \notin I^{\mathbf{s}}$. But the converse need not hold.

Many $u \in U$ may have no annihilators. However, if $m(u) > n$ then there necessarily exists an annihilator for u. This is a consequence of our next somewhat more general result. For ease of notation, we let $|I|$ denote the number of elements in I for any set of integers I.

Lemma 4.7. *Let* $u \in U, u \neq 0$ *and* $K \backslash Z(u) = \bigcup_{i=1}^{m(u)} A_i$. *Let* J *be a subset of* $\{1, \dots, m(u)\}$ *and set*

$$U_J = \{v : v \in U(u), v = 0 \text{ on } A_j \text{ for all } j \notin J\}.$$

If $\dim U_J < |J|$ *there exists a non-zero vector* $\mathbf{s} = (s_1, \dots, s_{m(u)})$ *such that* $s_j = 0$ *for all* $j \notin J$ *and if* $v \in U_J$ *satisfies* $s_j v \geq 0$ *on* A_j, $j = 1, \dots, m(u)$, *then* $s_j v = 0$ *on* A_j, $j = 1, \dots, m(u)$.

Before we prove the above lemma note that by taking $J = \{1, \dots, m(u)\}$ we have $U_J = U(u)$ and, as an immediate application:

Corollary 4.8. *If* $u \in U, u \neq 0$, *and* $m(u) > \dim U(u)$, *then there exists an annihilator for* u.

Proof of Lemma 4.7. Let $u_1, \dots, u_r$ be any basis for U_J. Set

$$c_{ij} = \int_{A_j} u_i dx, \ i = 1, \dots, r; \ j = 1, \dots, m(u).$$

Since $r = \dim U_J < |J|$, there exists an $\mathbf{s} = (s_1, \dots, s_{m(u)}) \neq \mathbf{0}$ with $s_j = 0, j \notin J$, satisfying

$$\sum_{j=1}^{m(u)} c_{ij} s_j = 0, \ i = 1, \dots, r.$$

Thus

$$\sum_{j=1}^{m(u)} s_j \int_{A_j} v \, dx = 0$$

for all $v \in U_J$. Assume $v \in U_J$ and $s_j v \geq 0$ on A_j, all j. If $s_j v \neq 0$ on A_j for some j, then

$$\sum_{j=1}^{m(u)} s_j \int_{A_j} v \, dx > 0.$$

This contradiction implies the lemma. □

Assume that we have $u \in U$, $u \neq 0$, with $m(u) > n$. From Corollary 4.8 there exists an annihilator $\mathbf{s}$ for u. Furthermore, from Property A we see that $U(u; \mathbf{s}) \neq \{0\}$. (Take $\varepsilon_i = \operatorname{sgn} s_i$ if $s_i \neq 0$, and $\varepsilon_i \in \{-1, 1\}$ arbitrary for $s_i = 0$.) From the definition of $I^{\mathbf{s}}$, every $v \in U(u; \mathbf{s})$ must vanish identically on A_i if $i \notin I^{\mathbf{s}}$. In particular $v \in U(u; \mathbf{s})$ vanishes identically on the A_i for which $s_i \neq 0$. Thus $u \notin U(u; \mathbf{s})$, and therefore $1 \leq \dim U(u; \mathbf{s}) < n$. (If we could find an annihilator $\mathbf{s}$ for the above u with $s_i \neq 0$, all i, then we would immediately contradict Property A. This we were unable to do.)

For u and $\mathbf{s}$ as above, set

$$K^{\mathbf{s}} = K \backslash \mathrm{int}\left(\overline{\bigcup_{i \notin I^{\mathbf{s}}} A_i}\right).$$

Notice that $K^{\mathbf{s}} = \overline{\mathrm{int}\, K^{\mathbf{s}}}$, $[K^{\mathbf{s}} \backslash Z(u)] = |I^{\mathbf{s}}|$ and $\bigcup\{K \backslash Z(v) : v \in U(u; \mathbf{s})\} \subseteq K^{\mathbf{s}}$.

Let d denote the *minimal* value of $\dim U(u; \mathbf{s})$ as u varies over all $u \in U$ with $m(u) > n$, and all annihilators $\mathbf{s}$ for u. Such a d, u and $\mathbf{s}$ exist. (We are minimizing in the set of values $\{1, \dots, n-1\}$.)

Lemma 4.9. *If $m(u) > n$, $\mathbf{s}$ is an annihilator for u, and $\dim U(u; \mathbf{s}) = d$, then*

$$[K^{\mathbf{s}} \backslash Z(u)] \leq d.$$

Proof. Assume $[K^{\mathbf{s}} \backslash Z(u)] > d$. Then $K \backslash Z(u) = \bigcup_{i=1}^{m(u)} A_i$, and $K^{\mathbf{s}} \backslash Z(u) = \bigcup_{i \in I^{\mathbf{s}}} A_i$, with $|I^{\mathbf{s}}| > d$. Apply Lemma 4.7 to u with $J = I^{\mathbf{s}}$. We obtain a non-zero sequence $\mathbf{t} = (t_1, \dots, t_{m(u)})$ with $t_i = 0$ for all $i \notin I^{\mathbf{s}}$, and such that, for $v \in U(u)$ satisfying $v = 0$ on A_i, all $i \notin I^{\mathbf{s}}$, if $t_i v \geq 0$ on A_i, all i, then $t_i v = 0$ on A_i, all i. Thus, if $v \in U(u; \mathbf{s})$ and $t_i v \geq 0$ on A_i, all i, then $t_i v = 0$ on A_i, all i. Change t_i for $i \notin I^{\mathbf{s}}$ by setting $t_i = s_i$ thereon. This new $\mathbf{t}$ is also an annihilator for u since $\mathbf{s}$ was. Furthermore, $U(u; \mathbf{t}) \subset U(u; \mathbf{s})$. The inclusion is a simple consequence of the definition. The two sets are not equal since $t_i \neq 0$ for at least one $i \in I^{\mathbf{s}}$, i.e., $I^{\mathbf{t}} \subset I^{\mathbf{s}}$. We have therefore contradicted the minimality of d. $\quad\square$

Thus from Lemma 4.9, if $m(u) > n$, $\mathbf{s}$ is an annihilator for u and $\dim U(u; \mathbf{s}) = d$, then $[K^{\mathbf{s}} \backslash Z(u)] \leq d$. Among all such u and $\mathbf{s}$, we now choose a u^* and $\mathbf{s}$, as above, for which $[K^{\mathbf{s}} \backslash Z(u^*)]$ is *maximal*.

Lemma 4.10. *Let u^* and $\mathbf{s}$ be as above. If $v \in U(u^*; \mathbf{s})$, then*

$$[K^{\mathbf{s}} \backslash Z(u^* - v)] \leq [K^{\mathbf{s}} \backslash Z(u^*)].$$

Proof. Set $u = u^* - v$, and assume that $[K^{\mathbf{s}} \backslash Z(u)] > [K^{\mathbf{s}} \backslash Z(u^*)]$. Since $v \in U(u^*; \mathbf{s})$ we have $v = 0$ on all A_i with $i \notin I^{\mathbf{s}}$. Thus $u = u^*$ on A_i, $i \notin I^{\mathbf{s}}$. Set $K^{\mathbf{s}} \backslash Z(u) = \bigcup_{j \in J} B_j$. Then $K \backslash Z(u) = (\bigcup_{j \in J} B_j) \cup (\bigcup_{i \notin I^{\mathbf{s}}} A_i)$. Since $|J| = [K^{\mathbf{s}} \backslash Z(u)] > [K^{\mathbf{s}} \backslash Z(u^*)] = |I^{\mathbf{s}}|$, it follows that $m(u) > n$. We now define an annihilator $\mathbf{t}$ for u by setting $t_i = s_i$ for all $i \notin I^{\mathbf{s}}$, and $t_j = 0$ for all $j \in J$. $\mathbf{t}$ is an annihilator for u since $\mathbf{s}$ is an annihilator for u^*. Furthermore, $I^{\mathbf{t}}$ is necessarily a subset of J, and $U(u; \mathbf{t}) \subseteq U(u^*; \mathbf{s})$. From the minimality property of $U(u^*; \mathbf{s})$ we obtain $U(u; \mathbf{t}) = U(u^*; \mathbf{s})$. Thus, in particular, $v \in U(u; \mathbf{t})$.

We now assert that $I^{\mathbf{t}} = J$, i.e., for each $j \in J$ there exists an element of $U(u; \mathbf{t})$ which does not vanish identically on B_j. Assume to the contrary that there exists a $k \in J$ such that every element of $U(u; t)$ vanishes identically on B_k. Thus $v = 0$ on B_k. Since $u = u^* - v$, and B_k is an open connected component of $K \backslash Z(u)$, we have that u^* vanishes identically on the boundary of B_k, and $B_k = A_i$ for some $i \in I^s$. Therefore every element of $U(u; t)$ vanishes identically on this A_i. From the definition of I^s, it now follows that $U(u; t) \subset U(u^*; s)$. This contradicts the minimality of the dimension of $U(u^*; s)$. Therefore $I^t = J$.

Since $I^t = J$ we have, by definition, $K^t = K^s$ and $K^t \backslash Z(u) = K^s \backslash Z(u)$. Furthermore, since $U(u; t) = U(u^*; s)$ the maximality of $[K^s \backslash Z(u^*)]$ now implies

$$[K^s \backslash Z(u)] = [K^t \backslash Z(u)] \leq [K^s \backslash Z(u^*)].$$

This contradiction proves the lemma. □

Let u^* and s be as defined earlier. From Property A there must exist a $v \in U(u^*)$, $v \neq 0$, for which $s_i v \geq 0$ on A_i, $i \notin I^s$, and $u^* v \geq 0$ on A_i, $i \in I^s$. The first of these two conditions implies that $v \in U(u^*; s)$. Thus in particular $v = 0$ on A_i for all $i \notin I^s$.

Lemma 4.11. *Let u^*, s and v be as above. For each $i \in I^s$ there exists a constant $\alpha_i \geq 0$ such that $\alpha_i u^* = v$ on A_i.*

Proof. We first prove that for all $x \in \overline{A}_k \cap \overline{A}_j$, $k \neq j$ (if such x exist), we have $v(x) = 0$. Assume to the contrary that there exist k and j, $k \neq j$, and an $x_0 \in \overline{A}_k \cap \overline{A}_j$ for which $v(x_0) \neq 0$. Since $v = 0$ on A_i for all $i \notin I^s$, we necessarily have $k, j \in I^s$. Choose $y_i \in A_i$, $i \in I^s$. By definition, $u^*(y_i) \neq 0$ for each such i. Because $|I^s| \leq d (< n)$ by Lemma 4.9, there exists a $\delta > 0$ such that $|\delta v(y_i)| < |u^*(y_i)|$ for all $i \in I^s$. From Lemma 4.10, $[K^s \backslash Z(u^* - \delta v)] \leq [K^s \backslash Z(u^*)] = |I^s|$. We contradict this inequality by showing that each of the points $\{y_i\}_{i \in I^s}$ and x_0 all lie in distinct open connected components of $K^s \backslash Z(u^* - \delta v)$.

Set $\varepsilon_i = \text{sgn } u^*$ on A_i, $i \in I^s$. From our construction of v, we have $\varepsilon_i v \geq 0$ on A_i, $i \in I^s$. Since $\varepsilon_i (u^* - \delta v)(y_i) > 0$, while $\varepsilon_i (u^* - \delta v)(x) = -\varepsilon_i \delta v(x) \leq 0$ on the (relative) boundary of A_i, it follows that the open connected component of $K^s \backslash Z(u^* - \delta v)$ containing y_i is itself contained in A_i. Now x_0 is not contained in any A_i, and $(u^* - \delta v)(x_0) = -\delta v(x_0) \neq 0$. Thus x_0 belongs to yet another component of $K^s \backslash Z(u^* - \delta v)$. This contradicts Lemma 4.10.

We now assume that v is not proportional to u^* on A_j for some $j \in I^s$. There then exists an $\alpha > 0$ such that $u^* - \alpha v$ takes on both positive and negative values on A_j, and $u^* - \alpha v$ does not vanish identically on any A_i, $i \in I^s$. Since v and u^* vanish on the (relative) boundary of A_i for each $i \in I^s$, we obtain

$$[K^s \backslash Z(u^* - \alpha v)] > [K^s \backslash Z(u^*)]$$

which again contradicts Lemma 4.10. □

On the basis of the above results, we may now conclude the existence of a $u^* \in U$ with $m = m(u^*) > \dim U(u^*) = n$, associated components A_i, $i = 1, \ldots, m$, a $v \in U(u^*)$, $v \neq 0$, and constants α_i, $i = 1, \ldots, m$, satisfying:

i) $v = \alpha_i u^*$ on A_i, $i = 1, \ldots, m$

ii) $\alpha_i \geq 0$ and all the α_i are zero except for at most some $d < n$.

We have now arrived at the final step in the proof of Theorem 4.6.

Proof of Theorem 4.6. The above α_i's take on the distinct values β_j, $j = 1, \ldots, k$, where $2 \leq k \leq n$ (by (ii)). Assume that β_j is taken on n_j times, $j = 1, \ldots, k$. Thus $\sum_{j=1}^{k} n_j = m > n$.

Since $\beta_j u^* - v$ vanishes identically on some A_i, it follows that $u^* \notin U(\beta_j u^* - v)$. Thus $\dim U(\beta_j u^* - v) < n$ and, from the induction hypothesis,

$$m - n_j = [K \backslash Z(\beta_j u^* - v)] \leq \dim U(\beta_j u^* - v), \quad j = 1, \ldots, k.$$

Since all but one of the n_j is bounded by d, this immediately implies that m is uniformly bounded above.

Now, consider $\bigcap_{j=1}^{k} U(\beta_j u^* - v)$. If $u \in \bigcap_{j=1}^{k} U(\beta_j u^* - v)$, then u must belong to $U(u^*)$ and vanish identically on each A_i, $i = 1, \ldots, m$, implying that $u = 0$. Thus $\dim \left(\bigcap_{j=1}^{k} U(\beta_j u^* - v) \right) = 0$. We shall now prove that $\dim \left(\bigcap_{j=1}^{k} U(\beta_j u^* - v) \right) > 0$. This contradiction then proves Theorem 4.6.

We shall prove by induction that $\dim \left(\bigcap_{j=1}^{r} U(\beta_j u^* - v) \right) > m - (n_1 + \cdots + n_r)$ for $r = 2, \ldots, k$. The case $r = k$ proves the desired result. For $r = 2$,

$$\dim \left(U(\beta_1 u^* - v) \cap U(\beta_2 u^* - v) \right) = \dim \left(U(\beta_1 u^* - v) \right) + \dim \left(U(\beta_2 u^* - v) \right)$$
$$- \dim \left(U(\beta_1 u^* - v) + U(\beta_2 u^* - v) \right)$$
$$\geq (m - n_1) + (m - n_2) - n$$
$$> m - (n_1 + n_2),$$

since $\dim(U(\beta_j u^* - v)) \geq m - n_j$, all j, and $\dim(U(\beta_1 u^* - v) + U(\beta_2 u^* - v)) \leq \dim U = n < m$. Assume that the desired inequality holds for $r - 1, 3 \leq r \leq k$. Then

$$\dim \left(\bigcap_{j=1}^{r} U(\beta_j u^* - v) \right) = \dim \left((\bigcap_{j=1}^{r-1} U(\beta_j u^* - v)) \cap U(\beta_r u^* - v) \right)$$
$$= \dim \left(\bigcap_{j=1}^{r-1} U(\beta_j u^* - v) \right) + \dim \left(U(\beta_r u^* - v) \right)$$
$$- \dim \left((\bigcap_{j=1}^{r-1} U(\beta_j u^* - v)) + U(\beta_r u^* - v) \right)$$
$$> (m - (n_1 + \cdots + n_{r-1})) + (m - n_r) - n$$
$$> m - (n_1 + \cdots + n_r).$$

Thus $\dim \left(\bigcap_{j=1}^{k} U(\beta_j u^* - v) \right) > m - (n_1 + \cdots + n_k) = m - m = 0$. This proves Theorem 4.6.   □

Remark. If $U(u) = U$ for all $u \in U, u \neq 0$ (which is the case, for example, if $Z(u)$ is a set of measure zero for all $u \in U, u \neq 0$), then the above arguments may be considerably simplified. In fact a contradiction will immediately ensue from Property A and Corollary 4.8.

There is one other implication of Property A which holds in this general setting. We first consider a simple example to explain and motivate the subsequent analysis.

Let $K = A_1 \cup A_2$, where A_1 and A_2 are disjoint compact connected subsets of $I\!\!R^d$, and $K = \overline{\text{int } K}$. Assume $\dim U = 2$, and $\dim U|_{A_1} = 2$, while $\dim U|_{A_2} \geq 1$. (By $U|_{A_i}$, we mean the restriction of the space U to the set A_i, $i = 1, 2$.) Now there necessarily exists a $u \in U$ which has a sign change on A_1. To see this, let $x_1, x_2 \in A_1$ where U is linearly independent on these two points, and let $u \in U$ satisfy $u(x_i) = (-1)^i$, $i = 1, 2$. Thus $[A_1 \backslash Z(u)] \geq 2$. If u identically vanishes on A_2, choose $v \in U$ which does not identically vanish on A_2, and replace u by $u + \varepsilon v$ with $\varepsilon \neq 0$, ε small. This new u still has a sign change on A_1 and does not vanish identically on A_2. Thus $[K \backslash Z(u)] \geq 3$, implying by Theorem 4.6 that U does not satisfy Property A. Of course there do exist U of dimension 2 on K which satisfy Property A. But for this to happen, it is necessary (but hardly sufficient) that $\dim U|_{A_1} + \dim U|_{A_2} = 2$. In other words, either all functions in U vanish identically on one of the two sets A_1 or A_2, or there exists a basis u_1, u_2 for U such that u_i vanishes identically off A_i, $i = 1, 2$.

As the above example indicates, if Property A is to hold, then there seems to be a restriction on U related in some way to the geometry (connectivity) of K. To explain more precisely what is occurring, we need to define some terms.

We remind the reader of some notation which has already been used, or which should be readily understood.

a) For $f \in C(K)$, supp $f = \{x : f(x) \neq 0\} = K \backslash Z(f)$. (This is an open set in K.)

b) For $U \subseteq C(K)$, and $B \subseteq K$, $U|_B$ will denote the set (or space) U restricted to the set B.

c) For $U \subseteq C(K)$, $Z(U) = \bigcap \{Z(u) : u \in U\}$.

d) For $U \subseteq C(K)$, supp $U = K \backslash Z(U)$ (i.e., supp $U = \bigcup \{\text{supp } u : u \in U\}$).

Definition 4.7. Assume U is a finite-dimensional subspace of $C(K)$. We shall say that U *decomposes* if there exist non-trivial subspaces V and W of U (i.e., each of positive dimension) such that

1) $U = V \oplus W$, i.e., $U = V + W$ and $V \cap W = \{0\}$;

2) supp $V \cap$ supp $W = \emptyset$.

In other words, U decomposes if there exist subsets B and C of K for which $B \cup C = K$ and $B \cap C = \emptyset$, and subspaces V and W of U (each of positive dimension) such that $V + W = U$, and every function in V vanishes identically off B, while every function in W vanishes identically off C.

Assume U decomposes with V, W, B and C as above. Why is this of interest to us? Let us go back to the original motivation for all our discussions, namely the approximation problem. Since U decomposes, we have for each $f \in C(K)$ and any measure μ,

$$\min_{u \in U} \int_K |f - u| d\mu = \min_{v \in V} \int_B |f - v| d\mu + \min_{w \in W} \int_C |f - w| d\mu \, .$$

That is, our approximation problem decomposes into two totally independent approximation problems. In particular, if U decomposes, then U is a unicity space for $C_1(K, \mu)$ if and only if V and W are unicity spaces for $C_1(B, \mu)$ and $C_1(C, \mu)$, respectively. Similarly U satisfies Property A if and only if both V and W satisfy Property A.

There is generally little reason why a given space should 'decompose'. However if we consider the last example, we see that, if U satisfies Property A, then either all functions in U have support on only one of the sets A_i, $i = 1, 2$, or U decomposes with $B = A_1$ and $C = A_2$. In general, we have:

Theorem 4.12. *Let U be a finite-dimensional subspace of $C(K)$ which satisfies Property A. If $[K \backslash Z(U)] \geq 2$, then U decomposes.*

In Example 4.6, $[K \backslash Z(U)] = 2$ and U does not decompose. This is one reason why the subspace U does not satisfy Property A. Theorem 4.12 is a necessary condition on U if it is to satisfy Property A. To be more precise, repeated application of Theorem 4.12 implies the following:

Assume U satisfies Property A. Let $K \backslash Z(U) = \bigcup_{i=1}^{r} A_i$, where the A_i are open (disjoint) connected components of K. Let $n_i = \dim U|_{A_i}$, $i = 1, \ldots, r$. Then $n_i \geq 1$, all i, and $\sum_{i=1}^{r} n_i = n$. That is, for each i, there exists a basis for $U|_{A_i}$, each of whose elements vanish identically off A_i. Furthermore, when U decomposes as above, then U satisfies Property A if and only if $U|_{A_i}$ satisfies Property A for each $i = 1, \ldots, r$.

The main tool to be used in the proof of Theorem 4.12 is the following proposition.

Proposition 4.13. *Let $W = \mathrm{span}\{w_1, \ldots, w_r\}$ be an r-dimensional subspace of $C(K)$. Assume there exists an M, finite, such that for all $w \in W$, $[K \backslash Z(w)] \leq M$. There then exists a $w^* \in W$ of the form*

$$w^* = w_1 + \sum_{i=2}^{r} a_i w_i$$

with the property that if $w \in W(w^*)$ satisfies $w(\mathrm{sgn}\, w^*) = |w|$ on $K \backslash Z(w^*)$, then $w = \alpha w^*$ for some $\alpha \geq 0$.

Note that w^* is not a totally arbitrary non-trivial element of W. We wrote w^* as a linear combination of the basis elements $\{w_1, \ldots, w_r\}$ with the coefficient of w_1 not zero (here normalized to be one). This is an important fact which will be used.

We divide the proof of the proposition into a series of lemmas. In each of these lemmas we assume that the conditions of the proposition hold.

Lemma 4.14. *Let $v_i \in W$, $v_i \neq 0$, $i = 1, \ldots, k$. Assume int $Z(v_i) \subset$ int $Z(v_{i+1})$, $i = 1, \ldots, k - 1$. Then the $\{v_i\}_{i=1}^{k}$ are linearly independent.*

Proof. The lemma is obviously true for $k = 1$ since $v_1 \neq 0$. We shall therefore assume that $v_1, \ldots, v_{k-1}$ are linearly independent, while $v_1, \ldots, v_k$ are linearly dependent. Thus $v_k = \sum_{i=1}^{k-1} a_i v_i$ with $(a_1, \ldots, a_{k-1}) \neq \mathbf{0}$. On int $Z(v_k)$, we have $\sum_{i=1}^{k-1} a_i v_i = 0$, while on int $Z(v_2) \subseteq$ int $Z(v_k)$, $0 = \sum_{i=1}^{k-1} a_i v_i = a_1 v_1$. But there exists an $x_1 \in$ int $Z(v_2) \backslash$int $Z(v_1)$ for which $v_1(x_1) \neq 0$. Thus $a_1 = 0$. Similarly, on int $Z(v_3)$, $0 = \sum_{i=2}^{k-1} a_i v_i = a_2 v_2$, while there exists an $x_2 \in$ int $Z(v_3) \backslash$ int $Z(v_2)$ for which $v_2(x_2) \neq 0$. Thus $a_2 = 0$. We continue in this fashion to obtain $a_1 = \ldots = a_{k-1} = 0$. This is a contradiction. □

Remark. For a different proof of this result, see the proof of Lemma 3.18.

Set

$$V = \left\{ w : w = w_1 + \sum_{i=2}^{r} a_i w_i \right\}$$

and

$$\widetilde{V} = \{ w : \alpha w \in V \text{ for some } \alpha \neq 0 \}.$$

Lemma 4.15. *There exists a $w^* \in V$ satisfying the following. If $w \in V$ and int $Z(w^*) \subseteq$ int $Z(w)$, then int $Z(w^*) =$ int $Z(w)$, and $[K \backslash Z(w^*)] \geq [K \backslash Z(w)]$.*

Proof. Choose an arbitrary $v_1 \in V$. If there exists a $v_2 \in V$ for which int $Z(v_1) \subset$ int $Z(v_2)$, then replace v_1 by v_2. Continue in this fashion. Since $V \subseteq W$ and dim $W = r$, it follows from Lemma 4.14 that this process must stop after at most r steps. Thus there exists a $\widetilde{w} \in V$ such that, if $w \in V$ and int $Z(\widetilde{w}) \subseteq$ int $Z(w)$, then int $Z(\widetilde{w}) =$ int $Z(w)$.

Among all $w \in V$ satisfying int $Z(\widetilde{w}) =$ int $Z(w)$, choose a $w^* \in V$ for which $[K \backslash Z(w^*)]$ is maximal. Such a choice is possible since $[K \backslash Z(w)] \leq M$ for all $w \in W$. The lemma is proved. □

We shall now prove that the above w^* satisfies the claim of the proposition.

Proof of Proposition 4.13. Let $K \backslash Z(w^*) = \bigcup_{i=1}^{k} A_i$, where the A_i are the open connected components of $K \backslash Z(w^*)$. Thus $k \leq M$. Set $\varepsilon_i = \operatorname{sgn} w^*$ on A_i, $i = 1, \ldots, k$. Assume, contrary to the statement of the proposition, that there exists a $w \in W(w^*)$ satisfying $\varepsilon_i w \geq 0$ on A_i, $i = 1, \ldots, k$, and such that $w \neq \alpha w^*$ for any $\alpha \geq 0$.

We divide the remaining argument into three parts.

Claim 1. *For all $x \in \overline{A}_i \cap \overline{A}_j$, $i, j \in \{1, \ldots, k\}$, $i \neq j$, we have $w(x) = 0$.*

Proof. The argument here is a repeat of an argument used in the proof of Lemma 4.11. Assume there exists an $x_0 \in \overline{A}_i \cap \overline{A}_j$, $i \neq j$, such that $w(x_0) \neq 0$. Let $y_m \in A_m$, $m = 1, \ldots, k$. From the definition of the A_m, there exists a $\delta > 0$, δ small, such that $\varepsilon_m (w^* - \delta w)(y_m) > 0$, $m = 1, \ldots, k$. We can also easily choose δ as above so that $w^* - \delta w \in \widetilde{V}$. By construction int $Z(w^*) \subseteq$ int $Z(w^* - \delta w)$ (since $w \in W(w^*)$). Paralleling the arguments used in Lemma 4.11, it is easily seen that $[K \backslash Z(w^*)] < [K \backslash Z(w^* - \delta w)]$. Renormalizing $w^* - \delta w$ we contradict the extremal properties of w^* as given in Lemma 4.15.

Claim 2. *For each $i \in \{1, \ldots, k\}$ there exists an $\alpha_i \geq 0$ such that $w = \alpha_i w^*$ on A_i.*

Proof. This is again an argument contained in the proof of Lemma 4.11. If there exists an $i \in \{1, \ldots, k\}$ for which the claim does not hold, there then exist $x_1, x_2 \in A_i$ and $\beta > 0$ such that $(w^* - \beta w)(x_1)(w^* - \beta w)(x_2) < 0$. The value β may be perturbed somewhat while maintaining this strict inequality. We therefore assume that $w^* - \beta w \in \widetilde{V}$, and $w^* - \beta w$ does not identically vanish on any A_j, $j = 1, \ldots, k$. We then obtain $[K \backslash Z(w^*)] < [K \backslash Z(w^* - \beta w)]$ and again contradict Lemma 4.15 as in Claim 1.

Claims 1 and 2, together with our assumption on w, imply that $w = \alpha_i w^*$ on each A_i, $i = 1, \ldots, k$, but not all the α_i are equal. Assume $\alpha_i \neq \alpha_j$ for some $i, j \in \{1, \ldots, k\}$, $i \neq j$. In this case, either $\alpha_i w^* - w \in \widetilde{V}$ or $\alpha_j w^* - w \in \widetilde{V}$ (or both). Furthermore, int $Z(w^*) \subset$ int $Z(\alpha w^* - w)$ for both $\alpha = \alpha_i$ and $\alpha = \alpha_j$. This is a consequence of the fact that $\alpha_i w^* - w$ vanishes identically on A_i, and $\alpha_j w^* - w$ vanishes identically on A_j. Again we contradict the choice of w^* as given in Lemma 4.15. This final contradiction proves our proposition.□

We shall now prove Theorem 4.12.

Proof of Theorem 4.12. Let $\dim U = n$, and assume that $[K \backslash Z(U)] \geq 2$. From Theorem 4.6, $[K \backslash Z(U)] \leq n$. Thus $K \backslash Z(U) = \bigcup_{i=1}^{k} A_i$, $2 \leq k \leq n$, where the A_i are the open connected components of $K \backslash Z(U)$. Set $B = A_1$ and $C = K \backslash A_1$. Let $\dim U|_B = m$ and $\dim U|_C = r$. Then $m, r \geq 1$, and $m + r \geq n$. The assertion of the theorem is equivalent to the claim that $m + r = n$. We therefore assume that $m + r > n$. We shall prove that this assumption contradicts Property A.

Let $l = m + r - n$. It is easily shown that there exists a basis $u_1, \ldots, u_l$, $v_{l+1}, \ldots, v_m, w_{l+1}, \ldots, w_r$ for U such that if $u|_B = u'$ and $u|_C = u''$, then

$$U|_B = \text{span}\{u_1', \ldots, u_l', v_{l+1}', \ldots, v_m'\}$$

$$U|_C = \text{span}\{u_1'', \ldots, u_l'', w_{l+1}'', \ldots, w_r''\}$$

while $v_i'' = 0, i = l+1, \ldots, m$, and $w_i' = 0, i = l+1, \ldots, r$.

Now, the conditions of Proposition 4.13 hold on $\overline{B}$. There therefore exists a function $v^* \in U$ of the form

$$v^* = u_1 + \sum_{i=2}^{l} a_i^* u_i + \sum_{i=l+1}^{m} b_i^* v_i$$

such that, if $v \in U$ satisfies

(4.1)
$$\begin{aligned} &a)\, v = 0 \text{ a.e. on } Z_{\overline{B}}(v^*) \\ &b)\, v(\operatorname{sgn} v^*) = |v| \text{ on } \overline{B} \backslash Z_{\overline{B}}(v^*), \end{aligned}$$

then $v = \alpha v^*$ on $\overline{B}$ for some $\alpha \geq 0$.

Set $u^* = u_1 + \sum_{i=2}^{l} a_i^* u_i$, and $W = \operatorname{span}\{u^*, w_{l+1}, \ldots, w_r\}$. We now apply Proposition 4.13 to $\overline{C}$ and W. Thus there exists a function $w^* \in W$ of the form

$$w^* = u^* + \sum_{i=l+1}^{r} c_i^* w_i$$

such that, if $w \in W$ satisfies

(4.2)
$$\begin{aligned} &a)\, w = 0 \text{ a.e. on } Z_{\overline{C}}(w^*) \\ &b)\, w(\operatorname{sgn} w^*) = |w| \text{ on } \overline{C} \backslash Z_{\overline{C}}(w^*), \end{aligned}$$

then $w = \beta w^*$ on $\overline{C}$ for some $\beta \geq 0$. Set

$$\widetilde{u} = u^* + \sum_{i=l+1}^{m} b_i^* v_i + \sum_{i=l+1}^{r} c_i^* w_i.$$

Then $\widetilde{u}|_B = v^*|_B$ and $\widetilde{u}|_C = w^*|_C$. As a consequence of Property A and the construction of B and C there necessarily exists a function $u \in U$, $u \neq 0$, satisfying

(4.3)
$$\begin{aligned} &i)\, u = 0 \text{ a.e. on } Z(\widetilde{u}) \\ &ii)\, u(\operatorname{sgn} \widetilde{u}) \geq 0 \text{ on } B \\ &iii)\, u(\operatorname{sgn} \widetilde{u}) \leq 0 \text{ on } C. \end{aligned}$$

From (4.1) ($\widetilde{u}|_B = v^*|_B$) and (4.3), (i) and (ii), it follows that $u = \alpha(u^* + \sum_{i=l+1}^{m} b_i^* v_i) + w$ for some $\alpha \geq 0$ and $w \in \operatorname{span}\{w_{l+1}, \ldots, w_r\}$. Thus $u \in W$ on C. From (4.2) ($\widetilde{u}|_C = w^*|_C$) and (4.3), (i) and (iii), it follows that $u = \beta(u^* + \sum_{i=l+1}^{r} c_i^* w_i) + v$ for some $\beta \leq 0$ and $v \in \operatorname{span}\{v_{l+1}, \ldots, v_m\}$. We have therefore proved that $u = \alpha \widetilde{u}$ and also that $u = \beta \widetilde{u}$. But $\alpha \geq 0$ and $\beta \leq 0$. This implies that $u = 0$, contradicting Property A. □

4. Characterizing Property A on $\mathbb{R}$

As noted in the previous section, Property A is generally difficult either to verify or disprove for a given finite-dimensional subspace U of $C(K)$. The two results of the previous section are necessary conditions which U must satisfy in order for Property A to hold. In this section we totally classify all U which satisfy Property A in the case where $K \subset \mathbb{R}$. (Recall that we always assume that K is compact and $K = \overline{\text{int}\, K}$.)

The first fact to be noted is that on the basis of Theorem 4.12 we can and shall make the following assumption, since we are only interested in U satisfying Property A.

Assumption I: $K = [a, b]$ and for each $x \in (a, b)$ there exists a $u \in U$ for which $u(x) \neq 0$.

For $K \subset \mathbb{R}$, $K = \overline{\text{int}\, K}$, the above assumption is equivalent to the demand that $[K \backslash Z(U)] = 1$ and K is 'minimal'. (By 'minimal' we mean that if $K = [c, d]$ and all $u \in U$ vanish identically on $[c, a] \cup [b, d]$, $c \leq a < b \leq d$, then we simply reduce K to $[a, b]$ with no loss of generality.)

In the characterization of the U satisfying Property A, we make use of Chebyshev (T-) systems, and weak Chebyshev (WT-) systems. Details and definitions concerning T- and WT-systems are to be found in Appendix A. In fact we shall use various results from Appendix A in this section, and now is the perfect time for the reader to skim through Appendix A. Let us just remind the reader that an n-dimensional subspace U of $C[a, b]$ is a T-system on $[a, b]$ if no $u \in U, u \neq 0$, vanishes at more than $n - 1$ points in $[a, b]$. It is a WT-system if no $u \in U$ has more than $n - 1$ sign changes on $[a, b]$. T-systems and, more importantly, WT-systems, and Property A are interrelated notions on $\mathbb{R}$ due to the inequality of Theorem 4.6, namely $[[a, b] \backslash Z(u)] \leq \dim U(u) \leq \dim U = n$. This inequality immediately implies that if U satisfies Property A then $u \in U$ has at most $n - 1$ sign changes, and therefore U is a WT-system. However the converse result need not hold. For example, let $K = [-2, 2]$ and $U = \text{span}\, \{u_1, u_2\}$, where $u_1(x) = 1$, and

$$u_2(x) = \begin{cases} x - 1, & 1 \leq x \leq 2 \\ 0, & -1 \leq x \leq 1 \\ x + 1, & -2 \leq x \leq -1 \, . \end{cases}$$

The space U is a WT-system on $[-2, 2]$. But U does not satisfy Property A since $[K \backslash Z(u_2)] = 2$, and $\dim U(u_2) = 1$. The problem here is that $U(u_2)$ is not a WT-system.

The subspaces U satisfying Property A are a subset of the set of WT-systems. The exact nature of this subset is the content of our main theorem. Before stating this theorem we present the following definition.

Definition 4.8. We say that $[c,d] \subseteq [a,b]$ is a *zero interval* of $u \in C[a,b]$ if $u = 0$ on $[c,d]$, and u does not vanish identically on $(c - \varepsilon, c)$ for any $\varepsilon > 0$ if $c > a$, while u does not vanish identically on $(d, d + \varepsilon)$ for any $\varepsilon > 0$ if $d < b$.

The main theorem of this section is the following.

Theorem 4.16. *Assume* U *is a finite-dimensional subspace of* $C[a,b]$ *and Assumption I holds. Then the following are equivalent.*

1) *U satisfies Property A.*
2) $[[a,b] \backslash Z(u)] \le \dim U(u)$ *for all* $u \in U$.
3) *(a) If* $[c,d]$, $a < c < d < b$, *is a zero interval of* $u \in U$, *then*
   (i) *there exists a* $v \in U$ *for which*

$$v(x) = \begin{cases} u(x), & a \le x \le c \\ 0, & c < x \le d, \end{cases}$$

   (ii) *there exists a* $w \in U$ *for which*

$$w(x) = \begin{cases} 0, & a \le x < d \\ u(x), & d \le x \le b. \end{cases}$$

(b) For $a \le c < d \le b$, *let*

$$V_{c,d} = \{u : u \in U, u = 0 \text{ on } [a,c) \cup (d,b]\} \,.$$

Then each $V_{c,d}$ *is a WT-system (if* $V_{c,d} \ne \{0\}$). *Furthermore, if* $v \in V_{c,d}$ *has no zero interval in* $[c,d]$, *then* v *has at most* $\dim V_{c,d} - 1$ *distinct zeros in* (c,d).

Note that (3)(a)(ii) is a totally redundant statement (take $w = u - v$). It has been included for æsthetic reasons. Before proving the theorem, we wish to consider (3) in considerably more detail. What has been listed under (3) is essentially the bare bones needed for proving its equivalence to (1) and (2). But (3) implies somewhat more about U then has been stated. The implications of (3) are considered below.

Proposition 4.17. *Assume* U *is an* n-*dimensional subspace of* $C[a,b]$, *Assumption I holds, and* U *satisfies* (3) *of Theorem 4.16.*

    A) *The collection of endpoints of zero intervals of all* $u \in U$ *is* $\{e_i\}_{i=0}^{k+1}$, *where* $a = e_0 < e_1 < \cdots < e_{k+1} = b$, *and* $k \le 2n - 2$.
    B) $U|_{(e_{i-1}, e_i)}$ *is a T-system on* (e_{i-1}, e_i), $i = 1, \ldots, k+1$.
Therefore U *satisfies* (3)(b) *if it satisfies* (3)(b) *only for* $c = e_i$ *and* $d = e_j$, *all* $0 \le i < j \le k+1$.

Proof. We claim that (A) follows from (3)(a). We shall prove that there exist at most n values $\{d_i\}_{i=1}^{r+1}$, $r + 1 \le n$, $d_0 = a < d_1 < \cdots < d_r < b = d_{r+1}$, such that d_i is a right endpoint of a zero interval of some $u \in U$. By totally analogous reasoning, it also follows that there exist $\{c_i\}_{i=0}^m$, $c_0 = a < c_1 <$

$\cdots < c_m < b = c_{m+1}$, $m + 1 \leq n$, such that some c_i is a left endpoint of any zero interval of some $u \in U$. Setting $\{e_i\}_{i=0}^{k+1} = \{c_i\}_{i=0}^{m+1} \cup \{d_i\}_{i=0}^{r+1}$ proves (A).

Assume $[c, d]$ is a zero interval of some $u \in U$, and $d < b$. From (3)(a), it follows that $[a, d]$ is also a zero interval of some other $u \in U$. Thus it suffices to prove that there do not exist $a < d_1 < \cdots < d_n < b$ such that $[a, d_i]$ is a zero interval of $u_i \in U$, $i = 1, \ldots, n$. Assume to the contrary that such $\{d_i\}_{i=1}^{n}$ and $\{u_i\}_{i=1}^{n}$ do exist. Let $d_0 = a$ and $d_{n+1} = b$. Choose $x_i \in (d_i, d_{i+1})$ such that $u_i(x_i) \neq 0$, $i = 1, \ldots, n$. From the definition of a zero interval, such x_i exist. Furthermore from Assumption I there exists an $x_0 \in (d_0, d_1)$ and a $u_0 \in U$ such that $u_0(x_0) \neq 0$. Now, consider the matrix $(u_i(x_j))_{i,j=0}^{n}$. This is an $(n + 1) \times (n + 1)$ triangular matrix with non-zero diagonal entries. Thus the $u_0, u_1, \ldots, u_n$ are linearly independent. But $\dim U = n$. This is a contradiction and (A) is proved.

To prove (B) we use various results from Appendix A. From (3)(b) with $c = a, d = b$, we have that U is a WT-system on $[a, b]$. Thus $U|_{[c,d]}$ is a WT-system for every $a \leq c < d \leq b$ from Proposition 7 of Appendix A. Since no $u \in U|_{(e_{i-1}, e_i)}$ vanishes identically on a subinterval of (e_{i-1}, e_i) unless $u = 0$ on all of (e_{i-1}, e_i) (see (A)), it now follows from Assumption I and Corollary 10 of Appendix A that $U|_{(e_{i-1}, e_i)}$ is a T-system on (e_{i-1}, e_i), $i = 1, \ldots, k + 1$. We have proved (B).

Now, consider $V_{c,d}$ as defined in (3)(b). If $e_{i-1} < c \leq e_i$, and $e_j \leq d < e_{j-1}$, and $u \in V_{c,d}$, then $u \in V_{e_i, e_j}$. This easily follows from (A). Thus, in verifying (3)(b) it suffices to verify it only for $c = e_i$ and $d = e_j$, where $0 \leq i < j \leq k + 1$.   □

Remark. The last sentence in (3)(b) also deserves additional comment. If $\dim V_{e_i, e_j} \neq 0$, and to each $x \in (e_i, e_j)$ there exists a $v \in V_{e_i, e_j}$ for which $v(x) \neq 0$, then the condition therein necessarily holds (see Corollary 9 of Appendix A).

We shall now prove Theorem 4.16.

Proof of Theorem 4.16. (1) $\Rightarrow$ (2). This is the content of Theorem 4.6.

(2) $\Rightarrow$ (3). We first prove (3)(b). Assume there exists a $c < d$ such that $\dim V_{c,d} = m \geq 1$, and $V_{c,d}$ is not a WT-system. Since $V_{c,d}$ is not a WT-system, there exists a $v \in V_{c,d}$ with at least m sign changes. Thus $[[a, b] \backslash Z(v)] \geq m + 1$. But from (2), $[[a, b] \backslash Z(v)] \leq \dim U(v)$. Since $U(v) \subseteq V_{c,d}$ (by definition), we have $\dim U(v) \leq m$. This contradiction implies that $V_{c,d}$ is a WT-system. Assume $v \in V_{c,d}$ has no zero interval in $[c, d]$. If v has at least $\dim V_{c,d}$ distinct zeros in (c, d), then $[[a, b] \backslash Z(v)] \geq \dim V_{c,d} + 1 \geq \dim U(v) + 1$, contradicting (2). Thus (3)(b) holds.

The more difficult part of the proof is the assertion that (2) implies (3)(a). Assume $u^* \in U$ has a zero interval $[c, d]$ where $a < c < d < b$. Then $U(u^*)$ is a subspace of U of dimension m, where $1 \le m < n$. Set $V = U(u^*)|_{[a,c]}$ and assume that $\dim V = r$, $1 \le r \le m$.

We first claim that $r < m$. If $r = m$, there then exist points $a < x_1 < \cdots < x_m < c$ and a $v \in U(u^*)$ for which $v(x_i) = (-1)^i$, $i = 1, \ldots, m$. Thus $[[a, c] \backslash Z(v)] \ge m$. Therefore $[[a, b] \backslash Z(v)] \ge m \ge \dim U(v)$ since $v \in U(u^*)$. For equality to hold, it is necessary that $U(v) = U(u^*)$, and also that v vanish identically on $[d, b]$. But then every $u \in U(v) = U(u^*)$ must necessarily vanish identically on $[d, b]$, contradicting the fact that $u^* \in U(u^*)$. Thus $1 \le r < m$.

Since $\dim V = r < m$, there exist $m - r$ linearly independent functions in $U(u^*)$ which identically vanish on $[a, c]$. Set

$$W = \{u : u \in U(u^*),\ u = 0 \text{ on } [a, c]\}.$$

W is a subspace of $U(u^*)$ of dimension $m - r$. Let $v_1, \ldots, v_r \in U(u^*)$ such that the $\{v_i|_{[a,c]}\}_{i=1}^r$ are a basis for V. We can and shall choose each v_i with at least $r - 1$ sign changes on $[a, c]$. We claim that for each $i \in \{1, \ldots, r\}$ there exists a $w_i \in W$ such that $v_i - w_i = 0$ on $[d, b]$. If this is true, then setting $v_i' = v_i - w_i$, $i = 1, \ldots, r$, we shall have found a basis for $U(u^*)|_{[a,c]}$, each of whose elements vanish identically outside $[a, c]$. Condition (3)(a) is an immediate consequence of this fact.

It thus remains to prove that $v_i|_{[d,b]} \in W$ for each $i \in \{1, \ldots, r\}$. Assume not. Since $\dim W = m - r$ and $v_i|_{[d,b]} \notin W$, there therefore exists a $w \in W$ and $\alpha \in \mathbb{R}$ such that $w + \alpha v_i$ has at least $m - r$ sign changes on the interval $[d, b]$. Furthermore, since these are sign changes, we can perturb α slightly without altering this fact. Thus we may assume that $\alpha \ne 0$, and $[[d, b] \backslash Z(w + \alpha v_i)] \ge m - r + 1$. On $[a, c]$, $w + \alpha v_i = \alpha v_i$. Thus $[[a, c] \backslash Z(w + \alpha v_i)] \ge r$. Because $w + \alpha v_i = 0$ on $[c, d]$, we have that $[[a, b] \backslash Z(w + \alpha v_i)] \ge (m - r + 1) + r = m + 1 > \dim U(u^*)$. Since $w + \alpha v_i \in U(u^*)$, this is a contradiction to (2), and we have proved (3)(a).

(3) $\Rightarrow$ (1). Let $u^* \in U$, $u^* \ne 0$. Set $[a, b] \backslash Z(u^*) = \bigcup_{i=1}^m A_i$. We must prove that for any choice of $\varepsilon_i \in \{-1, 1\}$, $i = 1, \ldots, m$, there exists a $u \in U(u^*)$, $u \ne 0$, such that $\varepsilon_i u \ge 0$ on A_i, $i = 1, \ldots, m$.

Since U satisfies (3), it also satisfies the conditions of Proposition 4.17. Let $a = e_0 < e_1 < \cdots < e_{k+1} = b$ be as in Proposition 4.17. From (A) of Proposition 4.17, u^* has at most a finite number of zero intervals and they are each of the form $[e_p, e_q]$ for some $0 \le p < q \le k + 1$. Let J denote the union of these intervals. It also follows from Proposition 4.17 that u^* has at most a finite number of zeros off J. Thus $m < \infty$, and $J = \overline{\text{int } Z(u^*)}$.

Let $[e_i, e_j]$ denote the closure of one of the connected components of the complement of J, $0 \le i < j \le k + 1$. That is, $(e_i, e_j) \cap J = \emptyset$, and if $i \ge 1$, then $[e_{i-1}, e_i] \subseteq J$, while if $j \le k$, then $[e_j, e_{j+1}] \subseteq J$. Applying (3)(a), we see that there exists a $v \in U$ satisfying

$$v(x) = \begin{cases} u^*(x), & e_i \le x \le e_j \\ 0, & x \notin [e_i, e_j]. \end{cases}$$

In order to prove our claim, it suffices to prove that, if $[[e_i, e_j]\backslash Z(v)] = \bigcup_{l=1}^{r} B_l$, then for any choice of $\varepsilon_l \in \{-1, 1\}$, $l = 1, \ldots, r$, there exists a $u \in U(v)$, $u \neq 0$, such that $\varepsilon_l u \geq 0$ on B_l, $l = 1, \ldots, r$. Since u^* has no zero interval in $[e_i, e_j]$, v has no zero interval in $[e_i, e_j]$. Furthermore $v \in V_{e_i, e_j}$ (= V) since $v = 0$ on $[a, e_i) \cup (e_j, b]$. From (3)(b), V is a WT-system. Let $\{x_l\}_{l=1}^{r-1}$, $e_i < x_1 < \cdots < x_{r-1} < e_j$, denote the zeros of v in (e_i, e_j). (If v has no zeros in (e_i, e_j), there is nothing to prove.) From the second half of (3)(b) we have that $r \leq s$, where $s = \dim V$. Set $x_0 = e_i$, $x_r = e_j$. Thus $[e_i, e_j]\backslash Z(v) = \bigcup_{l=1}^{r} B_l$, where $B_l = (x_{l-1}, x_l)$, $l = 2, \ldots, r - 1$, $B_1 = [x_0, x_1)$ or (x_0, x_1), and $B_r = (x_{r-1}, x_r]$ or (x_{r-1}, x_r). Choose $\varepsilon_l \in \{-1, 1\}$, $l = 1, \ldots, r$. From the sequence $(\varepsilon_1, \ldots, \varepsilon_r)$ form the sequence $(\delta_1, \ldots, \delta_p)$ where $\delta_i = \varepsilon_1 (-1)^{i+1}$, $i = 1, \ldots, p$, and the number of sign changes in the two sequences $(\varepsilon_1, \ldots, \varepsilon_r)$ and $(\delta_1, \ldots, \delta_p)$ is the same. Thus $p \leq r \leq s$. Set $C_1 = (x_0, x_{i_1})$ where $\varepsilon_1 = \ldots = \varepsilon_{i_1}$, $\varepsilon_{i_1} \varepsilon_{i_1 + 1} = -1$. Set $C_2 = (x_{i_1}, x_{i_2})$ where $\varepsilon_{i_1 + 1} = \ldots = \varepsilon_{i_2}$, $\varepsilon_{i_2} \varepsilon_{i_2 + 1} = -1$, etc., to obtain $C_1, \ldots, C_p$. We must prove the existence of a $u \in V$, $u \neq 0$, for which $\delta_i u \geq 0$ on C_i, $i = 1, \ldots, p$. Since V is a WT-system of dimension s, this follows from Corollary 12 of Appendix A. Thus U satisfies Property A.   □

Remark. In proving (3)(a) from (2), we essentially gave a different proof of Theorem 4.12 in this simpler case where $K = [a, b]$. It is worth noting that it is very easy to prove that (1) implies (3)(a). If (1) holds and u has a zero interval $[c, d]$, $a < c < d < b$, then $[[a, b]\backslash Z(U(u))] \geq 2$, and we can then immediately apply Theorem 4.12 to obtain (3)(a).

As an application of Theorems 4.16 and 4.12, we have the following corollary.

Corollary 4.18. *Let U be a finite-dimensional subspace of $C[a, b]$. Assume that no $u \in U$, $u \neq 0$, has zero intervals. Then U satisfies Property A if and only if U is a T-system on (a, b).*

Note that in the above corollary it is not necessary to assume that Assumption I holds. If the hypotheses of Assumption I do not hold then either from Theorem 4.12, or because $Z(U)$ itself contains an interval, it follows that there exists a $u \in U$, $u \neq 0$, which has zero intervals.

One additional consequence of Theorems 4.16 and 4.12 deserves mention. In Section 2, Definition 4.1, we defined Property A_μ where the measure μ was 'admissible'. On the basis of Theorems 4.12 and 4.16, it may be seen that if U satisfies Property A, then it also satisfies Property A_μ with respect to any such μ. This follows from the fact that, for any $u \in U$, $Z(u)\backslash Z(U)$ is the union of a finite number of intervals and points. Thus Property A on $K \subset \mathbb{R}$, K compact, satisfying $K = \overline{\text{int } K}$, is in fact μ independent.

We end this section by considering some examples of spaces satisfying Property A where $K = [a, b]$.

Example 4.7. U *is a T-system on* (a, b).

As noted in Corollary 4.18, U satisfies Property A. To be precise, this follows from Theorem 4.16 in any number of ways. For example, for every $u \in U, u \neq 0$, we have $U(u) = U$, while u has at most $\dim U - 1$ zeros in (a, b), and therefore $[[a, b] \backslash Z(u)] \leq \dim U = \dim U(u)$. Thus (2) of Theorem 4.16 holds. Condition (3) of Theorem 4.16 holds since the situation in (3)(a) can never occur, while (3)(b) is essentially vacuous as well ($V_{c,d} = \{0\}$ unless $c = a$ and $d = b$).

It has been known for many years that, if U is a T-system on (a, b), then U is a unicity space for $C_1([a, b], \mu)$ for all 'admissible' μ. Let us consider two other methods of proving this result without recourse to the theory developed in connection with Property A.

Method 1. In Theorem 3.1 it was proved that U is a unicity space for $C_1([a, b], \mu)$ if and only if there does *not* exist an $h \in L^{\infty}[a, b]$, and a $u^* \in U, u^* \neq 0$, for which

1) $|h| = 1$ on $[a, b]$

2) $\int_a^b hu\, d\mu = 0$, all $u \in U$

3) $h|u^*| \in C[a, b]$.

Let us assume that U is *not* a unicity space for $C_1([a, b], \mu)$ and h, u^* are as above. For (3) to hold it is necessary that u^* vanish at each point of discontinuity (jump) of h on (a, b). Since $u^* \in U$, and U is a T-system on (a, b), u^* has at most $\dim U - 1$ zeros on (a, b). Thus h can have no more than $\dim U - 1$ points of discontinuity in (a, b). However in order for (2) to hold, it is necessary, since U is a T-system on (a, b) (see Proposition 3 of Appendix B), that h have at least $\dim U$ sign changes and hence points of discontinuity in (a, b). This contradiction implies that U is a unicity space for $C_1([a, b], \mu)$.

Method 2. We recall (Theorem 2.1) that $u^* \in U$ is a best approximant to $f \in C[a, b]$ in the $L^1([a, b], \mu)$ norm if and only if

$$\left| \int_a^b \operatorname{sgn}(f - u^*) u\, d\mu \right| \leq \int_{Z(f-u^*)} |u| d\mu$$

for all $u \in U$.

Assume that U is not a unicity space for $C_1([a, b], \mu)$. There then exists an f with at least two best approximants v and w. That is, $v, w \in U, v \neq w$, and $\|f - v\|_1 = \|f - w\|_1 = E(f; U)$. The set of best approximants is convex (Proposition 1.13) and thus $(v + w)/2$ is also a best approximant to f from U. In fact, since

$$2|(f - (v + w)/2)(x)| \leq |(f - v)(x)| + |(f - w)(x)|$$

for all $x \in [a, b]$, while $\|f - v\|_1 = \|f - w\|_1 = \|f - (v + w)/2\|_1$, and $f, v, w \in$ $C[a, b]$, it follows that $2|(f - (v + w)/2)(x)| = |(f - v)(x)| + |(f - w)(x)|$ for all $x \in [a, b]$. In particular, if $(f - (v + w)/2)(x) = 0$, then $(f - v)(x) =$ $(f - w)(x) = 0$, implying that $v(x) = w(x)$.

Now, if $f - (v + w)/2$ has at least $\dim U$ zeros in (a, b), then $v - w$ has at least $\dim U$ zeros in (a, b) and it follows, since U is a T-system, that $v = w$. We therefore assume that $f - (v + w)/2$ has at most $\dim U - 1$ zeros in (a, b). Thus $\mu(Z(f - (v + w)/2)) = 0$ and, from our characterization,

$$\int_a^b \mathrm{sgn}\,(f - (v + w)/2)\, u\, d\mu = 0$$

for all $u \in U$. We again apply Proposition 3 of Appendix B which states that since U is a T-system on (a, b), $\mathrm{sgn}(f - (v + w)/2)$ must have at least $\dim U$ sign changes, (and therefore zeros) in (a, b). Thus $v = w$ and U is a unicity space for $C_1([a, b], \mu)$.

Example 4.8. *Pieceing together T-systems.*

Let $a = e_0 < e_1 < \cdots < e_k < e_{k+1} = b$. On each interval $I_i = [e_{i-1}, e_i]$, let U^i be a T-system of dimension n_i, $i = 1, \ldots, k + 1$. For convenience we shall assume that $n_i \geq 2$ for all i. U will denote the subspace of $C[a, b]$ satisfying $U|_{I_i} = U_i$, $i = 1, \ldots, k + 1$. Note the important fact that $U \subset C[a, b]$. This imposes continuity constraints at $e_1, \ldots, e_k$.

We shall show that U satisfies Property A by proving that it satisfies (3) of Theorem 4.16. Rather obviously U satisfies (3)(a). It is (3)(b) which needs checking. First note that the above $\{e_i\}_{i=0}^{k+1}$ are exactly the $\{e_i\}_{i=0}^{k+1}$ of Proposition 4.17. It is therefore necessary to prove that the $V_{i,j} = V_{e_i, e_j}$ of (3)(b) of Theorem 4.16 are WT-systems for every $0 \leq i < j \leq k + 1$. The last statement of (3)(b) is immediate. (See the remark after Proposition 4.17.)

With this goal in mind, let us first calculate the dimension of U. Choose n_1 points in the interior of I_1, and $n_i - 1$ points in the interior of I_i for each $i = 2, \ldots, k + 1$. Specify arbitrary data at these $n_1 + \cdots + n_{k+1} - k$ points. We claim that there exists a unique $u \in U$ interpolating the given data at the associated points. This then implies that $\dim U = n_1 + \cdots + n_{k+1} - k$. To prove this claim consider any $u \in U$. Since n_1 points and associated data are specified in I_1, $\dim U^1 = n_1$, and U^1 is a T-system on I_1, this uniquely defines the interpolating u on I_1. Consider all such u restricted to I_2. Only $n_2 - 1$ points and data are specified in I_2 (in its interior). However at the left endpoint e_1, u is already fixed since it is fixed on I_1. This uniquely determines u on I_2 since U^2 is a T-system on I_2 of dimension n_2. We continue in this manner to obtain the desired result. Although this may all seem a bit heuristic, we have indeed formally proved that

$$\dim U = n_1 + \cdots + n_{k+1} - k \;.$$

In a totally analogous way we can calculate $\dim V_{i,j}$ for $0 \le i < j \le k+1$. Recall that

$$V_{i,j} = \{u : u \in U, u = 0 \text{ on } [a, e_i) \cup (e_j, b]\}.$$

The above was the computation of $\dim V_{0,k+1}$, since $V_{0,k+1} = U$. Analogous arguments prove the following:

    i) for $1 \le j \le k$, $\dim V_{0,j} = n_1 + \cdots + n_j - j$,

    ii) for $1 \le i \le k$, $\dim V_{i,k+1} = n_{i+1} + \cdots + n_{k+1} - (k+1-i)$,

    iii) for $1 \le i < j \le k$, $\dim V_{i,j} = n_{i+1} + \cdots + n_j - (j-i+1)$.

The different cases occur because of the zero interpolatory conditions, or lack of them, at the endpoints.

To prove (3)(b) we must show that no $u \in V_{i,j}$ has more than $\dim V_{i,j} - 1$ sign changes. Since U^i is a T-system on I_i of dimension n_i, it follows that $u \in U^i$ has at most $n_i - 1$ sign changes (zeros) on I_i. However, if u vanishes at one of the endpoints of I_i, then this bound is reduced by one while, if u vanishes at both endpoints, it is reduced by two (except if $n_i = 2$, in which case it follows that $u = 0$ on all of I_i). Furthermore if $u \in U$ has a sign change at e_i, then obviously $u(e_i) = 0$. A simple counting argument now proves that $V_{i,j}$ is a WT-system for each $0 \le i < j \le k+1$ (unless $V_{i,j} = \{0\}$).

In an attempt to be somewhat complete, we shall run through this counting argument in one case. For example, consider $V_{0,j}$ where $1 \le j \le k$. Thus $\dim V_{0,j} = n_1 + \cdots + n_j - j$. Assume $u \in V_{0,j}$ and u has no zero intervals in $[e_0, e_j]$. (Otherwise either perturb u slightly so that it has no zero intervals and no fewer sign changes, or 'decompose' u and reduce it to other cases with this same assumption.) Now $u \in V_{0,j}$ has at most $n_i - 1$ sign changes on I_i for each $i = 1, \ldots, j-1$, and at most $n_j - 2$ sign changes on I_j since $u(e_j) = 0$. If $u(e_i) \ne 0$ for each $i = 1, \ldots, j-1$, then we are finished. If $u(e_i) = 0$ for some $i \in \{1, \ldots, j-1\}$, then u may have an additional sign change at e_i. But at the same time the possible number of sign changes on each of I_i and I_{i-1} is now one less. This demonstrates the fact that $u \in V_{0,j}$ has at most $n_1 + \cdots + n_j - j - 1 = \dim V_{0,j} - 1$ sign changes. Thus $V_{0,j}$ is a WT-system. As previously mentioned, the proofs in the other cases are totally analogous.

Example 4.9. *Polynomial splines with simple fixed knots.*

Given $[a, b]$ and $m \ge 2$, let $\xi_0 = a < \xi_1 < \cdots < \xi_r < b = \xi_{r+1}$. Consider the space $\mathcal{S}_{m-1,r}$ of functions s which on each of the intervals $[\xi_{i-1}, \xi_i]$, $i = 1, \ldots, r+1$, are polynomials of degree at most $m-1$, and such that $s \in C^{(m-2)}[a, b]$, i.e., there are $m-1$ continuity constraints at each ξ_i in (a, b). $\mathcal{S}_{m-1,r}$ is referred to as the space of *splines* of degree $m-1$ with the fixed *simple knots* $\{\xi_i\}_1^r$. (The term 'simple' is used to denote the fact that the only discontinuity permitted between consecutive intervals is that of the highest, i.e., $(m-1)$st, derivative.) This space is an important and useful subspace of the space of piecewise polynomials.

It is easy to see that $\mathcal{S}_{m-1,r}$ has dimension $m + r$. This fact may be discerned from the evident result that a natural basis for $\mathcal{S}_{m-1,r}$ is

$$1, x, \ldots, x^{m-1}, (x - \xi_1)_+^{m-1}, \ldots, (x - \xi_r)_+^{m-1}$$

where

$$(x - \xi_i)_+^{m-1} = \begin{cases} (x - \xi_i)^{m-1}, & x \geq \xi_i \\ 0, & x < \xi_i . \end{cases}$$

There is yet another basis for $\mathcal{S}_{m-1,r}$ which is more useful in practice (i.e., computationally) as well as in theory. This different basis (actually there are many) is called a *B-spline* basis. Since this is an important example, and we shall in the next chapter return to this space, we now consider the construction and certain relevant properties of B-splines in some detail.

Assume that we are given a set of knots $\xi_i < \cdots < \xi_j$. How do we choose $j - i$ minimal so that there exists a non-trivial spline s of degree $m - 1$ on all of $I\!\!R$, with the simple knots $\{\xi_i, \xi_{i+1}, \ldots, \xi_j\}$, and such that $s(x) = 0$ for all $x \in (-\infty, \xi_i] \cup [\xi_j, \infty)$?

Well, since s is a spline of degree $m - 1$ with the above simple knots and vanishes identically on $(-\infty, \xi_i]$, it follows that s must necessarily be of the form

$$s(x) = \sum_{k=i}^{j} c_k (x - \xi_k)_+^{m-1}$$

with $\mathbf{c} = (c_i, \ldots, c_j) \neq \mathbf{0}$. Since $(x - \xi_k)_+^{m-1} = (x - \xi_k)^{m-1}$ on (ξ_j, ∞) for every $k \in \{i, \ldots, j\}$, and $s = 0$ on (ξ_j, ∞), we also have that the polynomial

$$p(x) = \sum_{k=i}^{j} c_k (x - \xi_k)^{m-1}$$

vanishes identically on (ξ_j, ∞), and thus on all of $I\!\!R$. This is equivalent to the statement that $p^{(l)}(0) = 0$, $l = 0, 1, \ldots, m - 1$, or

$$\sum_{k=i}^{j} c_k \xi_k^l = 0, \quad l = 0, 1, \ldots, m - 1 .$$

The matrix $(\xi_k^l)_{k=i, l=0}^{j, \ m-1}$ has full rank equal to $\min \{j - i + 1, m\}$. Since $\mathbf{c} \neq \mathbf{0}$, a necessary and sufficient condition for the above to hold is that $j - i \geq m$. Thus if s is any non-trivial spline as above and vanishes on $(-\infty, \xi_i] \cup [\xi_j, \infty)$, then necessarily $j - i \geq m$. Furthermore for $i = j - m$ there exists a unique (up to multiplication by a constant) non-trivial spline, which we denote by B_j, of degree $m - 1$ with the simple knots $\xi_{j-m}, \ldots, \xi_j$, and with the property that $B_j = 0$ on $(-\infty, \xi_{j-m}] \cup [\xi_j, \infty)$. This B_j is called the jth B-spline (of degree $m - 1$).

What else may be said about the B-spline B_j? We claim that B_j has no zero in (ξ_{j-m}, ξ_j). Assume B_j has a zero in (ξ_{j-m}, ξ_j). Since B_j does not identically vanish on any subinterval of (ξ_{j-m}, ξ_j) (since $j - (j - m) = m$), and $B_j(\xi_{j-m}) = B_j(\xi_j) = 0$, it follows that B_j' has at least two sign changes on (ξ_{j-m}, ξ_j). Because $B_j'(\xi_{j-m}) = B_j'(\xi_j) = 0$, it follows that B_j'' has at least three sign changes on (ξ_{j-m}, ξ_j). Continuing in this fashion we obtain that $B_j^{(m-2)}$ has at least $m - 1$ sign changes on (ξ_{j-m}, ξ_j). However $B_j^{(m-2)}$ is a continuous piecewise linear function with the knots $\{\xi_k\}_{k=j-m}^{j}$, and satisfies $B_j^{(m-2)}(\xi_{j-m}) = B_j^{(m-2)}(\xi_j) = 0$. It is easily checked (draw a picture) that such a function has at most $m - 2$ sign changes on (ξ_{j-m}, ξ_j). This contradiction proves the claim.

The above argument proves somewhat more. Because $B_j^{(l)}(\xi_{j-m}) = B_j^{(l)}(\xi_j) = 0, l = 0, 1, \ldots, m - 2$, it follows, as above, that $B_j^{(m-2)}$ has at least $m - 2$ sign changes on (ξ_{j-m}, ξ_j). This can only occur if $B_j^{(m-2)}$ is alternately strictly increasing and strictly decreasing on consecutive (ξ_k, ξ_{k+1}), $k = j - m, \ldots, j - 1$, and has a sign change in (ξ_k, ξ_{k+1}) for each $k = j - m + 1, \ldots, j - 2$. Thus $B_j^{(m-1)}$ does not vanish identically on (ξ_k, ξ_{k+1}) for any $k \in \{j-m, \ldots, j-1\}$, and in fact strictly changes sign on consecutive intervals thereof.

We summarize as follows:

Proposition 4.19. *Given* $\cdots < \xi_i < \xi_{i+1} < \cdots < \xi_j < \cdots$, *let s be a nontrivial spline of degree $m - 1$ with the simple knots $\{\xi_k\}$. If supp $s \subseteq [\xi_i, \xi_j]$, then $j - i \geq m$. Furthermore if $i = j - m$ there then exists a unique (up to multiplication by a constant) non-trivial s as above, denoted B_j and called the jth B-spline. B_j has no zero in (ξ_{j-m}, ξ_j).*

Let us return to the space $\mathcal{S}_{m-1,r} = \mathrm{span}\{1, x, \ldots, x^{m-1}, (x-\xi_1)_+^{m-1}, \ldots, (x - \xi_r)_+^{m-1}\}$ defined on $[a, b]$, where $\xi_0 = a < \xi_1 < \cdots < \xi_r < b = \xi_{r+1}$. This space is of dimension $m+r$. If we use only the knots $\{\xi_i\}_{i=0}^{r+1}$, we can construct at most $r - m + 2$ B-splines (assuming $r + 2 \geq m$). This is certainly insufficient for constructing a B-spline basis for $\mathcal{S}_{m-1,r}$. What we shall therefore do is add an additional $m - 1$ knots to the left of a, and $m - 1$ knots to the right of b. Choose $\{\xi_i\}$, $i = -m + 1, \ldots, -1, r + 2, r + m$, such that

$$\xi_{-m+1} < \cdots < \xi_{-1} < \xi_0 = a < b = \xi_{r+1} < \xi_{r+2} < \cdots < \xi_{r+m}.$$

Construct the B-spline B_j of degree $m - 1$, using the knots $\{\xi_i\}_{i=-m+1}^{m+r}$, and such that B_j is positive on (ξ_{j-m}, ξ_j) and vanishes identically off (ξ_{j-m}, ξ_j), $j = 1, \ldots, m + r$. Note that supp $B_j \cap [a, b] \neq \emptyset$ for each j. We now have the correct count (i.e., $m + r$) of functions, and $B_j|_{[a,b]} \in \mathcal{S}_{m-1,r}$ for each j.

Proposition 4.20. *The B-splines $B_1, \ldots, B_{m+r}$, as defined above, span the space $\mathcal{S}_{m-1,r}$ on $[a, b]$.*

Proof. Since $\dim \mathcal{S}_{m-1,r} = m + r$, and there are $m + r$ B_j's, we must simply prove that the $\{B_j\}_{j=1}^{m+r}$ are linearly independent on $[a, b]$.

To this end, assume that

$$s = \sum_{j=1}^{m+r} a_j B_j = 0$$

on $[a, b]$. Consider s on all of $\mathbb{R}$. Since $\operatorname{supp} B_j = (\xi_{j-m}, \xi_j)$ for each j, we have that $s = 0$ on $(-\infty, \xi_{-m+1}] \cup [\xi_{m+r}, \infty)$. Now s vanishes identically on $(-\infty, \xi_{-m+1}]$ and on $[\xi_0, \xi_{r+1}]$. If s is not identically zero on (ξ_{-m+1}, ξ_0), then we contradict the first part of Proposition 4.19 (since $0 - (-m+1) = m-1 < m$). Thus s is identically zero on $(-\infty, \xi_{r+1}]$ (and by the similar argument on all of $\mathbb{R}$). On the interval (ξ_{-m+1}, ξ_{-m+2}), $s = a_1 B_1 = 0$ and $B_1 \neq 0$. Thus $a_1 = 0$. On (ξ_{-m+2}, ξ_{-m+3}), $s = a_1 B_1 + a_2 B_2 = a_2 B_2 = 0$ and $B_2 \neq 0$. Thus $a_2 = 0$. Continuing in this fashion we prove that $a_j = 0$, $j = 1, \ldots, m+r$. This proves that the $\{B_j\}_{j=1}^{m+r}$ are linearly independent on $[a, b]$.   □

To prove that $\mathcal{S}_{m-1,r}$ satisfies Property A, we need two further properties of B-splines. Much more is true than will be proved here, but we shall make do with these two facts.

Proposition 4.21.
1) If $s = \sum_{j=1}^{m+r} a_j B_j$ on $[a, b]$, and $s = 0$ on $[\xi_{k-1}, \xi_k]$ for some $k \in \{1, \ldots, r+1\}$, then $a_j = 0$, $j = k, \ldots, k+m-1$, i.e., $a_j = 0$ for all j satisfying $(\xi_{k-1}, \xi_k) \subseteq \operatorname{supp} B_j$.
2) For every $1 \le i \le j \le m+r$, the functions $B_i, B_{i+1}, \ldots, B_j$ span a WT-system on $[a, b]$.

Proof. We have actually already proved (1). To see this, simply set $a' = \xi_{k-1}$ and $b' = \xi_k$, and apply Proposition 4.20 to $\mathcal{S}_{m-1,r}$ restricted to the interval $[a', b']$. Alternatively $s|_{[\xi_{k-1}, \xi_k]} = \sum_{j=k}^{k+m-1} a_j B_j$ and we can reargue as in the proof of Proposition 4.20.

To establish (2), we must prove that every $s = \sum_{k=i}^{j} a_k B_k$ has at most $j - i$ sign changes on $[a, b]$. For $j = i$, this follows from Proposition 4.19. The general proof is also along these same lines.

Assume that there exists an $s = \sum_{k=i}^{j} a_k B_k$ with at least $j - i + 1$ sign changes on $[a, b]$. Since any such s, considered on all of $\mathbb{R}$, has support in (ξ_{i-m}, ξ_j), the function s has at least $j - i + 1$ sign changes in (ξ_{i-m}, ξ_j). Because $s^{(k)}(\xi_{i-m}) = s^{(k)}(\xi_j) = 0$, $k = 0, 1, \ldots, m-2$, it follows that $s^{(m-2)}$ has at least $j - i + m - 1$ sign changes on (ξ_{i-m}, ξ_j). $s^{(m-2)}$ is a continuous piecewise linear function with the knots $\{\xi_k\}_{k=i-m}^{j}$, and vanishes at ξ_{i-m} and ξ_j. Such an $s^{(m-2)}$ has at most $j - i + m - 2$ sign changes on (ξ_{i-m}, ξ_j). This contradiction proves (2).   □

We can now finally return to Property A.

Proposition 4.22. $\mathcal{S}_{m-1,r}$ satisfies Property A on $[a, b]$.

Proof. We shall prove that $\mathcal{S}_{m-1,r}$ satisfies (3) of Theorem 4.16. Condition (3)(a) is an immediate consequence of the definition of $\mathcal{S}_{m-1,r}$. Furthermore the $\{e_i\}_{i=0}^{k+1}$ of Proposition 4.17 are simply the knots $\{\xi_i\}_{i=0}^{r+1}$. We claim that, for each $0 \leq i < j \leq r+1$,

$$V_{i,j} = \{s : s \in \mathcal{S}_{m-1,r},\ s = 0 \text{ on } [a,\xi_i) \cup (\xi_j, b]\}$$

is a WT-system.

From (1) of Proposition 4.21,

   i) $V_{0,r+1} = \mathrm{span}\{B_1, \ldots, B_{r+m}\}$,

   ii) $V_{0,j} = \mathrm{span}\{B_1, \ldots, B_j\}$ for $1 \leq j \leq r$,

   iii) $V_{i,r+1} = \mathrm{span}\{B_{i+m}, \ldots, B_{r+m}\}$ for $1 \leq i \leq r$,

   iv) $V_{i,j} = \{0\}$ if $1 \leq i < j \leq r$ and $j - i \leq m - 1$,

   v) $V_{i,j} = \mathrm{span}\{B_{i+m}, \ldots, B_j\}$ if $1 \leq i < j \leq r$ and $j - i \geq m$.

From (2) of Proposition 4.21, $V_{i,j}$ is a WT-system in all these cases. Furthermore if $V_{i,j} \neq \{0\}$ then for each $x \in (\xi_i, \xi_j)$ there exists an $s \in V_{i,j}$ satisfying $s(x) \neq 0$ (s can be taken to be one of the B-splines in the basis for $V_{i,j}$). The last sentence of condition (3)(b) holds (see the remark after the proof of Proposition 4.17). Thus $\mathcal{S}_{m-1,r}$ satisfies Property A.   □

Before leaving this example, we wish to record some other properties of B-splines as they pertain to Property A.

In (2) of Proposition 4.21, we proved that $B_i, B_{i+1}, \ldots, B_j$ spans a WT-system on $[a, b]$ for any $1 \leq i < j \leq m + r$. In fact it is true (and this has little to do with splines) that the above implies that $B_{i_1}, B_{i_2}, \ldots, B_{i_k}$ spans a WT-system on $[a, b]$ for every choice of $\{i_1, \ldots, i_k\} \subseteq \{1, \ldots, m+r\}$. This fact together with (1) of Proposition 4.21 implies that $\mathrm{span}\{B_{i_1}, \ldots, B_{i_k}\}$ satisfies Property A on $[a, b]$ for any $\{i_1, \ldots, i_k\}$ as above. (It can actually be shown that

$$\det(B_{i_l}(x_j))_{l,j=1}^{k} \geq 0$$

for any $x_1 < \cdots < x_k$ and $1 \leq i_1 < \cdots < i_k \leq m + r$, and the above determinant is strictly positive if and only if $x_l \in \mathrm{supp}\, B_{i_l}$, $l = 1, \ldots, k$.)

All this can also be generalized in a somewhat different direction. We have considered splines with simple knots. One can also talk about splines with multiple knots. We say that the knot ξ has multiplicity p, $1 \leq p \leq m-1$, if we only demand that $s \in C^{(m-p-1)}$ in a neighborhood of ξ. If $p = 1$ then ξ is a simple knot. The space of splines of degree $m - 1$ with the r knots $a < \xi_1 < \cdots < \xi_r < b$ of multiplicity $p_1, \ldots, p_r$, respectively, $1 \leq p_i \leq m - 1$, $i = 1, \ldots, r$, also satisfies Property A. A proof thereof involves a refinement of the above proof. The case $p_i = m - 1$, $i = 1, \ldots, r$, is a special case of Example 4.8.

5. Property A on $I\!R^d$, $d \geq 2$

The problem of characterizing finite-dimensional subspaces $U \subset C(K)$ satisfying Property A where $K \subset I\!R^d$, $d \geq 2$, and K satisfies our usual assumptions, i.e., K compact, $K = \overline{\text{int}\, K}$, appears to be not at all easy. To date we know of no such characterization. In fact, little is known in this case. In this section we present a few examples (see also Examples 4.3, 4.4 and the exercises at the end of this chapter) and some (somewhat incomplete) results. The general feeling among researchers in the field is that in $I\!R^d$, $d \geq 2$, subspaces satisfying Property A are rare, aside from certain 'trivial' cases.

We consider two main examples. One general and one more specific.

A. Tensor Products

One classic set of subspaces in $I\!R^d$, $d \geq 2$, are those obtained by tensoring subspaces defined on lower dimensional domains. Let $K \subset I\!R^{d_1}, L \subset I\!R^{d_2}$, and assume that V is a finite-dimensional subspace of $C(K)$ while W is a finite-dimensional subspace of $C(L)$. By the tensor product of V and W, denoted $U = V \otimes W$, we mean the subspace of $C(K \times L)$, $K \times L \subset I\!R^{d_1+d_2}$, spanned by all functions of the form $v \cdot w$, where $v \in V$ and $w \in W$. (That is, each $u \in U$ is defined on (x,y), $x \in K$, $y \in L$.) Note that we do not mean merely the functions of the form $v \cdot w$, but all possible linear combinations thereof. Thus if $v_1, \ldots, v_{n_1}$, is a basis for V, and $w_1, \ldots, w_{n_2}$ is a basis for W, then

$$U = V \otimes W = \Big\{ \sum_{i=1}^{n_1} \sum_{j=1}^{n_2} a_{ij} v_i w_j : a_{ij} \in I\!R \Big\}.$$

It is easily seen that the $v_i w_j$, $i = 1, \ldots, n_1$; $j = 1, \ldots, n_2$, form a basis for U, and therefore $\dim U = n_1 n_2$.

We now consider when U satisfies Property A. To this end we assume that, in addition to the above conditions, we also have that K and L are compact, and $K = \overline{\text{int}\, K}, L = \overline{\text{int}\, L}$. Then $K \times L$ inherits these same properties. We also note some general facts.

Let $u = v \cdot w$ where $v \in V, w \in W$. Set $K \backslash Z(v) = \bigcup_{i=1}^{p} A_i$, and $L \backslash Z(w) = \bigcup_{j=1}^{q} B_j$, where the A_i and B_j are as in Definition 4.2. It is then easily seen that

$$(K \times L) \backslash Z(u) = \bigcup_{i=1}^{p} \bigcup_{j=1}^{q} (A_i \times B_j).$$

It therefore follows that

1) $[(K \times L) \backslash Z(u)] = [K \backslash Z(v)] \cdot [L \backslash Z(w)]$
2) $Z(u) = (Z(v) \times L) \cup (K \times Z(w))$.

With these facts, we can now prove:

Proposition 4.23. *Let $U = V \otimes W$ be as above. Assume that U satisfies Property A. Then both V and W satisfy Property A.*

Proof. Let $v \in V, v \neq 0$, and $K \backslash Z(v) = \bigcup_{i=1}^{p} A_i$. Choose $\varepsilon_i \in \{-1, 1\}$, $i = 1, \ldots, p$. We must prove the existence of a $v^* \in V(v)$, $v^* \neq 0$, satisfying $\varepsilon_i v^* \geq 0$ on A_i, $i = 1, \ldots, p$.

With this goal in mind, choose any $w \in W, w \neq 0$. Let $L \backslash Z(w) = \bigcup_{j=1}^{q} B_j$. Set $u = v \cdot w$, and $\varepsilon_{ij} = \varepsilon_i$, $i = 1, \ldots, p$; $j = 1, \ldots, q$. Since U satisfies Property A, there exists a $u^* \in U(u), u^* \neq 0$, such that $\varepsilon_{ij} u^* \geq 0$ on $A_i \times B_j$, all i, j.

Let $\widetilde{y} \in L$, and set $\widetilde{v}(\cdot) = u^*(\cdot, \widetilde{y})$, i.e., fix $\widetilde{y}$. Then $\widetilde{v} \in V$. Since u^* vanishes a.e. on $Z(v) \times L$ ($\subseteq Z(u)$), it follows from Fubini's Theorem that, for almost all $\widetilde{y} \in L$, $\widetilde{v}$ vanishes a.e. on $Z(v)$. But u^* is continuous and thus $\widetilde{v}$ vanishes a.e. on $Z(v)$ for all $\widetilde{y} \in L$. Now choose a $y \in L$ for which $v^*(\cdot) = u^*(\cdot, y)$ is not identically zero. Then $v^* \in V(v), v^* \neq 0$, and from the properties of u^*, we have $\varepsilon_i v^* \geq 0$ on A_i, $i = 1, \ldots, p$. Therefore V satisfies Property A. This same argument proves that W satisfies Property A. □

We do not claim that the converse result is valid. In fact it is generally not valid. However in one simple case it is certainly true.

Proposition 4.24. *Let $U = V \otimes W$ as above. Assume that $\dim V = 1$ and/or $\dim W = 1$. Then U satisfies Property A if and only if both V and W satisfy Property A.*

Proof. One half of this result is given by Proposition 4.23. We shall therefore assume that both V and W satisfy Property A and $\dim V = 1$. We shall prove that U satisfies Property A.

Let $v \in V, v \neq 0$. Then, from Theorem 4.6 $[K \backslash Z(v)] = 1$. Since V is one-dimensional, every $u \in U$ is necessarily of the form $u = v \cdot w$, $w \in W$, v fixed. Let $K \backslash Z(v) = A$. Then $(K \times L) \backslash Z(u) = \bigcup_{j=1}^{q} C_j$ if and only if $L \backslash Z(w) = \bigcup_{j=1}^{q} B_j$, and $C_j = A \times B_j$, $j = 1, \ldots, q$. The remaining steps in the proof of the proposition are immediate consequences of the fact that W satisfies Property A. □

On the basis of Theorem 4.12, we can generalize Proposition 4.24 to the case where $[K \backslash Z(V)] = \dim V$. ($V$ has a basis of functions with disjoint support.) But this is an uninteresting result. Because of Theorem 4.12, we should only consider V and W satisfying $[K \backslash Z(V)] = 1$ and $[L \backslash Z(W)] = 1$. Assume therefore that this is the case, and $\dim V$ and $\dim W$ are both at least two. Does the second half of Proposition 4.24 (i.e., the converse to Proposition 4.23) then hold? The conjecture is that it never ever holds in this setting. That is, U will not satisfy Property A. We end this subsection by proving a special case of this conjecture.

Proposition 4.25. *Let $U = V \otimes W$. Assume that $[K \backslash Z(V)] = [L \backslash Z(W)] = 1$ and both $\dim V$ and $\dim W$ are at least two. If $V(v) = V$ for all $v \in V$, $v \neq 0$, and $W(w) = W$ for all $w \in W$, $w \neq 0$, then U does not satisfy Property A.*

Proof. From Proposition 4.23, U does not satisfy Property A if either V or W does not satisfy Property A. We therefore assume that both V and W satisfy Property A.

We first claim that there exists a $v^* \in V, v^* \neq 0$, satisfying $[K \backslash Z(v^*)] \geq 2$, with the property that if $v \in V(v^*)$, and $v(\text{sgn } v^*) \geq 0$ on $K \backslash Z(v^*)$, then $v = \alpha v^*$ for some $\alpha \geq 0$. The stipulation $v \in V(v^*)$ is meaningless since $V(v^*) = V$. We included it for convenience and also in the hope that the above conditions would awaken in the reader a recollection of Proposition 4.13 (please look!). There we proved the existence of a v^* satisfying the above, but without the demand that $[K \backslash Z(v^*)] \geq 2$. Look back at the proof of Proposition 4.13, and especially at Lemma 4.15. The condition therein on int $Z(v^*)$ is meaningless because of the assumption $V(v) = V$ for all $v \in V, v \neq 0$. Therefore v^* can be taken to be any $v \in V$ which maximizes $[K \backslash Z(v)]$ over V. Since dim $V \geq 2$ we see that max $\{[K \backslash Z(v)] : v \in V\} \geq 2$. Any such v^* necessarily satisfies our claim. Analogously there exists a $w^* \in W, w^* \neq 0$, satisfying $[L \backslash Z(w^*)] \geq 2$ and such that if $w \in W(w^*)$, and $w(\text{sgn } w^*) \geq 0$ on $L \backslash Z(w^*)$, then $w = \beta w^*$ for some $\beta \geq 0$.

Set $u^* = v^* \cdot w^*$. Let $K \backslash Z(v^*) = \bigcup_{i=1}^{p} A_i$ and $L \backslash Z(w^*) = \bigcup_{j=1}^{q} B_j$. Thus $2 \leq p \leq \dim V$ and $2 \leq q \leq \dim W$. We have $(K \times L) \backslash Z(u^*) = \bigcup_{i=1}^{p} \bigcup_{j=1}^{q} (A_i \times B_j)$. Let $\varepsilon_i = \text{sgn } v^*$ on A_i, $i = 1, \dots, p$, and $\eta_j = \text{sgn } w^*$ on B_j, $j = 1, \dots, q$. Set $\sigma_{ij} = \varepsilon_i \eta_j$ for all $i = 1, \dots, p; j = 1, \dots, q; (i, j) \neq (1, 1)$, and $\sigma_{11} = -\varepsilon_1 \eta_1$.

We claim that there does *not* exist a $u \in U(u^*)$, $u \neq 0$, for which $\sigma_{ij} u \geq 0$ on $A_i \times B_j$, all (i, j). Assume $u \in U, u \neq 0$, exists satisfying these conditions. Let $x^* \in A_i$ for some $i \in \{2, \dots, p\}$. Then as a function of y, $\varepsilon_i \eta_j u(x^*, y) \geq 0$ on B_j, $j = 1, \dots, q$. From the construction of w^* it follows $\varepsilon_i \eta_j u(x^*, y) = \varepsilon_i \beta(x^*) w^*(y)$, where $\beta(x^*) \geq 0$. Similarly, for $y^* \in B_j$, some $j \in \{2, \dots, q\}$, we have $\varepsilon_i \eta_j u(x, y^*) = \eta_j v^*(x) \alpha^*(y^*)$, where $\alpha(y^*) \geq 0$. Thus on $\bigcup_{i=2}^{p} \bigcup_{j=2}^{q} (A_i \times B_j)$, $u = \gamma v^* w^*$ for some $\gamma \geq 0$.

We claim that $u = \gamma v^* w^*$ on all of $K \times L$. If not, set $\tilde{u} = u - \gamma v^* w^*$. Then $\tilde{u} \in U, \tilde{u} \neq 0$, and $U(\tilde{u}) \neq U$ (since $\bigcup_{i=2}^{p} \bigcup_{j=2}^{q} (A_i \times B_j) \subseteq Z(\tilde{u})$). Because $V(v) = V$ for all $v \in V, v \neq 0$, and $W(w) = W$ for all $w \in W, w \neq 0$, it is easily seen that $U(u) = U$ for all $u \in U, u \neq 0$. This contradiction implies that $u = \gamma v^* w^*$ on all of $K \times L$, $\gamma \geq 0$. Since u must satisfy $\sigma_{11} u \geq 0$ on $A_1 \times B_1$, we see that $\gamma = 0$, and thus $u = 0$. This contradiction proves the proposition.   □

Remark. As a consequence of Proposition 4.25, we have that the U of Example 3.4 does not satisfy Property A. In addition $P_{n,m}$ of Exercise 3 of Chapter 3 does not satisfy Property A for any $n, m \geq 1$.

B. A 'Non-Trivial' Example.

Let K be any polygonal bounded connected region of $\mathbb{R}^2$. (Its boundary is composed of straight lines.) We define any finite triangulation of K with the property that no vertex lies on the interior of a side of any other triangle. Such a triangulation is said to be *regular*. For example figure (a) is a regular triangulation, while figure (b) is not

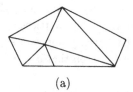   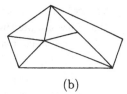

(a)   (b)

Each triangle of the regular triangulation will be called a *triangular cell*.

We consider the subspace U of $C(K)$ where each $u \in U$, restricted to each triangular cell, is a linear function, i.e., is in the span of $1, x, y$. The space U is generally referred to as the space of continuous piecewise linear functions on the given triangulation. The dimension of U is easily determined. It is simply the number of vertices. More in fact is true. Namely if the $(\mathbf{x}^i)_{i=1}^n$ are the vertices of the triangulation, then given any n data $(\gamma_i)_{i=1}^n$ there exists a unique $u \in U$ satisfying $u(\mathbf{x}^i) = \gamma_i$, $i = 1, \ldots, n$. The proof of this claim is a simple consequence of the fact that on any triangular cell there exists a unique linear function interpolating given data at its three vertices. In addition, the value of the linear function on any side of the triangular cell is completely determined by its values at the two endpoints thereof, and is independent of the value at the third vertex. Patching together triangular cells easily proves the claim.

We shall prove that U satisfies Property A. In proving this result, first recall from Section 2 that

$$U^* = \big\{ g : g \in C(K), \, |g| = |u| \text{ for some } u \in U \big\}.$$

From Definition 4.1, Property A holds if to each $g \in U^*$, $g \neq 0$, there exists a $v \in U, v \neq 0$, satisfying

$$a)\, v = 0 \text{ a.e. on } Z(g)$$

$$b)\, gv \geq 0 \text{ on } K.$$

Choose $g \in U^*$, $g \neq 0$. Let $v \in U$ satisfy $v(\mathbf{x}^i) = g(\mathbf{x}^i)$, $i = 1, \ldots, n$. v is uniquely defined. We shall prove that v satisfies all the above conditions. In the following arguments, $u \in U$ is such that $|g| = |u|$.

I. $v \neq 0$.

If $v = 0$, then $g(\mathbf{x}^i) = 0$, $i = 1, \ldots, n$. Therefore $u(\mathbf{x}^i) = 0$, $i = 1, \ldots, n$, implying $u = 0$. Thus $g = 0$, contradicting our assumption.

II. $v = 0$ a.e. on $Z(g)$.

$Z(g) = Z(u)$. Let Δ be a given triangular cell. Then $Z(u) \cap \Delta$ is either empty, all of Δ, or only one straight line in Δ. The collection of all such straight lines has measure zero. We shall therefore assume that $\Delta \subseteq Z(u) = Z(g)$. But then g vanishes at each of the vertices of Δ, and therefore v vanishes at each of the vertices of Δ, implying that $v = 0$ on Δ. Thus $v = 0$ a.e. on $Z(g)$.

III. $gv \geq 0$ on $K \backslash Z(g)$.

Let Δ be a given triangular cell. Consider the sign structure of g at the vertices of Δ. If g vanishes at all three vertices, then $\Delta \subseteq Z(g)$. We therefore assume that g does not vanish at all three vertices.

i) There exist two vertices where g has strictly opposite sign: Since $|g| = |u|$ on Δ and $g \in C(K)$, it is easily seen that $g = u$ or $g = -u$ on Δ. Thus $v = g$ on Δ.

ii) g is non-negative at each of the vertices of Δ: If g does not vanish at any interior point of Δ, then $g = u$ or $g = -u$ on Δ, and thus $v = g$ on Δ. Assume g vanishes at some interior point of Δ. Since $|g| = |u|$, it follows that $Z(g) = Z(u)$ is a straight line in Δ, u has one sign change in Δ (i.e., $[\Delta \backslash Z(u)] = 2$), and $g = |u| \geq 0$ on all of Δ. Since v is non-negative at each of the three vertices of Δ, $v \geq 0$ on Δ. Thus $vg \geq 0$ on Δ.

iii) g is non-positive at each of the vertices of Δ: The argument here is the exact analogue of the argument in (ii)

Thus $v \neq 0$ satisfies (a) and (b). Therefore U satisfies Property A.

Before ending this section, two general remarks are in order concerning 'trivial' examples.

Remark. Examples of U satisfying Property A for $K \subset \mathbb{R}^d$, $d \geq 2$, may also be constructed from subspaces satisfying Property A on lower dimensional domains. As an example, assume V satisfies Property A on $[a, b]$, where $0 < a < b$. Let $K = \{(x, y) : a^2 \leq x^2 + y^2 \leq b^2\}$, and

$$U = \{u : u(x, y) = v(\sqrt{x^2 + y^2}), v \in V\} .$$

Then U satisfies Property A on K. Other simple examples of this type may be found. Proposition 4.24 presents another set of such examples (see also Exercise 8).

Remark. Let $K_1, K_2 \subset \mathbb{R}^d$. A function h from K_1 to K_2 is said to be a C^1-*homeomorphism* if h is a one-to-one map from K_1 onto K_2, and both h and h^{-1} are C^1 functions. Assume K_1 is compact and $K_1 = \overline{\text{int } K_1}$. Since h is a homeomorphism, K_2 inherits these same properties. Because h and h^{-1} are C^1 functions, if $A \subseteq K_1$ has (Lebesgue) measure zero, then $h(A) \subseteq K_2$ has measure zero, and vice-versa. Let $U \subset C(K_1)$, and set

$$V = \{u(h^{-1}(\cdot)) : u \in U\}.$$

Then $V \subset C(K_2)$. From the above, we now see that the finite-dimensional subspace U of $C(K_1)$ satisfies Property A on K_1 if and only if V satisfies Property A on K_2.

6. Best Approximation under Interpolation: An Example

In Section 3 of Chapter 3 we considered the unicity set property for a specific family of closed convex subsets of finite-dimensional subspaces U of $C(K)$. In this section we study the unicity set property of best approximation under interpolatory constraints, and also obtain an analogue of Theorem 4.5.

Our problem is as follows. We are given an n-dimensional subspace U of $C(K)$, and m distinct points $t_1, \ldots, t_m$ in K. We wish to approximate $f \in C(K)$ in the $L^1(K, \mu)$ norm from the set of $u \in U$ satisfying $u(t_i) = f(t_i)$, $i = 1, \ldots, m$. In order that this problem be well defined, we shall assume that U is of dimension m over $\{t_1, \ldots, t_m\}$. In other words, to each $f \in C(K)$ there exists a $u \in U$ for which $u(t_i) = f(t_i)$, $i = 1, \ldots, m$. Equivalently, we can restate this condition as follows. Set

$$V = \{u : u \in U, \, u(t_i) = 0, \, i = 1, \ldots, m\}.$$

Then V is a subspace of U of dimension $n - m$. If $m = n$, then the interpolatory conditions uniquely determine the admissible $u \in U$, and there is no approximation problem. As such, we shall always assume that $1 \le m < n$.

For notational ease, set

$$M(f) = \{u : u \in U, \, u(t_i) = f(t_i), \, i = 1, \ldots, m\}.$$

Note that $M(f)$ is a convex subset of U which depends on f. In fact, as is readily seen, taking any fixed $\widetilde{u} \in M(f)$, we have

$$M(f) = \{\widetilde{u} + v : v \in V\}.$$

We easily characterize the best $L^1(K, \mu)$ approximants to f from $M(f)$ using the above representation of $M(f)$ and Theorems 2.1 and 2.3.

Proposition 4.26. *Let $f \in C(K)$. Then the following are equivalent:*

a) u^* *is a best $L^1(K, \mu)$ approximant to f from $M(f)$.*

b) *For all $v \in V$*

$$\left| \int_K \operatorname{sgn}(f - u^*) v \, d\mu \right| \le \int_{Z(f-u^*)} |v| \, d\mu.$$

c) *There exists an $h \in L^\infty(K, \mu)$ satisfying*

1) $|h| = 1$ *on K*

2) $\displaystyle \int_K h v \, d\mu = 0$ *all $v \in V$*

3) $\displaystyle \int_K h(f - u^*) d\mu = \|f - u^*\|_1$.

We can also characterize the unicity sets for our problem using Theorem 3.1.

Proposition 4.27. _There exists a unique best $L^1(K, \mu)$ approximant to each $f \in C(K)$ from $M(f)$ if and only if V is a unicity space for $C_1(K, \mu)$._

Proof. ($\Leftarrow$). For given $f \in C(K)$, let $\tilde{u} \in M(f)$. Since V is a unicity space for $C_1(K, \mu)$, there exists a unique best approximant $v^* \in V$ to $f - \tilde{u}$. The function $u^* = \tilde{u} + v^*$ is necessarily the unique best approximant to f from $M(f)$.

($\Rightarrow$). If V is not a unicity space for $C_1(K, \mu)$, then from Theorem 3.1 there exists an $h \in L^\infty(K, \mu)$ and a $v^* \in V$, $v^* \neq 0$, such that αv^* is a best approximant to $f = h|v^*| \in C(K)$ from V, for all $\alpha \in [-1, 1]$. Since $v^*(t_i) = 0$, $i = 1, \ldots, m$, we have that $M(f) = V$. We have constructed an $f \in C(K)$ without a unique best approximant from $M(f)$. □

The latter half of the proof of Proposition 4.27 is a reformulation of one of the principles implied by both Theorems 3.1 and 3.3. Namely, to check the unicity space property of a finite-dimensional subspace V of $C(K)$, it suffices to check it only for those $f \in C(K)$ which vanish on $Z(V)$.

As a result of Proposition 4.27 and Theorem 4.5, we have the following analogue of Theorem 4.5.

Proposition 4.28. _For each $\mu \in \mathcal{A}$ there exists a unique best $L^1(K, \mu)$ approximant to every $f \in C(K)$ from $M(f)$ if and only if V satisfies Property A._

The consequences of this result are generally negative in character for the following reason. Assume $K = [a, b]$, and among the points $\{t_i\}_{i=1}^m$ is at least one point in (a, b). Then $[K \backslash Z(V)] \geq 2$ or all functions in V have their support between two consecutive t_i. In the first case, we have from Theorem 4.12 that, if V satisfies Property A then it decomposes. If, for example, U is a T-system on (a, b), at least one of the points $\{t_i\}_{i=1}^m$ is in (a, b), and $1 \leq m < n$, then the associated V does not decompose and therefore does not satisfy Property A.

7. Property A^k and k-Convexity

In Section 4 of Chapter 3, we characterized the finite-dimensional subspaces U of $C_1(K, \mu)$ which are k-convex. To remind the reader (see Definitions 3.2 and 3.3), U is said to be k-convex if $\dim P_U(f) \leq k$ for every $f \in C(K)$ with equality for some element of $C(K)$. Thus the terms unicity space and 0-convex are equivalent. The characterization of k-convexity (Theorem 3.12) was based on Theorem 3.1. U is at most k-convex if and only if for every

$h \in L^\infty(K, \mu)$ satisfying

$$1) \quad |h| = 1 \text{ on } K$$

$$2) \quad \int_K hu \, d\mu = 0, \text{ all } u \in U,$$

the subspace $W_h = \{u : u \in U, hu \in C(K)\}$ satisfies $\dim W_h \leq k$.

We shall now characterize the $U \subset C(K)$ which are at most k-convex in $C_1(K, \mu)$ with respect to *every* $\mu \in \mathcal{A}$. For $k = 0$, this is exactly the problem studied in the last few sections.

In our search for such a characterization, we are naturally led to a generalization of Property A. First some notation. For $u_1, \ldots, u_r \in U$, set

$$a) \quad Z(u_1, \ldots, u_r) = \bigcap_{i=1}^{r} Z(u_i)$$

$$b) \quad U(u_1, \ldots, u_r) = \{u : u \in U, u = 0 \text{ a.e. on } Z(u_1, \ldots, u_r)\}.$$

Definition 4.9. Let $U \subset C(K)$, $\dim U = n$. We say that U satisfies *Property A^k*, $0 \leq k \leq n - 1$, if for every choice of $k + 1$ linearly independent elements $u_1, \ldots, u_{k+1}$ of U, with $K \backslash Z(u_1, \ldots, u_{k+1}) = \bigcup_{i=1}^{m} A_i$, and for every $\varepsilon_i \in \{-1, 1\}$, $i = 1, \ldots, m$, there exists a $u \in U(u_1, \ldots, u_{k+1})$, $u \neq 0$, satisfying $\varepsilon_i u \geq 0$ on A_i, $i = 1, \ldots, m$.

Note that if U satisfies Property A^k, $0 \leq k \leq n - 2$, then it also satisfies Property A^{k+1}. This follows from the definition (and also from this next characterization result). We have also abused our notation in that Property $A \equiv$ Property A^0.

Theorem 4.29. Let $U \subset C(K)$, $\dim U = n$, and $0 \leq k \leq n - 1$. Then U is at most k-convex in $C_1(K, \mu)$ for every $\mu \in \mathcal{A}$ if and only if U satisfies Property A^k.

Proof. ($\Rightarrow$). Assume that U does not satisfy Property A^k. We can therefore find $u_1, \ldots, u_{k+1}$ linearly independent in U with $K \backslash Z(u_1, \ldots, u_{k+1}) = \bigcup_{i=1}^{m} A_i$, and some choice of $\varepsilon_i \in \{-1, 1\}$, $i = 1, \ldots, m$, such that if $u \in U(u_1, \ldots, u_{k+1})$, and $\varepsilon_i u \geq 0$ on A_i, $i = 1, \ldots, m$, then $u = 0$.

Paralleling, almost word for word, the proof of Theorem 4.5, we prove the existence of an $h \in L^\infty(K)$ for which $|h| = 1$ on K, $h = \varepsilon_i$ on A_i, $i = 1, \ldots, m$, and such that $V = \{hu : u \in U\}$ contains no non-negative non-trivial function. From Corollary 4.4 there exists a $\mu \in \mathcal{A}$ for which $\int_K hu \, d\mu = 0$ for all $u \in U$. Since $h = \varepsilon_i$ on A_i, $i = 1, \ldots, m$, we see that $hu_i \in C(K)$, $i = 1, \ldots, k + 1$. (The u_i vanish at each point of discontinuity of h.) As a consequence of Theorem 3.12, we have that U is not k-convex in $C_1(K, \mu)$.

($\Leftarrow$). Assume U is not at most k-convex in $C_1(K, \mu)$ for some $\mu \in \mathcal{A}$. From Theorem 3.12 there exists an $h \in L^\infty(K)$ for which

$$1) \quad |h| = 1 \text{ on } K$$

$$2) \quad \int_K hu \, d\mu = 0, \text{ all } u \in U$$

and $W_h = \{u : u \in U, \, hu \in C(K)\}$ satisfies dim $W_h \geq k + 1$.

Let $u_1, \ldots, u_{k+1}$ be linearly independent elements of W_h. Set $K \backslash Z(u_1, \ldots, u_{k+1}) = \bigcup_{i=1}^m A_i$. Since each $u \in W_h$ must vanish at the points of discontinuity of h, we have that h is constant on each A_i. Set $\varepsilon_i = \mathrm{sgn}\, h$ on A_i, $i = 1, \ldots, m$. If U satisfies Property A^k, there necessarily exists a $u \in U(u_1, \ldots, u_{k+1})$ satisfying $\varepsilon_i u \geq 0$ on A_i, $i = 1, \ldots, m$. Thus $hu = |u|$ on A_i, $i = 1, \ldots, m$. But then $\int_K hu \, d\mu = \int_K |u| d\mu > 0$. This contradicts condition (2) on h. Thus U cannot satisfy Property A^k.   □

Before considering some examples, we shall prove two general results.

Proposition 4.30. *Let* $U \subset C(K)$, dim $U = n$, *and* $n > r > l \geq 0$. *Assume* U *contains an* $(n-r+l)$-*dimensional subspace* V *satisfying Property* A^l. *Then* U *satisfies Property* A^r.

Proof. Let $u_1, \ldots, u_{r+1}$ be any $r + 1$ linearly independent elements of U. Since $V \subseteq U$ and dim $V = n - r + l$, there exist $v_1, \ldots, v_{l+1} \in V$, linearly independent, such that $v_i \in \mathrm{span}\{u_1, \ldots, u_{r+1}\}$, $i = 1, \ldots, l + 1$. Therefore $Z(u_1, \ldots, u_{r+1}) \subseteq Z(v_1, \ldots, v_{l+1})$, and $V(v_1, \ldots, v_{l+1}) \subseteq U(u_1, \ldots, u_{r+1})$. Set $K \backslash Z(u_1, \ldots, u_{r+1}) = \bigcup_{i=1}^m A_i$, and $K \backslash Z(v_1, \ldots, v_{l+1}) = \bigcup_{j=1}^M B_j$. There exist disjoint subsets $J_1, \ldots, J_m$ of $\{1, \ldots, M\}$ such that $A_i \subseteq \bigcup_{j \in J_i} B_j$. Let $\varepsilon_i \in \{-1, 1\}$, $i = 1, \ldots, m$. Set $\eta_j = \varepsilon_i$ if $j \in J_i$, $i = 1, \ldots, m$, and otherwise choose $\eta_j \in \{-1, 1\}$, arbitrary. Since V satisfies Property A^l, there exists a $v \in V(v_1, \ldots, v_{l+1}) \subseteq U(u_1, \ldots, u_{r+1})$, $v \neq 0$, such that $\eta_j v \geq 0$ on B_j, $j = 1, \ldots, M$. Thus $\varepsilon_i v \geq 0$ on A_i, $i = 1, \ldots, m$.   □

We shall use Proposition 4.30 in the case where $l = 0$. Our next result is a partial analogue of Theorem 4.6.

Proposition 4.31. *Let* $U \subset C(K)$, dim $U = n$. *Assume that* U *satisfies Property* A^k, *and that* $U(u) = U$ *for every* $u \in U, u \neq 0$. *Then for any linearly independent* $u_1, \ldots, u_{k+1}$ *in* U,

$$[K \backslash Z(u_1, \ldots, u_{k+1})] \leq n \, .$$

Proof. Let $u_1, \ldots, u_{k+1}$ be as above, and assume that

$$K \backslash Z(u_1, \ldots, u_{k+1}) = \bigcup_{i=1}^m A_i \, ,$$

where $m > n$. Let $\{u_i\}_{i=1}^n$ be a basis for U, and set

$$c_{ij} = \int_{A_j} u_i \, dx, \qquad i = 1, \ldots, n; \, j = 1, \ldots, m \, .$$

Since $m > n$, there exists an $\mathbf{s} = (s_1, \ldots, s_m) \neq \mathbf{0}$ such that

$$\sum_{j=1}^m c_{ij} s_j = 0, \qquad i = 1, \ldots, n \, .$$

Thus

$$\sum_{j=1}^{m} s_j \int_{A_j} u\,dx = 0$$

for all $u \in U$. Therefore if $s_j u \geq 0$ on A_j, all j, then $s_j u = 0$ on A_j, all j. In the terminology of Definition 4.6, $\mathbf{s}$ is an annihilator. Let $\varepsilon_j = \mathrm{sgn}\, s_j$ if $s_j \neq 0$, and $\varepsilon_j \in \{-1, 1\}$ be otherwise arbitrarily chosen. If U satisfies Property A^k, then there exists a $u \in U$, $u \neq 0$, such that $\varepsilon_j u \geq 0$ on A_j, all j. From the above, any such u must vanish identically on some A_j (where $s_j \neq 0$), contradicting our assumption that $U(u) = U$.   □

The above proof is simply a repeat of the proof of Lemma 4.7 (see Corollary 4.8). As a simple consequence of Proposition 4.31, we have:

Corollary 4.32. *Assume $U(u) = U$ for every $u \in U, u \neq 0$, and $[K \backslash Z(U)] > \dim U$. Then U does not satisfy Property A^k for any k.*

Let us now consider some examples.

Example 4.10. $U = \mathrm{span}\{1, x, y, xy\}$ on $K = [-1, 1] \times [-1, 1]$.

For various and sundry reasons, e.g. Proposition 4.25, U does not satisfy Property A. However $V = \mathrm{span}\{1, x, y\} \subseteq U$ satisfies Property A (Example 4.3). Thus from Proposition 4.30, U satisfies Property A^1.

Example 4.11. *Multiplying T-systems.*

Let V be an n-dimensional T-system on (a, b). Assume $\varphi \in C[a, b]$, and φ has exactly k distinct zeros $\{x_i\}_{i=1}^{k}$ in (a, b), $1 \leq k \leq n - 1$. Set

$$U = \{\varphi v : v \in V\}\,.$$

Claim. *U satisfies Property A^k, but not Property A^{k-1} on $[a, b]$.*

Proof. Note that $\dim U = n$, and every $u \in U, u \neq 0$, vanishes on at most $n - 1 + k$ points in (a, b). Thus in particular $U(u) = U$.

We first prove that U does not satisfy Property A^{k-1}. To this end, let $x_{k+1}, \ldots, x_n$ be any $n - k$ points in $(a, b) \backslash \{x_1, \ldots, x_k\}$. Since $\dim V = n$, there exist k linearly independent functions $v_1, \ldots, v_k$ in V, all of which vanish at $x_{k+1}, \ldots, x_n$. Set $u_i = \varphi v_i$, $i = 1, \ldots, k$. The $u_1, \ldots, u_k$ are linearly independent in U, and each u_i vanishes at the $\{x_j\}_{j=1}^{n}$. (In fact $Z(u_1, \ldots, u_k) \cap (a, b) = \{x_1, \ldots, x_n\}$.) Thus

$$[K \backslash Z(u_1, \ldots, u_k)] = n + 1\,.$$

From Proposition 4.31, we see that U does not satisfy Property A^{k-1}.

It remains to prove that U satisfies Property A^k. Let $u_i = \varphi v_i$, $i = 1, \ldots, k + 1$, be linearly independent elements of U. The $v_1, \ldots, v_{k+1}$ are then $k + 1$ linearly independent elements of V. Since V is a T-system on

(a, b), $Z(v_1, \ldots, v_{k+1})$ contains at most $n - k - 1$ points in (a, b). Thus $Z(u_1, \ldots, u_{k+1})$ contains at most $n - 1$ points in (a, b). Let $Z(u_1, \ldots, u_{k+1}) \cap (a, b) = \{y_i\}_{i=1}^{s}$, where $a = y_0 < y_1 < \cdots < y_s < y_{s+1} = b$. Then $k \le s \le n - 1$, and $\{x_i\}_{i=1}^{k} \subseteq \{y_i\}_{i=1}^{s}$.

In addition,

$$K \backslash Z(u_1, \ldots, u_{k+1}) = \bigcup_{i=1}^{s+1} A_i ,$$

where $A_i = (y_{i-1}, y_i)$, $i = 1, \ldots, s + 1$, (except that A_1 and A_{s+1} may contain their left and right endpoints, respectively). Choose $\varepsilon_i \in \{-1, 1\}$, $i = 1, \ldots, s+1$. Let $\delta_i = \varepsilon_i \operatorname{sgn} \varphi(x)$, for $x \in A_i$, $i = 1, \ldots, s+1$. (φ does not vanish in any A_i.) Since V is a T-system, there exists a $v \in V, v \ne 0$, such that $\delta_i v \ge 0$ on A_i, $i = 1, \ldots, s+1$. Set $u = \varphi v$. Then $u \in U, u \ne 0$, and $\varepsilon_i u \ge 0$ on A_i, $i = 1, \ldots, s+1$. U satisfies Property A^k.   □

Example 4.12. $U = \operatorname{span}\{1, x, \ldots, x^m, x^{m+2}, \ldots, x^{m+2k}\}$ on $K = [-1, 1]$, and $m \ge 0$, $k \ge 1$.

Claim. U satisfies Property A^k, but not Property A^{k-1}.

Proof. Set $V = \operatorname{span}\{1, x, \ldots, x^m\}$. V is an $(m + 1)$-dimensional subspace of U satisfying Property A. Since $\dim U = m + k + 1$, it follows from Proposition 4.30 that U satisfies Property A^k.

It remains to prove that U does not satisfy Property A^{k-1}. From Proposition 7 of Appendix B, there exists a unique (up to multiplication by -1) function h with the property that $|h| = 1$, h has exactly $m + 1$ points of discontinuity (jumps), and

$$\int_{-1}^{1} h(x)v(x)dx = 0$$

for all $v \in V$. (The points of discontinuity of h are the zeros of the Chebyshev polynomial of the second kind of degree $m + 1$.) From the uniqueness of h, it follows that h is even if $m + 1$ is even, and h is odd if $m + 1$ is odd. Thus

$$\int_{-1}^{1} h(x)x^{m+2j}dx = 0, \qquad j = 1, \ldots, k ,$$

implying that

$$\int_{-1}^{1} h(x)u(x)dx = 0$$

for all $u \in U$.

Let $\{\xi_i\}_{i=1}^{m+1}$ denote the points of discontinuity of h. Since $\dim U = m + k + 1$, there exist at least k linearly independent functions $u_1, \ldots, u_k$ of U which vanish at these $\{\xi_i\}_{i=1}^{m+1}$. Thus $hu_i \in C(K)$, $i = 1, \ldots, k$. From Theorem 3.12, U is not $(k - 1)$-convex for $C_1(K, \mu)$, where μ is Lebesgue measure. Thus U does not satisfy Property A^{k-1}.   □

Exercises

1. In Definition 4.5, we demanded that $\mathcal{B}$ be a convex cone. Show that Corollary 4.4 holds without the restriction that $\mathcal{B}$ be a cone.

2. Let K be a compact convex subset of $\mathbb{R}^d$ satisfying $K = \overline{\operatorname{int} K}$. Set

$$U = \left\{ u(x_1, \ldots, x_d) = \sum_{i=1}^{d} a_i x_i + a_{d+1} : a_i \in \mathbb{R} \right\},$$

i.e., the set of affine linear functions. Prove that U satisfies Property A on K.

3. Let $K = [0,1] \times [0,1]$, and $U = \operatorname{span}\{u_1, u_2\}$ where $u_1(x, y) = 1$, and

$$u_2(x, y) = \begin{cases} x + y, & x + y \le 1 \\ 1, & x + y \ge 1. \end{cases}$$

Prove that U satisfies Property A.

4. Let $K = [0,1] \times [0,1]$. For given $0 = x_0 < x_1 < x_2 = 1$, and $0 = y_0 < y_1 < y_2 = 1$, set

$$U = \{u : u \in C(K), u \text{ linear on } [x_{i-1}, x_i] \times [y_{j-1}, y_j], i, j = 1, 2\}.$$

Prove that U does not satisfy Property A.

5. Let $K = [0,1] \times [0,1]$. For given $0 < x_1 < \cdots < x_m < 1$, set

$$U = \operatorname{span}\{1, y, (x - x_1)_+, \ldots, (x - x_m)_+\}.$$

Prove that U satisfies Property A on K.

6. Let $K = [0,1] \times [0,1]$. For given $0 < x_1 < \cdots < x_m < 1$, and $0 < y_1 < \cdots < y_k < 1$, set

$$U = \operatorname{span}\{1, (x - x_1)_+, \ldots, (x - x_m)_+, (y - y_1)_+, \ldots, (y - y_k)_+\}.$$

Prove that U satisfies Property A on K if and only if $\min\{m, k\} \le 1$. (Hint: Use Exercises 4 and 5.)

7. For $x_1, y_1 \in (0, 1)$, set

$$V = \operatorname{span}\{1, x, (x - x_1)_+\}$$

and

$$W = \operatorname{span}\{1, y, (y - y_1)_+\}$$

on $[0, 1]$. Set $U = V \otimes W$. Prove that U does not satisfy Property A on $[0,1] \times [0,1]$.

8. Let

$$U = \operatorname{span}\{r, r \sin\theta, r \cos\theta, \ldots, r \sin n\theta, r \cos n\theta\}$$

for $(r, \theta) \in [0, 1] \times [0, 2\pi]$. Prove that U satisfies Property A.

9. Let

$$P_n = \Big\{ \sum_{i+j \le n} a_{ij} x^i y^j : a_{ij} \in I\!R \Big\}.$$

Let K be any compact subset of $I\!R^2$ satisfying $K = \overline{\text{int}\,K}$. Prove that, for $n \ge 2$, P_n does not satisfy Property A.

10. Let $K = \{(x,y) : x^2 + y^2 \le 1, y \ge 0\}$. Prove that

$$\widetilde{P}_n = \Big\{ \sum_{i+j=n} a_{ij} x^i y^j : a_{ij} \in I\!R \Big\}$$

satisfies Property A on K.

11. Let $K = [0,1] \times [0,1]$, and set

$$U = \text{span}\{p_m(x) + q_k(y) : p_m \in \pi_m, \, q_k \in \pi_k\}.$$

Prove that U does not satisfy Property A if $m \ge 2$ and $k \ge 1$.

12. Let V and W be finite-dimensional subspaces of $C[0,1]$ containing the constant function. Set

$$U = \{v(x) + w(y) : v \in V, \, w \in W\},$$

where U is defined on $[0,1] \times [0,1]$. Prove that if U satisfies Property A on $[0,1] \times [0,1]$, then both V and W satisfy Property A on $[0,1]$.

13. Construct functions $v, w \in C[0,1]$ such that both $V = \text{span}\{v\}$ and $W = \text{span}\{w\}$ satisfy Property A on $[0,1]$, but

$$U = \{av(x) + bw(y) : a, b \in I\!R\}$$

does not satisfy Property A on $[0,1] \times [0,1]$.

14. Assume U is an n-dimensional T-system on $[0,1]$. Let $k + 1 \le n$, and

$$0 \le a_1 < b_1 < a_2 < b_2 < \cdots < a_{k+1} < b_{k+1} \le 1.$$

Prove that on $\bigcup_{i=1}^{k+1} [a_i, b_i]$, U satisfies Property A^k, but not Property A^{k-1}.

Notes and References

Property A was introduced by Strauss [1984] for $K = [a,b]$ and μ Lebesgue measure. It should be noted that this paper was actually written in 1977. Strauss proved therein Proposition 4.1 under the assumption that every $u \in U$ has at most a finite number of separated zeros (see Definition 7 of Appendix A). No notice was given to the fact that the unicity space property then carried over to a large class of measures. For a proof of Proposition 4.1 based on Theorem 3.1, see the proof of Theorem 4.29. Strauss used Property A to give a simpler proof of the fact that splines with simple fixed knots are a unicity space for $C_1([a,b], \mu)$ (see below). Kroó [1984] noted that if Property

A is satisfied then U is a unicity space for $C_1([a,b], \mu)$ for all μ of the form $d\mu = w(x)\,dx$ where $w(x)\,dx \in \mathcal{B}$, and

$$\mathcal{B} = \{w(x)\,dx \, : \, w \in L^\infty[a,b], \, \underset{x \in [a,b]}{\text{ess inf}}\; w(x) > 0\},$$

(see Example 1 after Definition 4.5). He further conjectured the converse result. Namely, if U is a unicity space for $C_1([a,b], \mu)$ for all μ as above, then U satisfies Property A. For dim $U = 1$ he proved his conjecture. Sommer [1983b], based on Kroó [1984], proved this conjecture with an additional constraint on U. Namely, if $|Z(u)| > 0$ for any $u \in U\backslash\{0\}$, then dim $U(u) = $ dim $U - 1$. In this same paper Sommer also proved that, if U satisfies Property A on $[a,b]$, then U is necessarily a WT-system. Finally Kroó [1985a] proved his above-mentioned conjecture. Independently, Pinkus [1986] proved that U is a unicity space for $C_1([a,b], \mu)$ for all μ of the form $d\mu = w(x)\,dx$, where w is strictly positive and continuous, if and only if U satisfies Property A. However this result was proved under the restriction that $|Z(u)| = |\text{int}\,Z(u)|$ for all $u \in U$. In Sommer [1985] and Kroó [1987b], Kroó's result from [1985a] was generalized to $K \subset \mathbb{R}^d$ satisfying $K = \overline{\text{int}\,K}$. Kroó [1987b] actually further extended this result to smooth strictly convex real Banach space valued functions. Finally Kroó [1987a] and Schmidt [1987] removed the restriction $|Z(u)| = |\text{int}\,Z(u)|$ imposed in Pinkus [1986]. In fact Schmidt [1987] stated and proved Theorem 4.5 in its present form.

Theorem 4.6 was first proved for $K = [a,b]$ by Pinkus [1986]. The proof given here is taken from Pinkus, Wajnryb [1988], and is actually less complicated than that found in Pinkus [1986]. Theorem 4.12 is also taken from Pinkus, Wajnryb [1988]. Theorem 4.16 may be essentially found in Pinkus [1986] and in this form corrects an oversight therein. If no $u \in U\backslash\{0\}$ vanishes on a subinterval of $[a,b]$, and U is a unicity space for $C_1([a,b], \mu)$ for all $d\mu = w(x)\,dx$, $w(x)\,dx \in \mathcal{B}$, then it was shown by Havinson [1958] that U is a T-system on $[a,b]$ (see Corollary 4.18).

If μ^* is Lebesgue measure, then the fact that π_n is a unicity space for $C_1([a,b], \mu^*)$ was first proved by Jackson [1921] (see Example 3.2). Krein [1962] (originally published in Russian in 1938) proved that if U is a T-system on (a,b), then U is a unicity space for $C_1([a,b], \mu^*)$. Example 4.8 was shown to be a unicity space for $C_1([a,b], \mu^*)$ by Carroll, Braess [1974]. Splines with simple fixed knots (Example 4.9) were shown to be unicity spaces for $C_1([a,b], \mu^*)$ independently by Galkin [1974] and Strauss [1975]. The approach taken for dealing with B-splines (Propositions 4.19–4.21) is essentially to be found in Chapter 19 of Powell [1981] (see also de Boor [1978] and Schumaker [1981]). Sommer in [1983a] and [1979] (the former was written in 1978) proved that both the above examples satisfy Property A. In fact, in these same papers, Sommer constructed a large class of WT-systems satisfying Property A on

$[a, b]$. In this he was not very far from proving that (3) implies (1) in Theorem 4.16.

Sommer [1987] obtained the result contained in Part B of Section 5 for what are called uni-diagonal triangulations. This is a triangulation of a rectangle in $I\!R^2$ by straight lines parallel to the two axes, plus diagonals in one direction on each of the small rectangles. Sommer [1988] also proved an analogue of Theorem 11 of Appendix A. Namely, every n-dimensional subspace $(n \geq 2)$ of $C[a, b]$ satisfying Property A contains an $(n-1)$-dimensional subspace satisfying Property A.

The results of Section 6 are new. They are of interest in the light of the fact that, for this same problem in the uniform norm, the uniqueness property is not lost for a T-system on $K = [a, b]$, see e.g. Paszkowski [1957].

The results of Section 7 are all taken from Kroó [1986a].

Exercise 2 is due to Kroó [1982], while Exercise 4 may be found in Sommer [1983b]. Exercise 10 is stated in Kroó [1985b]. Exercise 14 is a special case of an example given in Kroó [1986a].

5
One-Sided L^1-Approximation

1. Introduction

In the previous three chapters we concerned ourselves with various aspects of the basic problem of best L^1-approximation from a finite-dimensional subspace. Some of the topics dealt with were perhaps non-standard, but the setting (i.e., approximation from a finite-dimensional subspace) was a classic one. In this chapter we consider the problem of best one-sided L^1-approximation from below, which has been much studied of late. There is an essential difference between these two problems. We are here dealing with approximation from a convex subset of a finite-dimensional subspace which depends on the function being approximated. Moreover the results, i.e., characterization, uniqueness, etc., are different in nature from those of the previous chapters.

To be more precise, let U be a finite-dimensional subspace of $C(K)$. For each $f \in C(K)$, we denote by $\mathcal{U}(f)$ the convex subset of all $u \in U$ satisfying $u \leq f$. In Sections 2 and 3 we discuss questions of existence, characterization, and uniqueness in the problem of best approximating f from $\mathcal{U}(f)$ in the $L^1(K, \mu)$-norm. (We always assume that μ is an 'admissible' measure in the sense of Chapter 3.) Somewhat surprisingly, for most 'reasonable' U there exist $f \in C(K)$ with more than one best approximant (Theorems 5.13 and 5.14).

In Section 4 we ask for exact conditions on the subspace U which imply that there exists a unique best one-sided $L^1(K, \mu)$ approximant to each $f \in C(K)$ from $\mathcal{U}(f)$, for a large class of 'admissible' μ. It transpires (Theorem 5.17) that such U may be totally characterized. They satisfy the following. There is a basis $u_1, \ldots, u_n$ for U such that each u_i is non-negative, and the u_i have distinct support, i.e., $\operatorname{supp} u_i \cap \operatorname{supp} u_j = \emptyset$ for all $i \neq j$. This second property simply says that the one-sided approximation problem from an n-dimensional subspace is equivalent to n independent one-sided approximation problems, each from a one-dimensional subspace. In this sense the subspace U is said to be 'trivial'.

In Sections 5 and 6 we impose additional constraints on both the approximating and approximated functions. We demand that they be continuously differentiable. We further demand that the approximating subspace contain a strictly positive function. (There is also a restriction on K.) These seemingly innocuous assumptions alter the whole nature of the problem. Uniqueness is now a more common occurrence. It is proven (Theorem 5.22) that a subspace

U satisfies a condition called Property B (in somewhat of an analogy to Property A of Chapter 4) if and only if for each 'admissible' measure μ there exists a unique best one-sided $L^1(K, \mu)$ approximant from U to every $f \in C^1(K)$. Many subspaces satisfy Property B. In Section 6 we study various aspects of Property B, and consider a fair number of examples.

2. Best One-Sided Approximation in $C(K)$: Characterization

For each $f \in C(K)$, set

$$\mathcal{U}(f) = \{u : u \in U, u \le f\},$$

where U is a finite-dimensional subspace of $C(K)$. Let μ be a fixed 'admissible' measure on K in the sense of Chapter 3. We are interested in the problem

$$(5.1) \qquad \inf_{u \in \mathcal{U}(f)} \|f - u\|_1 = \inf_{u \in \mathcal{U}(f)} \int_K |f - u| \, d\mu.$$

Since $f - u \ge 0$ for all $u \in \mathcal{U}(f)$, the problem of finding, or characterizing, a $u \in \mathcal{U}(f)$ attaining the above infimum is equivalent to that of finding, or characterizing, a $u \in \mathcal{U}(f)$ attaining the supremum in

$$(5.2) \qquad \sup_{u \in \mathcal{U}(f)} \int_K u \, d\mu.$$

The values in (5.1) and (5.2) differ, but the problem is the same.

By definition $\mathcal{U}(f)$ is a closed convex subset of U. Thus if $\mathcal{U}(f) \ne \emptyset$, then the infimum in (5.1) (and the supremum in (5.2)) is attained (see Theorem 1.2). Set

$$P_{\mathcal{U}(f)}(f) = \{u^* : u^* \in \mathcal{U}(f), \|f - u^*\|_1 \le \|f - u\|_1 \text{ for all } u \in \mathcal{U}(f)\}.$$

From the above, if $\mathcal{U}(f) \ne \emptyset$ then $P_{\mathcal{U}(f)}(f) \ne \emptyset$. Each $u^* \in P_{\mathcal{U}(f)}(f)$ will be called a *best one-sided approximant to* f.

When is $\mathcal{U}(f)$ not the empty set? If $f \ge 0$ on all of K, then $0 \in \mathcal{U}(f)$, and $\mathcal{U}(f) \ne \emptyset$. Setting $f = -1$ we see that $\mathcal{U}(f) \ne \emptyset$ if U contains a strictly negative function. On the other hand, if U contains a strictly negative function (or equivalently a strictly positive function), then $\mathcal{U}(f) \ne \emptyset$ for every $f \in C(K)$. We have therefore proven that $\mathcal{U}(f) \ne \emptyset$ for all $f \in C(K)$ if and only if U contains a strictly positive function. Even if U does not contain a strictly positive function, we shall still consider problem (5.1) (or (5.2)). We shall, however, restrict ourselves to those $f \in C(K)$ with $\mathcal{U}(f) \ne \emptyset$.

Now that we have dealt with the problem of existence, let us turn to the question of characterizing best one-sided approximants. In Chapter 1 we have two general results, Theorems 1.6 and 1.11, which are applicable. Unfortunately both these results give us no new information. Both Theorems

1.6 and 1.11 applied to our problem simply state that $u^* \in P_{\mathcal{U}(f)}(f)$ if and only if

$$\int_K u^* d\mu \geq \int_K u\, d\mu$$

for all $u \in \mathcal{U}(f)$, i.e., (5.2) holds. This is a meaningless tautology since it is totally uninformative. (All characterization results are tautologies, but most are not meaningless.)

The fact of the matter is that there seems to be no 'good' characterization result, unless we presuppose additional assumptions on U or on f. This we now do. First however let us recall that, for each $f \in C(K)$,

$$Z(f) = \{x : f(x) = 0\}.$$

Theorem 5.1. *Let U be a finite-dimensional subspace of $C(K)$, and assume that $\int_K u\, d\mu \neq 0$ for some $u \in U$. Let $f \in C(K)$ be such that there exists a $v \in U$ for which $v < f$ on all of K. Then $u^* \in P_{\mathcal{U}(f)}(f)$ if and only if $u^* \in \mathcal{U}(f)$ and for any $w \in U$ satisfying $w \leq 0$ on $Z(f - u^*)$, it follows that $\int_K w\, d\mu \leq 0$.*

Proof. ($\Leftarrow$). Let $u \in \mathcal{U}(f)$. Then $u(x) \leq f(x) = u^*(x)$ for $x \in Z(f - u^*)$. By assumption $\int_K (u - u^*)d\mu \leq 0$, i.e., $\int_K u\, d\mu \leq \int_K u^* d\mu$. Thus $u^* \in P_{\mathcal{U}(f)}(f)$.

($\Rightarrow$). Suppose that $u^* \in P_{\mathcal{U}(f)}(f)$. There exists a $\tilde{u} \in U$ for which $\int_K \tilde{u}\, d\mu > 0$. If $Z(f - u^*) = \emptyset$, then $u^* + \varepsilon \tilde{u} \in \mathcal{U}(f)$ for some $\varepsilon > 0$ and $\int_K (u^* + \varepsilon \tilde{u})d\mu > \int_K u^* d\mu$, contradicting our assumption that $u^* \in P_{\mathcal{U}(f)}(f)$. We therefore have $Z(f - u^*) \neq \emptyset$.

Suppose there exists a $w \in U$ satisfying $w \leq 0$ on $Z(f - u^*)$, and $\int_K w\, d\mu > 0$. By assumption there exists a $v \in U$ with $v < f$ on K. This implies that $\tilde{v} = u^* - v$ is strictly positive on $Z(f - u^*)$. We therefore have a $\delta > 0$, δ small, such that $\tilde{w} = w - \delta \tilde{v}$ satisfies $\tilde{w} < 0$ on $Z(f - u^*)$, and $\int_K \tilde{w}\, d\mu > 0$. Now K is compact and $Z(f - u^*)$ is closed. Furthermore $f - u^* > 0$ on $K \backslash Z(f - u^*)$, $\tilde{w} < 0$ on $Z(f - u^*)$, and f, u^* and $\tilde{w}$ are all continuous functions on K. It therefore follows that for some $\varepsilon > 0$, ε small, $u = u^* + \varepsilon \tilde{w} \in \mathcal{U}(f)$, i.e., $u \leq f$. By construction

$$\int_K u\, d\mu > \int_K u^* d\mu.$$

This contradicts our assumption that $u^* \in P_{\mathcal{U}(f)}(f)$. □

The characterization result of Theorem 5.1, while elegant, is not easy to use. With a little work we can rewrite Theorem 5.1 in the following more amenable form.

Theorem 5.2. *Let U be an n-dimensional subspace of $C(K)$, and assume that $\int_K u\, d\mu \neq 0$ for some $u \in U$. Let $f \in C(K)$ be such that there exists a $v \in U$ satisfying $v < f$ on all of K. Then $u^* \in P_{\mathcal{U}(f)}(f)$ if and only if*

$u^* \in \mathcal{U}(f)$ and there exist distinct points $\{x_i\}_{i=1}^k$ in K, $1 \le k \le n$, and positive numbers $\{\lambda_i\}_{i=1}^k$ for which

$$a) \ (f - u^*)(x_i) = 0, \qquad i = 1, \dots, k$$

$$b) \int_K u \, d\mu = \sum_{i=1}^k \lambda_i u(x_i), \quad \text{all } u \in U.$$

We first separate out the main idea in the proof of Theorem 5.2, namely the integration formula. This we do for two reasons. Firstly we wish to highlight this fact, and secondly we shall use it again in a later section.

Lemma 5.3. *Let U be an n-dimensional subspace of $C(K)$, and assume that $\int_K u \, d\mu \ne 0$ for some $u \in U$. Let A be a closed subset of K, with the property that if $u \in U$ satisfies $u(x) \le 0$, all $x \in A$, then $\int_K u \, d\mu \le 0$. There then exist points $\{x_i\}_{i=1}^k$ in A, $1 \le k \le n$, and positive numbers $\{\lambda_i\}_{i=1}^k$ such that*

$$\int_K u \, d\mu = \sum_{i=1}^k \lambda_i u(x_i)$$

for all $u \in U$.

Proof. Let $u_1, \dots, u_n$ be any basis for U and

$$P = \{(u_1(x), \dots, u_n(x)) : x \in A\}.$$

Let Q denote the closed convex cone generated by P. Set

$$\mathbf{c} = \left(\int_K u_1 d\mu, \dots, \int_K u_n d\mu \right).$$

By assumption $\mathbf{c} \ne \mathbf{0}$. If $\mathbf{c} \in Q$, then since Q is a closed convex cone in $\mathbb{R}^n$, there exist distinct points $\{x_i\}_{i=1}^k$ in A, and positive numbers $\{\lambda_i\}_{i=1}^k$, $1 \le k \le n$, such that

$$\int_K u_j d\mu = \sum_{i=1}^k \lambda_i u_j(x_i), \qquad j = 1, \dots, n.$$

Thus

$$\int_K u \, d\mu = \sum_{i=1}^k \lambda_i u(x_i)$$

for all $u \in U$, proving the lemma. It remains to prove that $\mathbf{c} \in Q$.

Assume not. Since Q and $\{\mathbf{c}\}$ are closed convex subsets of $\mathbb{R}^n$, there exists a separating hyperplane. Since Q is a cone, there necessarily exists a separating hyperplane passing through the origin. That is, there exists a vector $\mathbf{a} = (a_1, \dots, a_n) \ne \mathbf{0}$ for which

$$\sum_{j=1}^n a_j q_j \le 0 < \sum_{j=1}^n a_j c_j$$

for all $\mathbf{q} = (q_1, \ldots, q_n) \in Q$, where $\mathbf{c} = (c_1, \ldots, c_n)$. Set $v = \sum_{j=1}^n a_j u_j$. Then $v(x) \le 0$ for all $x \in A$, and $\int_K v \, d\mu > 0$. This contradicts our assumption. Thus $\mathbf{c} \in Q$.   □

Remark. The assumption $\int_K u \, d\mu \ne 0$ for some $u \in U$ may be dropped if $k = 0$ is admitted.

Proof of Theorem 5.2. ($\Leftarrow$). Assume $u^* \in \mathcal{U}(f)$, and (a) and (b) hold. Then for every $u \in \mathcal{U}(f)$

$$\int_K u \, d\mu = \sum_{i=1}^k \lambda_i u(x_i) \le \sum_{i=1}^k \lambda_i f(x_i) = \sum_{i=1}^k \lambda_i u^*(x_i) = \int_K u^* d\mu,$$

i.e., $\int_K u \, d\mu \le \int_K u^* d\mu$ for all $u \in \mathcal{U}(f)$. Thus $u^* \in P_{\mathcal{U}(f)}(f)$. (Equivalently it follows from (a) and (b) that, if $u \le 0$ on $Z(f - u^*)$, then $\int_K u \, d\mu \le 0$, and we can apply Theorem 5.1.)

($\Rightarrow$). We now suppose that $u^* \in P_{\mathcal{U}(f)}(f)$. From Theorem 5.1, if $u \in U$ satisfies $u \le 0$ on $Z(f - u^*)$, then $\int_K u \, d\mu \le 0$. Apply Lemma 5.3 with $A = Z(f - u^*)$.   □

As an immediate consequence of this theorem, we have these next two results.

Corollary 5.4. *Let U be an n-dimensional subspace of $C(K)$, and assume that U contains a strictly positive function. Then $u^* \in P_{\mathcal{U}(f)}(f)$ if and only if $u^* \in \mathcal{U}(f)$ and conditions (a) and (b) of Theorem 5.2 hold.*

Corollary 5.5. *Let the conditions of Theorem 5.2 hold. Let $u^* \in P_{\mathcal{U}(f)}(f)$ with $\{x_i\}_{i=1}^k$ as in (a) and (b). Then for any $u \in P_{\mathcal{U}(f)}(f)$,*

$$u(x_i) = u^*(x_i), \qquad i = 1, \ldots, k.$$

When is the best one-sided approximant unique? A sufficient condition readily follows from the above corollary.

Corollary 5.6. *Let the conditions of Theorem 5.2 hold. Assume the $\{x_i\}_{i=1}^k$ are as in (a) and (b). If $u \in U$ and $u(x_i) = 0$, $i = 1, \ldots, k$, implies $u = 0$, then $P_{\mathcal{U}(f)}(f)$ is a singleton.*

Unfortunately, if U and f do not satisfy the two assumptions of Theorems 5.1 and 5.2, then the theorems are no longer valid. Let us consider each of these assumptions separately.

If $\int_K u \, d\mu = 0$ for all $u \in U$, then it is evident (see (5.2)) that $P_{\mathcal{U}(f)}(f) = \mathcal{U}(f)$. In this case, if f is a strictly positive function, then $Z(f - u) = \emptyset$ for some $u \in \mathcal{U}(f)$. This is a somewhat degenerate situation and does not quite contradict the statement of Theorem 5.1. What happens if $\int_K u \, d\mu \ne 0$ for some $u \in U$, $\mathcal{U}(f) \ne \emptyset$, but there exists no $v \in U$ satisfying $v < f$? We may then have a situation as exemplified by the following. Let $U = $ span $\{x\}$ on $K = [0, 1]$ with μ any 'admissible' measure. Set $f(x) = x^2$. Then $\mathcal{U}(f) = \{\alpha x : \alpha \leq 0\}$, and $P_{\mathcal{U}(f)}(f) = \{0\}$. It is easily checked that the necessary conditions of neither Theorem 5.1 nor Theorem 5.2 hold for this example.

As these examples indicate, there seems to be no alternative characterization to (5.2). However if we return to the first half of the proof of Theorem 5.2, we see that none of the assumptions were used therein. We have,

Corollary 5.7. *If* $u^* \in \mathcal{U}(f)$ *and there exist distinct points* $\{x_i\}_{i=1}^k$, $1 \leq k \leq n$, *and positive values* $\{\lambda_i\}_{i=1}^k$ *satisfying* (a) *and* (b) *of Theorem 5.2, then* $u^* \in P_{\mathcal{U}(f)}(f)$.

The formula (b) of Theorem 5.2 is called a *quadrature formula* for U. It has the additional property that all the coefficients $\{\lambda_i\}_{i=1}^k$ are positive numbers. If

$$(5.3) \qquad \int_K u \, d\mu = \sum_{i=1}^k \lambda_i u(x_i)$$

for all $u \in U$, where $\lambda_i > 0, i = 1, \ldots, k$ and $1 \leq k < \infty$, then we shall say that (5.3) is a *positive quadrature formula with* k *active points* (i.e., the $\{x_i\}_{i=1}^k$). In Theorem 5.2 we always had $k \leq n$. This is not really a restriction. It follows from this next lemma which is also to be used later.

Lemma 5.8. *Let* U *be an* n-*dimensional subspace of* $C(K)$, *and assume there exists a* $u \in U$ *with* $\int_K u \, d\mu \neq 0$. *Let* $u_1, \ldots, u_n$ *be any basis for* U. *If*

$$\int_K u \, d\mu = \sum_{i=1}^k \lambda_i u(x_i)$$

for all $u \in U$, *where* $1 \leq k < \infty$, $\lambda_i > 0$, $i = 1, \ldots, k$ *and*

$$\text{rank}(u_j(x_i))_{i=1, j=1}^{k, \ n} < k,$$

then there exists a positive quadrature formula for U *with* r *active points* $\{y_i\}_{i=1}^r$, *where* $1 \leq r < k$, *and* $\{y_i\}_{i=1}^r \subset \{x_i\}_{i=1}^k$.

Proof. Since rank $(u_j(x_i))_{i=1, j=1}^{k, \ n} < k$, there exists a vector $\boldsymbol{\sigma} = (\sigma_1, \ldots, \sigma_k) \neq \mathbf{0}$, such that

$$\sum_{i=1}^k \sigma_i u(x_i) = 0,$$

all $u \in U$. By assumption, $\sum_{i=1}^k \lambda_i u(x_i) \neq 0$ for some $u \in U$. Thus $\boldsymbol{\lambda} = (\lambda_1, \ldots, \lambda_k)$ and $\boldsymbol{\sigma}$ are linearly independent vectors.

For every $\delta \in \mathbb{R}$,

$$\int_K u \, d\mu = \sum_{i=1}^k (\lambda_i + \delta \sigma_i) u(x_i)$$

for all $u \in U$. Choose $\delta_0 \in \mathbb{R}$ such that $\lambda_i + \delta_0 \sigma_i \geq 0$, $i = 1, \ldots, k$, and $\lambda_j + \delta_0 \sigma_j = 0$ for some $j \in \{1, \ldots, k\}$. This proves the lemma. □

On the basis of this lemma, it suffices to consider positive quadrature formulae for U with at most n active points if there exists a $u \in U$ with $\int_K u \, d\mu \neq 0$. If $\int_K u \, d\mu = 0$ for all $u \in U$, then we can consider positive quadrature formulae with at most $n + 1$ active points.

3. Best One-Sided Approximation in $C(K)$: Uniqueness

We now turn to the question of when we have a unique best one-sided approximant from $\mathcal{U}(f)$ to f for every $f \in C(K)$ satisfying $\mathcal{U}(f) \neq \emptyset$. In other words, when does $P_{\mathcal{U}(f)}(f)$ contain at most one function for every $f \in C(K)$? If $P_{\mathcal{U}(f)}(f)$ contains at most one function for every $f \in C(K)$, then we shall, following the terminology of the previous chapters, say that U is a *unicity space* for $C_{1+}(K, \mu)$. (The + indicates one-sided approximation.) Obviously, if $\int_K u \, d\mu = 0$ for all $u \in U$, then U is not a unicity space for $C_{1+}(K, \mu)$. Equally obviously (see (5.2)), if dim $U = 1$ and $\int_K u \, d\mu \neq 0$ for a $u \in U$, then U is a unicity space for $C_{1+}(K, \mu)$.

We shall delineate two criteria for determining when U is a unicity space for $C_{1+}(K, \mu)$. The first is a consequence of Corollary 5.4 (and as such presupposes the existence of a strictly positive function in U). The second is totally general and more in the spirit of Theorem 3.3 and Corollary 3.4.

Theorem 5.9. *Let U be an n-dimensional subspace of $C(K)$, $n \geq 2$. Assume U contains a strictly positive function. Then U is a unicity space for $C_{1+}(K, \mu)$ if and only if each positive quadrature formula for U contains at least n active points.*

Proof. ($\Rightarrow$). Assume that

$$\int_K u \, d\mu = \sum_{i=1}^{k} \lambda_i u(x_i)$$

for all $u \in U$, where $\lambda_i > 0$, $i = 1, \ldots, k$, and $1 \leq k \leq n - 1$. Since dim $U = n$, there exists a $u^* \in U$, $u^* \neq 0$, satisfying $u^*(x_i) = 0$, $i = 1, \ldots, k$. Set $f = |u^*|$. Then $\pm u^* \in \mathcal{U}(f)$ and, from (a) and (b) of Theorem 5.2, $\pm u^* \in P_{\mathcal{U}(f)}(f)$. Thus U is *not* a unicity space for $C_{1+}(K, \mu)$.

($\Leftarrow$). Since dim $U = n$, it follows from Lemma 5.8 that we may assume that, for every positive quadrature formula for U with n active knots $\{x_i\}_{i=1}^{n}$, we have det $(u_j(x_i))_{i,j=1}^{n} \neq 0$, where $u_1, \ldots, u_n$ is any basis for U. Thus if $u \in U$ and $u(x_i) = 0$, $i = 1, \ldots, n$, then $u = 0$.

Assume that U is not a unicity space for $C_{1+}(K, d\mu)$. Let $f \in C(K)$ and $u^1, u^2 \in P_{\mathcal{U}(f)}(f)$ where $u^1 \neq u^2$. From Corollary 5.5 and our assumption, there exists a positive quadrature formula for U with n active points

$\{x_i\}_{i=1}^n$ and $u^1(x_i) = u^2(x_i)$, $i = 1, \ldots, n$. From the above, $u^1 = u^2$. This is a contradiction.   □

We shall shortly return to Theorem 5.9, since it begs the question of when every quadrature formula for U contains at least n active points. As we shall show, the answer is seldom.

We now turn to the next characterization of unicity spaces for $C_{1+}(K, \mu)$, without the restriction of U containing a strictly positive function. In Theorem 3.3 and Corollary 3.4 we characterized unicity spaces for $C_1(K, \mu)$ by looking at a set of 'test' functions, i.e., U^*. The basic idea here is the same. The set of test functions is considerably smaller.

Theorem 5.10. *A finite-dimensional subspace U of $C(K)$ is a unicity space for $C_{1+}(K, \mu)$ if and only if for every $u \in U$, $u \neq 0$, the zero function is not in $P_{\mathcal{U}(|u|)}(|u|)$.*

Proof. ($\Rightarrow$). Assume there exists a $u \in U$, $u \neq 0$, for which $0 \in P_{\mathcal{U}(|u|)}(|u|)$. But $\pm u \in \mathcal{U}(|u|)$, and therefore $\int_K u \, d\mu = 0$. Thus $\pm u \in P_{\mathcal{U}(|u|)}(|u|)$, and U is not a unicity space for $C_{1+}(K, \mu)$.

($\Leftarrow$). Assume that U is not a unicity space for $C_{1+}(K, \mu)$. Let $f \in C(K)$, and $u^1, u^2 \in P_{\mathcal{U}(f)}(f)$, $u^1 \neq u^2$. Set $u^* = (u^1 - u^2)/2$ and $g = f - (u^1 + u^2)/2$. Then $\pm u^* \in P_{\mathcal{U}(g)}(g)$. Since $|u^*| \leq g$, we also have $\pm u^* \in P_{\mathcal{U}(|u^*|)}(|u^*|)$. Thus the zero function is in $P_{\mathcal{U}(|u^*|)}(|u^*|)$.   □

Using the definition of one-sided approximants, we can rewrite Theorem 5.10.

Corollary 5.11. *A finite-dimensional subspace U of $C(K)$ is a unicity space for $C_{1+}(K, \mu)$ if and only if for each $u \in U$, $u \neq 0$, there exists a $v \in U$ satisfying*

$$a) \ v \leq |u|$$

$$b) \ \int_K v \, d\mu > 0.$$

Note that (a) and (b) trivially hold for $u \in U$ satisfying $\int_K u \, d\mu \neq 0$. Take $v = u$ or $v = -u$. Thus the conditions of Corollary 5.11 need be verified only for those $u \in U$ satisfying $\int_K u \, d\mu = 0$.

If U contains a strictly positive function, then Corollary 5.11 may be recast as follows.

Corollary 5.12. *Let U be a finite-dimensional subspace of $C(K)$, and assume U contains a strictly positive function. Then U is a unicity space for $C_{1+}(K, \mu)$ if and only if for each $u \in U$, $u \neq 0$, there exists a $v \in U$ satisfying*

$$a) \ v(x) \leq 0, \quad \text{all } x \in Z(u)$$

$$b) \ \int_K v \, d\mu > 0.$$

Proof. ($\Rightarrow$). If U is a unicity space for $C_{1+}(K, \mu)$, then (a) and (b) of Corollary 5.11 hold. Thus (a) and (b) of Corollary 5.12 hold.

($\Leftarrow$). Assume (a) and (b) of Corollary 5.12 hold. Let u and v be as in Corollary 5.12, and $\tilde{u} \in U$ satisfy $\tilde{u} > 0$. For $\varepsilon > 0$, ε sufficiently small, $w = v - \varepsilon\tilde{u}$ satisfies $w(x) < 0$ for all $x \in Z(u)$, and $\int_K w\, d\mu > 0$. For $\delta > 0$, δ sufficiently small, it is evident that δw satisfies (a) and (b) of Corollary 5.11. Thus U is a unicity space for $C_{1+}(K, \mu)$. $\quad\square$

Theorems 5.9 and 5.10, and Corollaries 5.11 and 5.12, are nice elegant results. But for a given subspace U, it is not at all easy to check whether U is a unicity space for $C_{1+}(K, \mu)$ using these results. We shall show that many spaces are not unicity spaces for $C_{1+}(K, \mu)$. In fact, while positivity is generally a good property vis-a-vis uniqueness, we prove that in this context it is not.

Theorem 5.13. *Let U be an n-dimensional subspace of $C(K)$, $n \geq 2$. Assume that $\mathrm{int}\, K$ is connected, and there exists a $u^* \in U$ such that $u^* > 0$ on $\mathrm{int}\, K$. Then U is not a unicity space for $C_{1+}(K, \mu)$.*

Proof. For convenience, set $u_1 = \alpha u^*$ where $\alpha > 0$ is chosen so that $\int_K u_1 d\mu = 1$. There exists a basis $u_1, u_2, \ldots, u_n$ for U, with $\int_K u_i d\mu = 0$, $i = 2, \ldots, n$. Let $V = \mathrm{span}\,\{u_2, \ldots, u_n\}$. Then $\int_K v\, d\mu = 0$ for all $v \in V$.

Set
$$V_1 = \{v : v \in V, \|v\|_1 = 1\}.$$

For each $v \in V_1$, define
$$J(v) = \{x : v(x) \leq 0\}$$

and let $\mu(J(v))$ denote the μ-measure of the set $J(v)$. Since V_1 is compact and equicontinuous, there exists a $v^* \in V_1$ such that $\mu(J(v^*)) \geq \mu(J(v))$ for all $v \in V_1$. Since $J(cv) = J(v)$ for all $c > 0$ and $v \in V$, $v \neq 0$, it therefore follows that $\mu(J(v^*)) \geq \mu(J(v))$ for all $v \in V$, $v \neq 0$.

Define
$$v_+^*(x) = \max\{v^*(x), 0\}.$$

Since $\int_K v^* d\mu = 0$, we have $v_+^* \neq 0$. Assume $u \in \mathcal{U}(v_+^*)$. Then $u = bu_1 + v$, $v \in V$, and $\int_K u\, d\mu = b$. We claim that $b \leq 0$. To this end, suppose that $b > 0$. For each $x \in J(v^*)$ we have $v_+^*(x) = 0$ and therefore $u(x) = bu_1(x) + v(x) \leq 0$. Thus $v(x) \leq -bu_1(x) \leq 0$, implying $J(v^*) \subseteq J(v)$. Since $\int_K v^* d\mu = 0$, and $\mathrm{int}\, K$ is connected, there exists an $x^* \in \mathrm{int}\, K$ with $v^*(x^*) = 0$, while v^* takes on strictly positive values in every neighborhood of x^*. Moreover $bu_1(x^*) + v(x^*) \leq 0$ and therefore $v(x^*) \leq -bu_1(x^*) < 0$ (by our assumption on u_1). Therefore $J(v)$ contains a neighborhood of x^*. Thus $J(v^*) \subset J(v)$, and more importantly $\mu(J(v^*)) < \mu(J(v))$. But $v \neq 0$ (since $v(x^*) < 0$) and we have contradicted our definition of v^*. Thus $b \leq 0$.

To summarize, for all $u \in \mathcal{U}(v_+^*)$ we have $\int_K u\, d\mu \leq 0$. But $0, v^* \in \mathcal{U}(v_+^*)$, and $\int_K v^* d\mu = 0$. Thus $0, v^* \in P_{\mathcal{U}(v_+^*)}(v_+^*)$, and U is *not* a unicity space for $C_{1+}(K, \mu)$. □

The condition that U contains a function, strictly positive on its interior, is essential. If $U = \text{span}\{x, |x|\}$ on $K = [-1, 1]$, then $|u| \in U$ for every $u \in U$, and U is therefore a unicity space for $C_{1+}(K, \mu)$ as a consequence of Theorem 5.10. The condition that int K be connected is also necessary. To see this, let U be as previously defined, and

$$K = \{(x, y) : |y| \leq |x| \leq 1\}.$$

Again Theorem 5.10 implies that U is a unicity space for $C_{1+}(K, \mu)$.

Our second negative result is based on Theorem 5.9. Recall that $[K]$ is the symbol for the number of connected components of K.

Theorem 5.14. *Let U be an n-dimensional subspace of $C(K)$, $n \geq 2$. Assume that U contains a strictly positive function, and also that $[K] \leq n - 1$. Then U is not a unicity space for $C_{1+}(K, \mu)$.*

Proof. On the basis of Theorem 5.9, it suffices to prove that under the above assumptions there exists a positive quadrature formula for U with k active points, $1 \leq k \leq n - 1$. To this end, let

$$\int_K u\, d\mu = \sum_{i=1}^{n} \lambda_i u(x_i)$$

be any positive quadrature formula for U where $\lambda_i \geq 0, i = 1, \ldots, n$. Such a quadrature formula necessarily exists. If $\lambda_i = 0$ for some i, then we are finished. Furthermore, if $u_1, \ldots, u_n$ is any basis for U, and $\det (u_j(x_i))_{i,j=1}^{n} = 0$, then we are also finished, as a consequence of Lemma 5.8. We therefore assume that we are given a positive quadrature formula for U with n active points $\{x_i\}_{i=1}^{n}$, and that $\det (u_j(x_i))_{i,j=1}^{n} \neq 0$.

Set

$$V = \{v : v \in U, \int_K v\, d\mu = 0\}.$$

Since V is a subspace of U, and since U contains a strictly positive function, $\dim V = n - 1$. Let $v_1, \ldots, v_{n-1}$ be any basis for V. Set

$$\mathbf{v}(x) = (v_1(x), \ldots, v_{n-1}(x))$$

for each $x \in K$. From the above,

$$\mathbf{0} = \sum_{i=1}^{n} \lambda_i \mathbf{v}(x_i).$$

If any $n - 1$ of the vectors $\mathbf{v}(x_1), \ldots, \mathbf{v}(x_n)$ are linearly dependent, then there are at least two linearly independent solutions to $\sum_{i=1}^{n} a_i \mathbf{v}(x_i) = \mathbf{0}$ (recall that $\lambda_i > 0$ for all i) and therefore rank $(v_j(x_i))_{i=1,j=1}^{n,\,n-1} < n-1$. But then det $(u_j(x_i))_{i,j=1}^{n} = 0$, contradicting our hypothesis.

Since $[K] \le n-1$, and $x_1, \ldots, x_n \in K$, there exists a connected component of K containing at least two distinct x_i. Assume A is a connected component of K, and $x_1, x_2 \in A$. Set

$$B = \Big\{ -\sum_{i=2}^{n} \alpha_i \mathbf{v}(x_i) : \alpha_i \ge 0 \Big\}.$$

Since the $\mathbf{v}(x_2), \ldots, \mathbf{v}(x_n)$ are linearly independent vectors in $\mathbb{R}^{n-1}$, B is a polygonal convex cone, and

$$\text{int } B = \Big\{ -\sum_{i=2}^{n} \alpha_i \mathbf{v}(x_i) : \alpha_i > 0, \; i = 2, \ldots, n \Big\}.$$

Because

$$\mathbf{v}(x_1) = -\sum_{i=2}^{n} \frac{\lambda_i}{\lambda_1} \mathbf{v}(x_i),$$

we have that $\mathbf{v}(x_1) \in \text{int } B$. On the other hand, if $\mathbf{v}(x_2) \in B$, then

$$\mathbf{v}(x_2) = -\sum_{i=2}^{n} \alpha_i \mathbf{v}(x_i)$$

for some $\alpha_i \ge 0$, $i = 2, \ldots, n$. Thus

$$(1 + \alpha_2)\mathbf{v}(x_2) + \sum_{i=3}^{n} \alpha_i \mathbf{v}(x_i) = \mathbf{0},$$

implying that the $\mathbf{v}(x_2), \ldots, \mathbf{v}(x_n)$ are linearly dependent and contradicting our earlier hypothesis. Thus $\mathbf{v}(x_2) \notin B$.

Now A is connected and $x_1, x_2 \in A$. Furthermore B is connected, $\mathbf{v}(x_1) \in \text{int } B$, $\mathbf{v}(x_2) \notin B$, and $\mathbf{v}$ is continuous on A. Following a path in A from x_1 to x_2, we obtain a $y \in A$ for which $\mathbf{v}(y) \in \partial B = B \backslash \text{int } B$. Thus

$$\mathbf{v}(y) = -\sum_{i=2}^{n} \alpha_i \mathbf{v}(x_i)$$

where $\alpha_i \ge 0$, $i = 2, \ldots, n$, and at least one of the α_i is zero. Assume for the sake of the argument that $\alpha_n = 0$. Then

$$\mathbf{v}(y) + \sum_{i=2}^{n-1} \alpha_i \mathbf{v}(x_i) = \mathbf{0},$$

implying that

$$\int_K v \, d\mu = v(y) + \sum_{i=2}^{n-1} \alpha_i v(x_i)$$

for all $v \in V$.

Let $u_1 \in U, u_1 > 0$ on K. Then $U = \text{span} \{u_1, v_1, \ldots, v_{n-1}\}$. It is now easily seen that there exists a $c > 0$ for which

$$\int_K u_1 d\mu = c\left(u_1(y) + \sum_{i=2}^{n-1} \alpha_i u_1(x_i)\right).$$

Thus

$$\int_K u \, d\mu = cu(y) + \sum_{i=2}^{n-1} c\alpha_i u(x_i)$$

for all $u \in U$. We have constructed a positive quadrature formula for U with k active points, $1 \le k \le n - 1$. U is not a unicity space for $C_{1+}(K, \mu)$. □

If $[K] \ge n$, then the above theorem simply does not hold. For example, assume that $K = \bigcup_{i=1}^n A_i$, where each A_i is a distinct component of K, and let u_i be strictly positive on A_i and vanish elsewhere. Set $U = \text{span} \{u_1, \ldots, u_n\}$. Then U is a unicity space for $C_{1+}(K, \mu)$. The reader familiar with the contents of the last chapter might justifiably think that this example is somewhat special. To nip possible conjectures in the bud, consider the following example. Let $U = \text{span} \{1, x\}$ on $K = [-2, -1] \cup [1, 2]$, where μ is Lebesgue measure. U is a unicity space since there does not exist a positive quadrature formula for U with one active point in K.

Since unicity spaces in $C_{1+}(K, \mu)$ are such rare birds, it is natural to ask about the size of the set of $f \in C(K)$ for which $P_{\mathcal{U}(f)}(f)$ is a singleton. If we measure distance in terms of the uniform norm, then this set need not be very big. As such we use the $L^1(K, \mu)$ norm. If U does not contain a strictly positive function, then given any $f \in C(K)$ and any $\varepsilon > 0$, there always exists a $g \in C(K)$ satisfying $\|f - g\|_1 < \varepsilon$ for which $\mathcal{U}(g) = \emptyset$. Furthermore, it is not difficult to construct U as above, an $f \in C(K)$ and $\varepsilon_0 > 0$, such that $\mathcal{U}(f) = \emptyset$ and $\mathcal{U}(g) = \emptyset$ for all $g \in C(K)$ satisfying $\|f - g\|_1 < \varepsilon_0$. As such it is important to note that, in this next proposition, we always restrict ourselves to those $f \in C(K)$ for which $\mathcal{U}(f) \ne \emptyset$.

Proposition 5.15. If U is a finite-dimensional subspace of $C(K)$, then the set $\{f : P_{\mathcal{U}(f)}(f) \text{ a singleton}\}$ is $L^1(K, \mu)$ dense in $\{f : \mathcal{U}(f) \ne \emptyset\}$.

Proof. Let $f \in C(K)$ with $\mathcal{U}(f) \ne \emptyset$. From Lemma 3.5, there exist points $\{x_i\}_{i=1}^m$, m finite, such that if $w \in U$ satisfies $w(x_i) \le 0$, $i = 1, \ldots, m$, and $\int_K w \, d\mu \ge 0$, then $w = 0$. Choose any $v \in \mathcal{U}(f)$. Given $\varepsilon > 0$, let $g \in C(K)$ satisfy $g \ge v$, $g(x_i) = v(x_i)$, $i = 1, \ldots, m$, and $\|f - g\|_1 < \varepsilon$. Such a g is easily constructed.

We claim that $P_{\mathcal{U}(g)}(g)$ is a singleton. By construction $v \in \mathcal{U}(g)$. If $u \in P_{\mathcal{U}(g)}(g)$, then $u(x_i) \le g(x_i)$, $i = 1, \ldots, m$, and $\int_K u \, d\mu \ge \int_K v \, d\mu$. Set $w = u - v$. Then $w(x_i) \le 0$, $i = 1, \ldots, m$, and $\int_K w \, d\mu \ge 0$. From Lemma 3.5, $w = 0$. Thus $u = v$ and $P_{\mathcal{U}(g)}(g) = \{v\}$. □

4. Unicity Spaces for all μ

In Chapter 4 we considered conditions on the subspace U which guaranteed that U was a unicity space for $C_1(K, \mu)$ for all $\mu \in \mathcal{A}$, i.e., Property A. In this section we tackle this same problem, but in our setting of best one-sided approximation.

Before stating our main result, let us consider some simple subspaces which are necessarily unicity spaces for $C_{1+}(K, \mu)$ for all 'admissible' μ. If $U = \mathrm{span}\{u\}$ is a one-dimensional subspace of $C(K)$, and $u \geq 0$ on all of K, then it easily follows from (5.2) that U is a unicity space for $C_{1+}(K, \mu)$ for every 'admissible' μ. A simple generalization of this example is the following. Let $U \subset C(K)$ and $\dim U = n$. Assume $U = \mathrm{span}\{u_1, \ldots, u_n\}$, where each u_i is non-negative on K, and the u_i have disjoint support, i.e., $\mathrm{supp}\ u_i \cap \mathrm{supp}\ u_j = \emptyset$ for all $i \neq j$. From (5.2) we see that U is a unicity space for $C_{1+}(K, \mu)$ for every 'admissible' μ. In fact, with the above assumptions, the n-dimensional problem has reduced itself to n independent one-dimensional problems (and the unique solution u^* is μ independent). Thus there do exist subspaces U of arbitrary dimension which are unicity spaces for $C_{1+}(K, \mu)$ for all 'admissible' μ. We claim that these subspaces are the only ones with this property. In fact this converse result holds if U is a unicity space for $C_{1+}(K, \mu)$ for all $\mu \in \mathcal{B}$, where $\mathcal{B}$ is a considerably smaller set than the set of all 'admissible' measures.

The following is very much the same as Condition B of Definition 4.5.

Definition 5.1. Let $\mathcal{B}$ be a subset of the set of 'admissible' measures. We shall say that $\mathcal{B}$ satisfies *Condition B'* if

1) $\mathcal{B}$ is a convex cone
2) $f \in C(K)$ and $\int_K f\, d\mu \geq 0$ for all $\mu \in \mathcal{B}$ implies $f \geq 0$ on K.

Analogous to Corollary 4.4 we have:

Lemma 5.16. *Assume $\mathcal{B}$ satisfies Condition B'. Let V be a finite-dimensional subspace of $C(K)$ which does not contain a non-negative non-trivial function. Then there exists a $\mu \in \mathcal{B}$ for which $\int_K v\, d\mu = 0$ for all $v \in V$.*

We can now state the main result of this section.

Theorem 5.17. *Let U be an n-dimensional subspace of $C(K)$, and assume $\mathcal{B}$ satisfies Condition B'. Then U is a unicity space for $C_{1+}(K, \mu)$ for all $\mu \in \mathcal{B}$ if and only if U has a basis of functions $u_1, \ldots, u_n$ satisfying*

$$a)\ u_i \geq 0, \quad i = 1, \ldots, n$$
$$b)\ \mathrm{supp}\ u_i \cap \mathrm{supp}\ u_j = \emptyset, \ \text{all}\ i \neq j.$$

Our proof of this theorem is based on the following proposition.

Proposition 5.18. *Assume that the conditions of Theorem 5.17 hold. If U is a unicity space for $C_{1+}(K,\mu)$ for all $\mu \in \mathcal{B}$, then given any $n-1$ distinct points $\{x_i\}_{i=1}^{n-1}$ in K, there exists a $u \in U$, $u \neq 0$, for which $u \geq 0$ and $u(x_i) = 0$, $i = 1,\ldots,n-1$.*

Proof. Our proof is via induction on the number of points. We prove that given any r distinct points $\{x_i\}_{i=1}^r$, $0 \leq r \leq n-1$, there exists a $u \in U, u \neq 0$, satisfying $u \geq 0$ and $u(x_i) = 0$, $i = 1,\ldots,r$. This statement for $r = 0$ should be taken to mean that U contains a non-negative, non-trivial function. If this is not the case, then from Lemma 5.16 there exists a $\mu \in \mathcal{B}$ for which $\int_K u \, d\mu = 0$ for all $u \in U$. This immediately implies that U is not a unicity space for $C_{1+}(K,\mu)$. Thus U contains a non-negative, non-trivial function.

Assume that our result is valid for $r-1$, where $1 \leq r \leq n-1$. Let $\{x_i\}_{i=1}^r$ be any r distinct points in K, and assume that there does *not* exist a non-negative, non-trivial function in U vanishing at the $\{x_i\}_{i=1}^r$.

From the induction hypothesis there exist non-negative, non-trivial $u_j \in U$ satisfying $u_j(x_i) = 0$, $i,j = 1,\ldots,r$; $i \neq j$. We may therefore assume that $u_j(x_i) = \delta_{ij}$, $i,j = 1,\ldots,r$. Set

$$V = \{v : v \in U, \; v(x_i) = 0, \; i = 1,\ldots,r\}.$$

V is a subspace of U, and since $r \leq n-1$, we have dim $V \geq n-r > 0$. Furthermore $u_j \notin V$ for each $j = 1,\ldots,r$, and the $u_1,\ldots,u_r$ are linearly independent. Thus dim $V = n - r$. By assumption V contains no non-negative, non-trivial function. From Lemma 5.16 there exists a $\mu \in \mathcal{B}$ for which $\int_K v \, d\mu = 0$ for all $v \in V$. Choose $v^* \in V$, $v^* \neq 0$, and set $v_+^*(x) = \max\{v^*(x), 0\}$. Then $v_+^* \in C(K)$, and $v_+^* \neq 0$. We claim that $0, v^* \in P_{\mathcal{U}(v_+^*)}(v_+^*)$ (with the above measure). To prove this assertion we must show that every $u \in \mathcal{U}(v_+^*)$ satisfies $\int_K u \, d\mu \leq 0$.

Let $u \in \mathcal{U}(v_+^*)$. Then $u = \sum_{j=1}^r \alpha_j u_j + v$, where $v \in V$. Now $\alpha_j = u(x_j) \leq v_+^*(x_j) = 0$, $j = 1,\ldots,r$. Furthermore $\int_K u_j \, d\mu > 0$ for each j since u_j is a non-negative, non-trivial function. Thus

$$\int_K u \, d\mu = \int_K \left(\sum_{j=1}^r \alpha_j u_j + v \right) d\mu = \sum_{j=1}^r \alpha_j \int_K u_j \, d\mu \leq 0.$$

This proves that U is not a unicity space for $C_{1+}(K,\mu)$, contradicting our hypothesis. The proposition is proved. □

Proof of Theorem 5.17. ($\Leftarrow$). If (a) and (b) hold, then on the basis of the remarks at the beginning of this section, U is a unicity space for $C_{1+}(K,\mu)$ for every $\mu \in \mathcal{B}$.

($\Rightarrow$). Assume U is a unicity space for $C_{1+}(K,\mu)$ for every $\mu \in \mathcal{B}$. We shall construct $u_1,\ldots,u_n$ satisfying (a) and (b).

Let $x_1, \ldots, x_n$ be any n distinct points in K with the property that if $u \in U$ satisfies $u(x_i) = 0$, $i = 1, \ldots, n$, then $u = 0$. From Proposition 5.18, there exist non-negative, non-trivial $u_1, \ldots, u_n$ in U satisfying $u_j(x_i) = 0$, $i, j = 1, \ldots, n$; $i \neq j$. If $u_j(x_j) = 0$, then $u_j = 0$. Thus $u_j(x_j) > 0$ for every $j = 1, \ldots, n$. For convenience we may therefore assume that $u_j(x_i) = \delta_{ij}$, $i, j = 1, \ldots, n$. The $\{u_j\}_{j=1}^n$ form a basis for U satisfying (a). It remains to prove that (b) holds.

Assume that $i, j \in \{1, \ldots, n\}$, $i \neq j$, satisfy supp $u_i \cap$ supp $u_j \neq \emptyset$. Let $y \in K$ be such that $u_i(y), u_j(y) > 0$. Obviously $y \notin \{x_1, \ldots, x_n\}$. From Proposition 5.18, there exists a non-negative, non-trivial $u^* \in U$ satisfying $u^*(y) = 0$, and $u^*(x_k) = 0$, $k \in \{1, \ldots, n\} \backslash \{i, j\}$. Since $\{u_1, \ldots, u_n\}$ is a basis for U, and $u_k(x_l) = \delta_{kl}$, $k, l = 1, \ldots, n$,

$$u^* = u^*(x_i)u_i + u^*(x_j)u_j.$$

Since u^* is non-negative, $u^*(x_i), u^*(x_j) \geq 0$. However $u^*(y) = 0$ while $u_i(y), u_j(y) > 0$. A contradiction now ensues since u^* was also assumed to be non-trivial.   □

5. Best One-Sided Approximation in $C^1(K)$

On the basis of the negative results of the previous two sections, the reader might suppose that any further investigation is futile. After all, unicity spaces for given $C_{1+}(K, \mu)$ are rare, and unicity spaces for $C_{1+}(K, \mu)$ for all 'admissible' μ are essentially trivial. This is very unfortunate. The unicity space property is a 'good' quality. However this situation is radically altered if we demand that both the approximating subspace U, and the approximated functions f, be continuously differentiable. We shall see that there are many unicity spaces in this setting, and also many 'non-trivial' spaces which are unicity spaces with respect to all 'admissible' measures.

As such, we shall in this section always assume that $U \subset C^1(K)$. Similarly we are only interested in approximating functions in $C^1(K)$. We shall presuppose two additional assumptions, one on K and the other on U. These conditions are assumed to hold throughout this section.

1. U contains a strictly positive function
2. K is a compact convex subset of $\mathbb{R}^d$, with piecewise smooth boundary (and $K = \overline{\text{int } K}$).

Beware! We shall generally not refer to these conditions in the statements of the results of this and the next section, but they are to be assumed. One more word of explanation is in order. We mean by $f \in C^1(K)$ that f has an extension such that it and its first partial derivatives are continuous on some open neighborhood of K.

In the problem of best one-sided L^1-approximation, the zero set of the error function $f - u^*$ is of paramount importance. Let $K = [a, b]$, $(f - u^*)(x) =$

0, and $x \in (a,b)$. If $f, u^* \in C^1(K)$, then since $f - u^* \geq 0$, we must also
have $(f - u^*)'(x) = 0$. This simple observation is the crucial property which
separates one-sided approximation in $C(K)$ from one-sided approximation in
$C^1(K)$. (Look back at the proofs of the central theorems of the previous two
sections to see why they are invalid in this new setting.)

For $f \in C^1[a,b]$, we define $Z_1(f)$ to be the set of zeros of f in $[a,b]$, with
the proviso that, if $x \in (a,b)$ then $f'(x) = 0$. For $f \in C^1(K)$, the definition of
$Z_1(f)$ is somewhat more complicated, but the basic notion is the same. We
let $Z_1(f)$ denote the set of zeros of f on K for which the following hold:

1) If $x \in \text{int } K$, then all first partial derivatives of f at x vanish (i.e.,
   the gradient of f at x is zero).
2) If $x \in \partial K$, then all directional derivatives to f at x vanish for all
   directions tangent to K at x.

Note that if $f \in C^1(K)$ and $f \geq 0$, then $Z_1(f) = Z(f)$. Furthermore, if
$f, g \in C^1(K)$ and $f \geq |g|$, then $Z_1(f) \subseteq Z_1(g)$.

For a finite-dimensional subspace $U \subset C^1(K)$ we shall say that U is
a *unicity space* for $C^1_{1+}(K, \mu)$ (terrible notation) if to each $f \in C^1(K)$ there
exists a unique best one-sided $L^1(K, \mu)$ approximant to f from U, i.e., $P_{\mathcal{U}(f)}(f)$
is a singleton. (From our assumption on U, $\mathcal{U}(f) \neq \emptyset$ for all f.) We also, as
previously, shall always restrict ourselves to 'admissible' measures μ.

Our first result totally characterizes unicity spaces for $C^1_{1+}(K, \mu)$.

Theorem 5.19. *Let U be a finite-dimensional subspace of $C^1(K)$. Then U is
a unicity space for $C^1_{1+}(K, \mu)$ if and only if to each $u \in U$, $u \neq 0$, there exists
a $v \in U$ satisfying*

$$a)\, v \leq 0 \text{ on } Z_1(u)$$

$$b) \int_K v \, d\mu > 0.$$

This result is similar to Corollary 5.12. There is, however, a fundamental
difference.

Proof. ($\Leftarrow$). We assume that (a) and (b) hold. Since U contains a strictly
positive function, we may further assume, by perturbation, that to each $u \in
U$, $u \neq 0$, there exists a $v \in U$ satisfying

$$a')\, v < 0 \text{ on } Z_1(u)$$

$$b') \int_K v \, d\mu > 0.$$

Now, assume that U is not a unicity space for $C^1_{1+}(K, \mu)$. Let $f \in C^1(K)$
be such that $u^1, u^2 \in P_{\mathcal{U}(f)}(f)$, $u^1 \neq u^2$. Set $u^* = (u^1 - u^2)/2$, and $g =
f - (u^1 + u^2)/2$. Then $g \geq |u^*|$, and $0, \pm u^* \in P_{\mathcal{U}(g)}(g)$. Since $g \in C^1(K)$, and
$g \geq |u^*|$, we have $Z(g) = Z_1(g) \subseteq Z_1(u^*)$. By assumption, there exists a $v \in U$
satisfying $v < 0$ on $Z_1(u^*)$, and $\int_K v \, d\mu > 0$. Thus $v < 0$ on $Z_1(g) = Z(g)$. It

now follows that for $\varepsilon > 0$, ε sufficiently small, $\varepsilon v \in \mathcal{U}(g)$. But $\int_K \varepsilon v \, d\mu > 0$ contradicting the fact that $0 \in P_{\mathcal{U}(g)}(g)$. Therefore U is a unicity space for $C^1_{1+}(K, \mu)$.

($\Rightarrow$). To prove the converse we assume that (a) and (b) do not hold. There therefore exists a $u^* \in U$, $u^* \neq 0$, with the property that if $v \in U$ and $v \leq 0$ on $Z_1(u^*)$, then $\int_K v \, d\mu \leq 0$. This implies, by Lemma 5.3, the existence of points $\{x_i\}_{i=1}^k$ in $Z_1(u^*)$, $1 \leq k \leq n = \dim U$, and positive numbers $\{\lambda_i\}_{i=1}^k$, for which

$$\int_K u \, d\mu = \sum_{i=1}^k \lambda_i u(x_i)$$

for all $u \in U$.

Suppose there exists an $f \in C^1(K)$ satisfying $f(x_i) = 0$, $i = 1, \dots, k$, and $f \geq |u^*|$. Then it easily follows from the above that $\pm u^* \in P_{\mathcal{U}(f)}(f)$, and therefore U is not a unicity space for $C^1_{1+}(K, \mu)$. It remains to prove the existence of such an f. This is the crux of the proof. Note that if $x_i \in Z(u^*) \backslash Z_1(u^*)$, then we would have no chance of constructing such an f. If u^* has at most a finite number of zeros, then this is easily done. (Let $f = |u^*|$ in a neighborhood of each x_i, and join these locally defined f in a continuously differentiable manner, maintaining the desired inequality.) However we can make no such assumption. We prove the existence of the desired f in a separate lemma. We shall further prove it under the hypothesis that $K \subset \mathbb{R}^2$. The basic idea is contained in this setting. The reader is urged to fill in the details in the general case.

Lemma 5.20. *Given* $\{(x_i, y_i)\}_{i=1}^k$ *in* $Z_1(u^*)$, $x_i, y_i \in \mathbb{R}$, $i = 1, \dots, k$, *there exists an* $f \in C^1(K)$ *satisfying*

$$1) \ f(x_i, y_i) = 0, \quad i = 1, \dots, k$$
$$2) \ f \geq |u^*|.$$

Proof. Recall that K is compact, convex, and has piecewise smooth boundary. If $(x, y) \in Z_1(u^*)$, then $u^*(x, y) = 0$, and one of the following hold.

  i) If $(x, y) \in \text{int } K$, then $u_x^*(x, y) = u_y^*(x, y) = 0$.
 ii) If $(x, y) \in \partial K$, and a tangent to K exists at (x, y), then the derivative of u^* in the tangent direction is zero.
iii) If $(x, y) \in \partial K$, and a tangent to K does not exist at (x, y), then there is no additional assumption.

Let (x_i, y_i), $i = 1, \dots, k$, be as in the statement of the lemma. To construct the desired f, it suffices to construct, in a neighborhood of each (x_i, y_i), a continuously differentiable g_i which vanishes at (x_i, y_i) and satisfies $g_i \geq |u^*|$ in this neighborhood. We then set

$$h_i(x, y) = g_i(x, y) + M[(x - x_i)^2 + (y - y_i)^2]$$

for some $M > 0$. h_i vanishes at (x_i, y_i), satisfies $h_i \geq |u^*|$ in the above neighborhood (with equality only at (x_i, y_i)) and is strictly positive on the boundary of the neighborhood. Since there are only a finite number of such h_i we can find an $f \in C^1(K)$ which agrees with h_i on each appropriate neighborhood, and which satisfies $f \geq |u^*|$ on all of K.

It therefore remains to construct the g_i. This we now do.

i) Assume $(x_i, y_i) \in \text{int } K$. Without loss of generality, let $(x_i, y_i) = (0,0)$. Then $u^*(0,0) = u_x^*(0,0) = u_y^*(0,0) = 0$. Choose $R > 0$ such that the circle of radius R with center zero is in K and contains no other (x_j, y_j). For $0 \leq r \leq R$, set

$$h(r) = \max_{0 \leq \theta \leq 2\pi} \left| \frac{\partial}{\partial r} u^*(r\cos\theta, r\sin\theta) \right|.$$

Since $u^* \in C^1(K)$, and $(0,0) \in Z_1(u^*)$, it follows that $h \in C[0,R]$, and $h(0) = 0$, For $x = r\cos\theta$, $y = r\sin\theta$, $0 \leq r \leq R$, set

$$g_i(x,y) = g_i(r\cos\theta, r\sin\theta) = \int_0^r h(s)\, ds.$$

Then $g_i(0,0) = 0$, $g_i \geq 0$, and g_i is continuously differentiable on its domain of definition. Furthermore,

$$g_i(x,y) = g_i(r\cos\theta, r\sin\theta) = \int_0^r h(s)\, ds \geq \int_0^r \left| \frac{\partial}{\partial s} u^*(s\cos\theta, s\sin\theta) \right| ds$$

$$\geq \left| \int_0^r \frac{\partial}{\partial s} u^*(s\cos\theta, s\sin\theta) ds \right| = |u^*(r\cos\theta, r\sin\theta)| = |u^*(x,y)|.$$

ii) $(x_i, y_i) \in \partial K$ and a tangent to K exists at (x_i, y_i). Without loss of generality, we assume that $(x_i, y_i) = (0,0)$, and $K \subseteq \{(x,y) : x \geq 0\}$. Then the tangent to K at $(0,0)$ is along the y-axis. Therefore $u^*(0,0) = u_y^*(0,0) = 0$. We apply Whitney's Extension Theorem. This theorem says that we may assume that $u^* \in C^1(\mathbb{R}^2)$, i.e., u^* has an extension to $\mathbb{R}^2$, and remains continuously differentiable on all of $\mathbb{R}^2$. Now let

$$\tilde{g}_i(y) = \int_0^y |u_s^*(0,s)|\, ds$$

for all $|y| \leq R$ (some $R > 0$). Then $\tilde{g}_i \in C^1[-R, R]$, $\tilde{g}_i(0) = 0$, and $\tilde{g}_i(y) \geq |u^*(0,y)|$ for all $|y| \leq R$. Let

$$g_i(x,y) = \tilde{g}_i(y) + Mx,$$

where $M \geq \|u_x^*\|_\infty$. For $x > 0$ and any fixed y, we have from Lagrange's Theorem

$$u^*(x,y) = u^*(0,y) + xu_x^*(\tilde{x}, y),$$

where $0 < \tilde{x} < x$. Thus, for $|y| \leq R$,

$$|u^*(x,y)| \leq |u^*(0,y)| + x|u_x^*(\tilde{x}, y)| \leq \tilde{g}_i(y) + Mx = g_i(x,y).$$

iii) $(x_i, y_i) \in \partial K$, and no tangent to K at (x_i, y_i) exists. Since ∂K is piecewise smooth, one-sided tangents to K exist at every $(x, y) \in \partial K$. Since K is convex, we see that the point (x_i, y_i) is a 'corner' of K with angle $< \pi$. We may therefore assume that $(x_i, y_i) = (0, 0)$, and K is a subset of a cone in $\mathbb{R}^2$ which lies in the union of the first and fourth quadrant and touches the y-axis only at $(0, 0)$. Set $g_i(x, y) = Mx$. It is not difficult to show that there exists an $M > 0$, M sufficiently large, such that $g_i \geq |u^*|$ in some neighborhood of $(0, 0)$ in K.

This proves the lemma, and also Theorem 5.19. □

From Theorem 5.19 and its proof, we also obtain the following characterization of unicity spaces.

Corollary 5.21. *An n-dimensional subspace U of $C^1(K)$ is a unicity space for $C^1_{1+}(K, \mu)$ if and only if there does not exist a $u^* \in U$, $u^* \neq 0$, points $\{x_i\}_{i=1}^k$ in $Z_1(u^*)$, $1 \leq k \leq n$, and positive values $\{\lambda_i\}_{i=1}^k$ satisfying*

$$\int_K u \, d\mu = \sum_{i=1}^k \lambda_i u(x_i)$$

for all $u \in U$.

Are there 'nice' subspaces U which are unicity spaces for $C^1_{1+}(K, \mu)$? Yes, in fact there are many. Let us consider one simple example.

Let $U = \text{span}\{1, x\}$ and $K = [a, b]$. (Recall that U is not a unicity space for $C_{1+}(K, \mu)$ for any 'admissible' μ.) For any $u \in U$, $u \neq 0$, it is easily seen that $Z_1(u)$ is either empty, the set $\{a\}$, or the set $\{b\}$. In all three cases, we can easily construct a $v \in U$ which vanishes on $Z_1(u)$ and such that $v > 0$ on (a, b). Thus $\int_a^b v \, d\mu > 0$ for every 'admissible' μ. This proves that U is a unicity space for $C^1_{1+}(K, \mu)$ for every 'admissible' μ.

We shall now characterize those U which are unicity spaces for $C^1_{1+}(K, \mu)$ for all 'admissible' μ. To this end, we define Property B.

Definition 5.2. *A finite-dimensional subspace U of $C^1(K)$ is said to satisfy Property B if to each $u \in U$, $u \neq 0$, there exists a $v \in U$, $v \neq 0$, for which*

$$a) \, Z_1(u) \subseteq Z_1(v)$$
$$b) \, v \geq 0.$$

That is, there exists a non-negative, non-trivial function in U vanishing at the points $Z_1(u)$.

If U satisfies Property B, then it follows from Theorem 5.19 that U is a unicity space for $C^1_{1+}(K, \mu)$ for all 'admissible' measures μ. We shall prove that the converse also holds. For the converse result we do not need *all* 'admissible' μ.

Let $\mathcal{D}$ denote the set of measures of the form $w(x)\,dx$, where w runs over the class of all strictly positive continuous functions on K. (The set $\mathcal{D}$ satisfies Condition B', see Definition 5.1.)

Theorem 5.22. *If a finite-dimensional subspace U of $C^1(K)$ is a unicity space for $C^1_{1+}(K,\mu)$ for all $\mu \in \mathcal{D}$, then U satisfies Property B.*

Proof. Assume that U does not satisfy Property B. There then exists a $u^* \in U$, $u^* \neq 0$, with the property that if $u \in U$ satisfies $Z_1(u^*) \subseteq Z_1(u)$ and $u \geq 0$ on K, then $u = 0$. Set

$$P = \{u : u \in U, \ u \leq 0 \text{ on } Z_1(u^*)\}.$$

P is rather large since U contains a strictly negative function. Furthermore, by hypothesis, P contains no non-negative, non-trivial function. Let

$$\widetilde{P} = \{u(x)\,dx : \ u \in P\},$$

and consider $\widetilde{P}$ as a subset of $M(K)$, the set of Borel measures of bounded total variation on K, endowed with the weak*-topology induced by $C(K)$. From the definition of P, we see that $\widetilde{P}$ is a closed convex cone in some finite-dimensional subspace of $M(K)$. Set

$$Q = \left\{\nu : \ \nu \in M(K), \ \nu \geq 0, \ \int_K d\nu = 1\right\}.$$

Thus Q is simply the set of probability densities. Q is convex and compact in this same topology. By assumption, $Q \cap \widetilde{P} = \emptyset$. There therefore exists a continuous linear functional w on $M(K)$ such that $w(u) \leq 0$ for all $u \in \widetilde{P}$, and $w(\nu) > 0$ for all $\nu \in Q$. Every continuous linear functional in this space may be represented as integration over K against a continuous function, which we again denote by w. Thus $w \in C(K)$,

$$\int_K u(x)w(x)dx \leq 0$$

for all $u \in P$, and

$$\int_K w(x)d\nu > 0$$

for all $\nu \in Q$. Since the point functionals are in Q, it follows that $w > 0$ on K, i.e., $w(x)dx \in \mathcal{D}$,

We have therefore constructed a $d\tilde{\mu} = w(x)dx \in \mathcal{D}$ for which

$$\int_K u\,d\tilde{\mu} \leq 0$$

for all $u \in P$. From Theorem 5.19, U is *not* a unicity space for $C^1_{1+}(K,\tilde{\mu})$. This proves the theorem.   □

6. Property B

In this section we present several examples of spaces satisfying Property B, and some examples of spaces which do not satisfy Property B. Scattered throughout are various general results. We start with a simple example.

Example 5.1. $U = \text{span}\{1, x^3\}$, $K = [-1, 1]$.

U does not satisfy Property B. This is a consequence of the fact that $Z_1(x^3) = \{0\}$, and there exists no non-negative non-trivial function in U which vanishes at 0.

Note that the above U is a T-system on $[-1, 1]$ since no non-trivial $u \in U$ has more than one zero. T-systems do not necessarily satisfy Property B. None the less a large subset of T-systems do satisfy Property B. To delineate this subset, we introduce the following notation.

Let A be a subset of $[a, b]$. A is said to have *index* $I(A)$, where $I(A)$ counts the number of points in A under the convention that points in $A \cap (a, b)$ are counted twice, and points in $A \cap \{a, b\}$ once.

As an immediate consequence of Theorem 4 of Appendix A, we have this next result.

Proposition 5.23. *If U is an n-dimensional T-system in $C^1[a, b]$, and for every $u \in U$, $u \neq 0$, we have $I(Z_1(u)) \leq n - 1$, then U satisfies Property B.*

There is a large class of T-systems which automatically satisfy the condition $I(Z_1(u)) \leq n - 1$ for all $u \in U$, $u \neq 0$. These are the ET_2-systems.

Definition 5.3. Let $U \subset C^1[a, b]$, $\dim U = n$. We say that U is an ET_2-system (extended Chebyshev of order two) if $|Z^*(u)| \leq n-1$ for all $u \in U$, $u \neq 0$, where $|Z^*(u)|$ is the count for the number of zeros of u in $[a, b]$, with a zero x counted twice if $u(x) = u'(x) = 0$.

In other words, we count the multiplicities of a zero, but up to multiplicity at most two. Since $I(Z_1(u)) \leq |Z^*(u)|$, we easily obtain:

Proposition 5.24. *If U is an ET_2-system on $[a, b]$, then U satisfies Property B.*

In fact we can somewhat weaken the condition defining ET_2-systems in that, even if u has a double zero at an endpoint, we need count it only once.

We know that U is a T-system of dimension n on $[a, b]$ if and only if for any n distinct points $\{x_i\}_{i=1}^n$ in $[a, b]$ and n given data $\{e_i\}_{i=1}^n$ there exists a unique $u \in U$ satisfying $u(x_i) = e_i$, $i = 1, \ldots, n$ (see Appendix A). There is a similar result for ET_2-systems. Let $\{x_i\}_{i=1}^k$ be any k distinct points in $[a, b]$. Let $w(x_i)$ be either 1 or 2. For each x_i, we define data e_i if $w(x_i) = 1$, and data e_i, f_i if $w(x_i) = 2$. Then U is an ET_2-system of dimension n on $[a, b]$ if and only if given any k distinct points $\{x_i\}_{i=1}^k$ with $\sum_{i=1}^k w(x_i) = n$, and n data as above, there exists a unique $u \in U$ satisfying $u(x_i) = e_i$, $i = 1, \ldots, k$, and $u'(x_i) = f_i$ for those i with $w(x_i) = 2$. While ET_2-systems satisfy Property B, nothing even remotely similar to the ET_2 property is necessary in order that the subspace U satisfy Property B.

Example 5.2. $U = \text{span}\{1, u\}$, $K = [a, b]$.

Let $M = \{x : u(x) = \max u\}$, $m = \{x : u(x) = \min u\}$, and $A = \{x : u'(x) = 0\} \cup \{a, b\}$. It is readily verified that U satisfies Property B on $[a, b]$ if and only if $A = M \cup m$. Note that u may well equioscillate many times so that U is not necessarily a T-system, or even close to being one.

Example 5.2 is a special case of a rather general result.

Proposition 5.25. *Let V be an $(n-1)$-dimensional ET_2-system on $[a, b]$ $(n \geq 2)$. Let $u^* \in C^1[a, b]$, and $U = \text{span}\{V, u^*\}$. Then U satisfies Property B if and only if there exists no $u \in U$ with a sign change and satisfying $I(Z_1(u)) \geq n - 1$.*

Proof. ($\Rightarrow$). Assume U satisfies Property B and there exists a $u \in U$ which has a sign change and such that $I(Z_1(u)) \geq n-1$. As a consequence of Property B, there exists a $w \in U$, $w \neq 0$, satisfying $w \geq 0$ and $Z_1(u) \subseteq Z_1(w)$. Since V is an ET_2-system of dimension $n-1$, it follows that $u, w \notin V$. Thus $u = cu^* + v_1$, and $w = du^* + v_2$, where $c, d \neq 0$, and $v_1, v_2 \in V$. Now $du - cw = dv_1 - cv_2 \in V$, and $du - cw \neq 0$ since u and w are linearly independent (u changes sign and w does not). Therefore $I(Z_1(du - cw)) \leq n-2$. However $Z_1(u) \subseteq Z_1(w)$, whence $Z_1(u) \subseteq Z_1(du - cw)$. Thus $I(Z_1(du - cw)) \geq n - 1$. This is a contradiction. ($\Leftarrow$). Assume no $u \in U$ has a sign change and satisfies $I(Z_1(u)) \geq n - 1$. To prove that U satisfies Property B, it suffices to check Property B only for those $u \in U$ which have sign changes on $[a, b]$. Assume $u \in U$ has a sign change on $[a, b]$. By hypothesis $I(Z_1(u)) \leq n - 2$. V is an ET_2-system of dimension $n - 1$ on $[a, b]$. From the previous remarks concerning ET_2-systems (see also Theorem 4 of Appendix A) there exists a $v \in V$, $v \neq 0$, such that $Z_1(u) \subseteq Z_1(v)$ and $v \geq 0$. Thus U satisfies Property B.   □

What possible u^* can be adjoined to the ET_2-system V so that the resulting subspace U satisfies Property B? This seems to be a difficult question to answer in general. However if n is odd, a partial result is contained in our next proposition.

Proposition 5.26. *Let V be an $(n-1)$-dimensional ET_2-system on $[a, b]$. Let $u^* \in C^1[a, b]$ and assume that $U = \text{span}\{V, u^*\}$ satisfies Property B. If n is odd $(n \geq 3)$, then U is a WT-system.*

An equivalent statement is that u^* lies in the convexity cone of V (see Definition 2 of Appendix B). Note that this condition is necessary, but not sufficient.

Proof. Assume that U is not a WT-system. There then exists a $u \in U$ with at least n sign changes on $[a, b]$. Since V is an ET_2-system of dimension $n-1$, $u \notin V$. Let $a < t_1 < \ldots < t_n < b$ satisfy

$$u(t_i) = 0, \qquad i = 1, \ldots, n,$$

and

$$u'(t_{2i}) \leq 0, \qquad i = 1, \ldots, (n-1)/2.$$

(Recall that n is odd so that $(n-1)/2$ is an integer.) Such points necessarily exist. (We may have to take $-u$ rather than u.) Since V is an ET_2-system, there exists a unique $v \in V$ satisfying $v(t_{2i}) = u(t_{2i})$ and $v'(t_{2i}) = u'(t_{2i})$ for $i = 1, \ldots, (n-1)/2$. Let $v_{kl} \in V$, $k = 1, \ldots, (n-1)/2$; $l = 0, 1$ where

$$v_{kl}^{(j)}(t_{2i}) = \begin{cases} 1, & k = i, l = j \\ 0, & \text{otherwise.} \end{cases}$$

The $\{v_{kl}\}$ are linearly independent, and constitute a basis for V. Since $u(t_{2i}) = 0$,

$$v = \sum_{i=1}^{(n-1)/2} [u(t_{2i})v_{i0} + u'(t_{2i})v_{i1}]$$

$$= \sum_{i=1}^{(n-1)/2} u'(t_{2i})v_{i1}.$$

Since u has a sign change and U satisfies Property B, it follows from Proposition 5.25 that $I(Z_1(u)) \leq n - 2$. Thus $u'(t_{2k}) < 0$ for at least one index $k \in \{1, \ldots, (n-1)/2\}$.

V is an ET_2-system of dimension $n - 1$ on $[a, b]$, and by construction $|Z^*(v_{i1})| \geq n - 2$ for each i. Thus $|Z^*(v_{i1})| = n - 2$ for each i, implying that v_{i1} has no zeros in $[a, b]$ other than those specified above. Each v_{i1} must therefore change sign at t_{2i} and nowhere else. Since $v_{i1}'(t_{2i}) = 1$, we see that $v_{i1}(x) < 0$ for $x < t_2$, while $v_{i1}(x) > 0$ for $x > t_{n-1}$. Since $u'(t_{2i}) \leq 0$, $i = 1, \ldots, (n-1)/2$, and $u'(t_{2k}) < 0$ for some k, it follows that $v(x) > 0$ for $x < t_2$, while $v(x) < 0$ for $x > t_{n-1}$. Therefore $(u - v)(t_1) = -v(t_1) < 0$, and $(u - v)(t_n) = -v(t_n) > 0$, i.e., $u - v$ has a sign change. By construction $I(Z_1(u-v)) \geq n-1$. From Proposition 5.25, we see that we have contradicted the fact that U satisfies Property B. □

In preparation for the next two examples, we need the following generalization of Theorem 4 of Appendix A.

Proposition 5.27. *Let $U \subset C^1[a, b]$ be a WT-system of dimension n. Let A be a subset of $[a, b]$ with $I(A) \leq n - 1$. There then exists a $u \in U$, $u \neq 0$, such that $u \geq 0$ and $A \subseteq Z_1(u)$.*

Proof. From Theorem 4 of Appendix A, we know that the above result is valid if U is a T-system on $[a, b]$. Let $u_1, \ldots, u_n$ be any basis for U. Smooth

$u_1, \ldots, u_n$ (see Proposition 6 of Appendix A). Thus, for each $\varepsilon > 0$, $U_\varepsilon = \operatorname{span}\{u_1^\varepsilon, \ldots, u_n^\varepsilon\}$ is a T-system on $[a, b]$, and $\lim_{\varepsilon \to 0^+} u_i^\varepsilon = u_i$, $i = 1, \ldots, n$, uniformly on $[a, b]$.

Let $u^\varepsilon = \sum_{i=1}^n a_i^\varepsilon u_i^\varepsilon$ satisfy $u^\varepsilon \geq 0$, $A \subseteq Z_1(u^\varepsilon)$, and $\sum_{i=1}^n (a_i^\varepsilon)^2 = 1$. Let $\varepsilon \to 0^+$. On a subsequence $a_i^\varepsilon \to b_i$, $i = 1, \ldots, n$. Set $u = \sum_{i=1}^n b_i u_i$. Then $u \geq 0$, $A \subseteq Z_1(u)$, and $\sum_{i=1}^n b_i^2 = 1$, i.e., $u \neq 0$.   □

Example 5.3. *Pieceing together ET_2-systems.*

Let $a = e_0 < e_1 < \cdots < e_k < e_{k+1} = b$. On each interval $I_i = [e_{i-1}, e_i]$, $i = 1, \ldots, k+1$, let U_i be an ET_2-system of dimension n_i, $n_i \geq 2$, with the additional properties that $U_i' = \{u' : u \in U_i\}$ is a T-system on I_i, and the constant function is in U_i. Let $U \subset C^1[a, b]$ be the subspace defined by $U|_{I_i} = U_i$, $i = 1, \ldots, k+1$. We claim that U satisfies Property B.

This example is similar to Example 4.8 of Chapter 4. There are, however, a number of important differences. One of these is the demand that $U \subset C^1[a, b]$, i.e., at each e_i, $i = 1, \ldots, k$, we demand both continuity of the function and its derivative.

To prove our claim, set

$$V_{i,j} = \{u : u \in U, \ u = 0 \text{ on } [a, e_i) \cup (e_j, b]\}$$

for each $0 \leq i < j \leq k + 1$. From the method of proof of Example 4.8, it is readily verified that

  i) $\dim V_{0,k+1} (= \dim U) = n_1 + \cdots n_{k+1} - 2k$,
  ii) for $1 \leq j \leq k$, $\dim V_{0,j} = n_1 + \cdots + n_j - 2j$,
  iii) for $1 \leq i \leq k$, $\dim V_{i,k+1} = n_{i+1} + \cdots + n_{k+1} - 2(k + 1 - i)$,
  iv) for $1 \leq i < j \leq k$, $\dim V_{i,j} = n_{i+1} + \cdots + n_j - 2(j - i + 1)$.

If any of these values is negative, then we understand it to mean that $\dim V_{i,j} = 0$.

For each $0 \leq i < j \leq k + 1$ we also set

$$V_{i,j}' = \{u' : u \in U, \ u' = 0 \text{ on } [a, e_i) \cup (e_j, b]\}.$$

Directly from Example 4.8, and since $\dim U_i' = \dim U_i - 1$ for each i, we obtain:

  i) $\dim V_{0,k+1}' = n_1 + \cdots + n_{k+1} - 2k - 1$,
  ii) for $1 \leq j \leq k$, $\dim V_{0,j}' = n_1 + \cdots + n_j - 2j$,
  iii) for $1 \leq i \leq k$, $\dim V_{i,k+1}' = n_{i+1} + \cdots + n_{k+1} - 2(k + 1 - i)$,
  iv) for $1 \leq i < j \leq k$, $\dim V_{i,j}' = n_{i+1} + \cdots + n_j - 2(j - i + 1) + 1$.

Note that the relationship between $\dim V_{i,j}$ and $\dim V_{i,j}'$ is not a simple one. For example, $\dim V_{0,k+1}' = \dim V_{0,k+1} - 1$, while $\dim V_{i,j}' = \dim V_{i,j} + 1$ if $1 \leq i < j \leq k$. (In this latter case there exist $u \notin V_{i,j}$ for which $u' \in V_{i,j}'$.)

Let $u \in U$, $u \neq 0$, and assume that $[e_r, e_s]$, $r < s$, is a zero interval of u. By construction, there exists a $w \in U$ such that $w = u$ on $[e_s, b]$, while

$w = 0$ on $[a, e_s]$. The analogous result holds in the other direction. In proving Property B, it suffices to consider only those $u \in V_{i,j}$, some $0 \le i < j \le k+1$, which have no zero intervals in $[e_i, e_j]$. For these u we must prove the existence of a $v \in V_{i,j}$, $v \ne 0$, $v \ge 0$, satisfying $Z_1(u) \subseteq Z_1(v)$.

For $u \in V_{i,j}$, we let $Z_1(u; i, j)$ denote the restriction of $Z_1(u)$ to (e_i, e_j) for $1 \le i < j \le k$; to $[a, e_j)$ for $i = 0 < j \le k$; to $(e_i, b]$ for $1 \le i < j = k+1$; and to $[a, b]$ if $i = 0$, $j = k+1$.

Claim 1. *Let $u \in V_{i,j}$ and assume that u has no zero interval in $[e_i, e_j]$. Then $I(Z_1(u; i, j)) \le \dim V_{i,j} - 1$.*

Proof. There are really four cases to consider depending on whether $i = 0$ and/or $j = k+1$. We shall only consider two and leave the other two to the reader. The analysis is very much the same.

Assume $i = 0$, $j = k+1$, and let $I(Z_1(u)) = m$. Then u' has at least $m - 1$ distinct zeros in (a, b). Assume u' has no zero interval in $[a, b]$. Then $u' \in V'_{0,k+1}$, and from Corollary 9 of Appendix A, $m - 1 \le \dim V'_{0,k+1} - 1 = \dim V_{0,k+1} - 2$, i.e., $m \le \dim V_{0,k+1} - 1$. If u' has zero intervals, then the inequality is even sharper (recall that by assumption u has no zero intervals). There is simply a bit more bookkeeping involved.

Assume $i = 0$, $1 \le j \le k$, and let $I(Z_1(u; 0, j)) = m$. Since $u(e_j) = 0$, u' has at least m distinct zeros in (a, e_j), and $u' \in V'_{0,j}$. Assume u' has no zero interval in $[a, e_j]$. Then from Corollary 9 of Appendix A, $m \le \dim V'_{0,j} - 1 = \dim V_{0,j} - 1$. Again if u' has zero intervals in $[a, e_j]$, there is a bit more work, but the counting works out (believe me).

The cases $1 \le i \le k$, $j = k+1$, and $1 \le i < j \le k$ are dealt with in a similar manner. □

Claim 2. *$V_{i,j}$ is a WT-system on $[e_i, e_j]$.*

Proof. Assume $V_{i,j}$ is not a WT-system on $[e_i, e_j]$. There then exists a $v \in V_{i,j}$ with at least $\dim V_{i,j}$ sign changes thereon.

Again there are four cases to consider, and we shall only discuss two. Assume $i = 0$ and $j = k+1$. Since v has at least $\dim V_{0,k+1}$ sign changes on $[a, b]$, v' has at least $\dim V_{0,k+1} - 1 = \dim V'_{0,k+1}$ sign changes on $[a, b]$. Since $V'_{0,k+1}$ is a WT-system (see Example 4.8), and $v' \in V'_{0,k+1}$, this is impossible. Thus $V_{0,k+1}$ is a WT-system.

Assume $i = 0$, $1 \le j \le k$. Let $v \in V_{0,j}$ have at least $\dim V_{0,j}$ sign changes on $[a, e_j]$. Since $v(e_j) = 0$, it follows that $v' \in V'_{0,j}$ has at least $\dim V_{0,j} = \dim V'_{0,j}$ sign changes on $[a, e_j]$. This is impossible since $V'_{0,j}$ is a WT-system. The remaining two cases are similarly dealt with. □

Claims 1 and 2, together with Proposition 5.27, immediately imply that U satisfies Property B.

Example 5.4. *Polynomial splines with simple fixed knots.*

We are given $m \geq 3$, an interval $[a, b]$ and knots $\xi_0 = a < \xi_1 < \cdots < \xi_r < b = \xi_{r+1}$. As in Example 4.9 of Chapter 4, we let $\mathcal{S}_{m-1,r}$ denote the space of splines of degree $m - 1$ with the fixed simple knots $\{\xi_i\}_{i=1}^r$. Note that $\mathcal{S}_{m-1,r} \subset C^1[a, b]$ since $m \geq 3$. (If $m = 3$, $\mathcal{S}_{m-1,r} = \mathcal{S}_{2,r}$ satisfies the conditions of the previous Example 5.3.) We shall prove that $\mathcal{S}_{m-1,r}$ satisfies Property B.

For $0 \leq i < j \leq r + 1$, set

$$V_{i,j} = \{s : s \in \mathcal{S}_{m-1,r}, \ s = 0 \text{ on } [a, \xi_i) \cup (\xi_j, b]\}.$$

From Example 4.9, each $V_{i,j}$ is a WT-system, and the following hold.

  i) For $i = 0, j = r + 1$, $V_{0,r+1} = \text{span}\{B_1, \ldots, B_{r+m}\}$,
  ii) for $i = 0, 1 \leq j \leq r$, $V_{0,j} = \text{span}\{B_1, \ldots, B_j\}$,
  iii) for $1 \leq i \leq r, j = r + 1$, $V_{i,r+1} = \text{span}\{B_{i+m}, \ldots, B_{r+m}\}$,
  iv) for $1 \leq i < j \leq r, \ j - i \leq m - 1$, $V_{i,j} = 0$,
  v) for $1 \leq i < j \leq r, \ j - i \geq m$, $V_{i,j} = \text{span}\{B_{i+m}, \ldots, B_j\}$.

On the basis of Proposition 5.27, it suffices to prove that if $s \in V_{i,j}$ has no zero interval on $[\xi_i, \xi_j]$, then $I(Z_1(s; i, j)) \leq \dim V_{i,j} - 1$, (where $Z_1(s; i, j)$ is defined as in the previous example). The proof of this inequality is totally analogous to the proof of Claim 1 of Example 5.3. Thus $\mathcal{S}_{m-1,r}$ satisfies Property B.

We now turn to the multi-dimensional setting.

Example 5.5. $U = \text{span}\{1, x, y\}$, $K = \{(x, y) : x^2 + y^2 \leq 1\}$.

It is readily verified that if $u \in U$, $u \neq 0$, and $(x_0, y_0) \in Z_1(u)$, then $(x_0, y_0) \in \partial K$, i.e., $x_0^2 + y_0^2 = 1$. Furthermore, since ∂K is smooth, the tangent line to K at (x_0, y_0) is necessarily the zero set of u. Thus u is of one sign on K, and therefore Property B holds on K.

Example 5.6. $U = \text{span}\{1, x, y\}$, $K = \{(x, y) : 0 \leq x, y \leq 1\}$.

The function $u(x, y) = x - y$ satisfies $Z_1(u) = \{(0, 0), (1, 1)\}$. If $v \in U$, and $v(0, 0) = v(1, 1) = 0$, then $v = \alpha u$ for some $\alpha \in \mathbb{R}$. Thus there does not exist a non-negative, non-trivial function v in U satisfying $Z_1(u) \subseteq Z_1(v)$. U does not satisfy Property B on K.

Example 5.7. $U = \text{span}\{1, x, y, x^2, y^2\}$, $K = \{(x, y) : x^2 + y^2 \leq 1\}$.

We shall prove that U satisfies Property B on K. Let $u \in U$, $u \neq 0$. If $Z_1(u)$ contains no interior points of K, then set

$$v(x, y) = 1 - (x^2 + y^2).$$

Since $Z_1(v) = \partial K$, and $v \geq 0$ on K, Property B holds for all such u.

We therefore assume that $Z_1(u)$ contains an interior point of K. Let $(x_0, y_0) \in Z_1(u) \cap \text{int } K$. Since $u(x_0, y_0) = u_x(x_0, y_0) = u_y(x_0, y_0) = 0$, it is readily verified that

$$u(x, y) = a(x - x_0)^2 + b(y - y_0)^2$$

for some constants a and b. If $ab \geq 0$, set $v = u$ or $v = -u$, as appropriate. Assume $ab < 0$. We claim that $Z_1(u)$ contains no other points of K, and we can then choose, for example, $v(x, y) = (x - x_0)^2 + (y - y_0)^2$. It remains to prove that if $ab < 0$, then $Z_1(u) = \{(x_0, y_0)\}$. Assume $(x_1, y_1) \in Z_1(u) \cap \text{int } K$. Then

$$u(x, y) = a'(x - x_1)^2 + b'(y - y_1)^2.$$

But, as may be easily checked, this is impossible for $(x_1, y_1) \neq (x_0, y_0)$, unless $ab = 0$ and $(x_1 - x_0)(y_1 - y_0) = 0$. Assume $(x_1, y_1) \in Z_1(u) \cap \partial K$. We then have $u(x_1, y_1) = 0$, and $y_1 u_x(x_1, y_1) - x_1 u_y(x_1, y_1) = 0$. Since $ab < 0$, it follows from the first equation that $x_1 \neq x_0$ and $y_1 \neq y_0$. These two equations, together with the equality $x_1^2 + y_1^2 = 1$, imply that

$$1 - x_1 x_0 - y_1 y_0 = 0.$$

Since $x_0^2 + y_0^2 < 1 = x_1^2 + y_1^2$, we have

$$x_1 x_0 + y_1 y_0 \leq (x_1^2 + y_1^2)^{1/2}(x_0^2 + y_0^2)^{1/2} < 1.$$

Thus $1 - x_1 x_0 - y_1 y_0 > 0$, and $Z_1(u) = \{(x_0, y_0)\}$. Thus U satisfies Property B on K.

Example 5.8. $U = \text{span}\{1, x, y, xy\}$, $K \in \mathbb{R}^2$ is arbitrary (with interior).

Let $(x_0, y_0) \in \text{int } K$, and set $u(x, y) = (x - x_0)(y - y_0)$. Then $(x_0, y_0) \in Z_1(u)$. If $v \in U$, and $(x_0, y_0) \in Z_1(v)$, then $v = au$ for some $a \in \mathbb{R}$. No such v is non-negative and non-trivial on K. Thus U does not satisfy Property B on K.

Example 5.9. *Tensor products.*

Let $V \subset C^1(K)$, $W \subset C^1(L)$, where $K \subseteq \mathbb{R}^{d_1}, L \subseteq \mathbb{R}^{d_2}$, $\dim V = n_1$, and $\dim W = n_2$. We assume that the assumptions of the previous section apply here. In particular K and L are convex.

Set $U = V \otimes W$, i.e., the tensor product of V and W. For a discussion of tensor products, see Section 5 of Chapter 4. Note that $U \subset C^1(K \times L)$. Two simple properties are easily checked, but hardly worth stating formally as lemmas.

1. U contains a strictly positive function if and only if both V and W contain strictly positive functions.

2. Assume U contains a strictly positive function and dim $V = 1$, i.e., $n_1 = 1$. Then U satisfies Property B on $K \times L$ if and only if W satisfies Property B on L.

Of course the analogous result to 2 holds if we interchange V and W. What happens however if $n_1, n_2 \geq 2$?

Theorem 5.28. *Let $U = V \otimes W$ be as defined above. Assume* dim $V = n_1 \geq 2$, dim $W = n_2 \geq 2$, *and U contains a strictly positive function. Then U does not satisfy Property B on $K \times L$.*

We shall use this next lemma in the proof of Theorem 5.28.

Lemma 5.29. *Let V be an n-dimensional subspace of $C^1(K)$ $n \geq 2$, containing a strictly positive function. There exist points $\{x_i\}_{i=1}^{n-1}$ in int K, and a function $v^* \in V$ for which*

1) $v^*(x_i) = 0$, $i = 1, \ldots, n-1$,

2) v^* *has a sign change on K,*

3) *if $v \in V$, $v(x_i) = 0$, $i = 1, \ldots, n-1$, then $v = \alpha v^*$ for some $\alpha \in \mathbb{R}$.*

Proof. Let $v_1, \ldots, v_n$ be any basis for V. Since $V \subset C^1(K)$, and $K = \overline{\text{int } K}$, there exist points $\{z_i\}_{i=1}^n$ in int K such that $\det(v_i(z_j))_{i,j=1}^n \neq 0$. Let $v^i \in V$ satisfy $v^i(z_j) = \delta_{ij}$, $i, j = 1, \ldots, n$. Then $v^1, \ldots, v^n$ is another basis for V. If v^i changes sign on K for some $i \in \{1, \ldots, n\}$, set $v^* = v^i$ and $\{x_1, \ldots, x_{n-1}\} = \{z_1, \ldots, z_n\} \backslash \{z_i\}$. The lemma then holds.

We therefore assume that $v^i \geq 0$ for all $i = 1, \ldots, n$. Since V contains a strictly positive function, for each $x \in K$ there exists an $i \in \{1, \ldots, n\}$ for which $v^i(x) > 0$. Since K is connected, it follows from continuity considerations that there exists a point $z^* \in \text{int } K$ and $i, j \in \{1, \ldots, n\}$, $i \neq j$, with $v^i(z^*), v^j(z^*) > 0$. Thus $z^* \notin \{z_1, \ldots, z_n\}$. Set $\{x_1, \ldots, x_{n-1}\} = \{z_1, \ldots, z_n, z^*\} \backslash \{z_i, z_j\}$, and

$$v^* = v^j(z^*)v^i - v^i(z^*)v^j.$$

By construction $v^*(x_i) = 0$, $i = 1, \ldots, n-1$. v^* has a sign change on K since $v^*(z_i) = v^j(z^*) > 0$, while $v^*(z_j) = -v^i(z^*) < 0$. Now, if $v \in V$ and $v(x_i) = 0$, $i = 1, \ldots, n-1$, then $v = v(z_i)v^i + v(z_j)v^j$, and v also satisfies $v(z^*) = 0$. This implies that $v = \alpha v^*$, where $\alpha = v(z_i)/v^j(z^*)$. □

Proof of Theorem 5.28. Since U contains a strictly positive function, so do V and W. From Lemma 5.29, there exist points $\{x_i\}_{i=1}^{n_1-1}$ in int K and a $v^* \in V$ satisfying (1)-(3) of Lemma 5.29. Similarly there exist points $\{y_j\}_{i=1}^{n_2-1}$ in int L and a $w^* \in W$ satisfying (1)-(3) of Lemma 5.29.

Set $u^*(x, y) = v^*(x)w^*(y)$. Then $u^* \in U$, $u^* \neq 0$. It easily follows that $(x_i, y_i) \in Z_1(u^*)$ for $i = 1, \ldots, n_1-1$; $j = 1, \ldots, n_2-1$. If U satisfies Property B, there exists a $u \in U$, $u \neq 0$, such that $u \geq 0$, and $u(x_i, y_j) = 0$, $i = 1, \ldots, n_1-1$; $j = 1, \ldots, n_2-1$. For each fixed $i \in \{1, \ldots, n_1-1\}$, $u(x_i, y) \in W$, and $u(x_i, y)$ vanishes at $y_1, \ldots, y_{n_2-1}$. Thus from Lemma 5.29, $u(x_i, y) = \alpha_i w^*(y)$. But $w^*(y)$ has a sign change and $u \geq 0$. Thus $u(x_i, y) = 0$ for all $y \in L$. Similarly $u(x, y_j) = 0$ for all $x \in K$, and $j = 1, \ldots, n_2 - 1$. Now for any fixed $y \in L$, consider $u(x, y)$ as a function of x, i.e., in V. From the above, it vanishes at each x_i, $i = 1, \ldots, n_1 - 1$, and must therefore be a multiple of v^*. Since v^* has a sign change, u is necessarily identically zero along this line. This implies that u is identically zero on all of $K \times L$, a contradiction. Thus U does not satisfy Property B. □

Exercises

1. Let $U \subset C(K)$, dim $U = n$. Assume that

$$\int_K u \, d\mu = \sum_{i=1}^{k} \lambda_i u(x_i)$$

for all $u \in U$, with $1 \leq k < \infty$, and $\lambda_i > 0$, $i = 1, \ldots, k$. Prove that there exists an $f \in C(K)$ with $Z(f - u) = \{x_i\}_{i=1}^{k}$ for all $u \in P_{\mathcal{U}(f)}(f)$.

2. Assume that $U \subset C(K)$, dim $U = n$, U contains a strictly positive function, and μ is an 'admissible' measure. For each $f \in C(K)$, set

$$\mathcal{U}_+(f) = \{u : u \in U, \ 0 \leq u \leq f\}.$$

Prove that if $f \in C(K)$ and $f > 0$, then $u^* \in P_{\mathcal{U}_+(f)}(f)$ if and only if there exist distinct points $\{x_i\}_{i=1}^{k}$ and $\{y_i\}_{i=1}^{m}$, $k \geq 1$, $m \geq 0$, $m + k \leq n$, and positive numbers $\{\lambda_i\}_{i=1}^{k}$ and $\{\sigma_i\}_{i=1}^{m}$, satisfying

$a)\ (f - u^*)(x_i) = 0, \quad i = 1, \ldots, k$

$b)\ u^*(y_i) = 0, \quad i = 1, \ldots, m$

$c)\ \int_K u \, d\mu = \sum_{i=1}^{k} \lambda_i u(x_i) - \sum_{i=1}^{m} \sigma_i u(y_i), \quad$ all $u \in U.$

3. Let $U \subset C(K)$, dim $U = n$, satisfy

$$\int_K u \, d\mu = \sum_{i=1}^{n} \lambda_i u(x_i)$$

for all $u \in U$, where $\lambda_i > 0$, $i = 1, \ldots, n$. Assume that U is linearly independent over the $\{x_i\}_{i=1}^{n}$, i.e., if $u \in U$ and $u(x_i) = 0$, $i = 1, \ldots, n$, then $u = 0$. Prove that there exists a $\gamma > 0$ such that

$$\int_K v \, d\mu \geq \gamma \int_K |v| d\mu$$

for all $v \in U$ satisfying $v(x_i) \geq 0$, $i = 1, \ldots, n$.

4. Assume that the hypotheses of Theorem 5.9 hold. Apply Exercise 3 to prove that for each $f \in C(K)$, there exists a $\gamma > 0$ (dependent on f) such that, if $P_{\mathcal{U}(f)}(f) = \{u^*\}$ then, for all $u \in \mathcal{U}(f)$,

$$\|f - u\|_1 \geq \|f - u^*\|_1 + \gamma \|u^* - u\|_1.$$

5. Let $U \subset C(K)$, dim $U = n$, and let μ be an 'admissible' measure. Prove that

$$\int_K u \, d\mu = 0$$

for all $u \in U$ if and only if, for every $f \in C(K)$ with $\mathcal{U}(f) \neq \emptyset$, $P_{\mathcal{U}(f)}(f) \subseteq P_U(f)$, i.e., every best one-sided $L^1(K, \mu)$ approximant to f is a best two-sided $L^1(K, \mu)$ approximant to f.

6. Construct U, K, and μ as in Exercise 5 for which there exists an $f \in C(K)$ with $\mathcal{U}(f) \neq \emptyset$ and $P_{\mathcal{U}(f)}(f) \subseteq P_U(f)$, but such that

$$\int_K u \, d\mu \neq 0$$

for some $u \in U$.

7. Assume that $U \subset C(K)$, dim $U = n$, and U contains a strictly positive function. Prove that every positive quadrature for U has at least $n - m$ active points if and only if dim $P_{\mathcal{U}(f)}(f) \leq m$ for all $f \in C(K)$.

8. Let $U \subset C(K)$, dim $U = n$, and μ be an 'admissible' measure. Prove that U is a unicity space for $C_{1+}(K, \mu)$ if and only if for every $u \in U$, $u \neq 0$, satisfying $\int_K u \, d\mu = 0$, the zero function is not in $P_{\mathcal{U}(u_+)}(u_+)$.

9. Tchakaloff [1957] essentially proved that for $K \subset \mathbb{R}^2$, μ an 'admissible' measure, and

$$U = \Big\{ \sum_{0 \leq i+j \leq n} a_{ij} x^i y^j \ : \ a_{ij} \in \mathbb{R} \Big\},$$

there exists a positive quadrature formula for U with at most $(n+1)(n+2)/2$ active points. Prove this result, and also that if K is connected then this bound can be lowered by one.

10. Prove Lemma 5.20 for $K \subset \mathbb{R}^d$, $d \geq 3$.

11. Generalize Theorems 5.19 and 5.22 to the case where K is the union of a finite number of disjoint compact convex subsets of $\mathbb{R}^d$, each with piecewise smooth boundary and non-empty interior.

12. Prove that Theorem 5.28 holds for K and L satisfying the conditions of Exercise 11, if $[K] \leq n_1 - 1$ and $[L] \leq n_2 - 1$.

13. Let U be an ET_2-system on $[0, 1]$. Let $0 \leq a < b < c < d \leq 1$. Prove that U does not satisfy Property B on $[a, b] \cup [c, d]$.

Notes and References

The study of best one-sided L^1-approximation does not have a long history. The qualitative theory seems to have been first considered by Bojanic, DeVore [1966]. They studied algebraic polynomials π_n on $[a, b]$. They proved, by looking at Gaussian-type quadrature formulae, the existence of $f \in C[a, b]$ with more than one best one-sided L^1-approximant from π_n. They also showed, for

f continuous on $[a, b]$ and differentiable on (a, b), the uniqueness of the best one-sided L^1-approximant from π_n. Additional results concerned f in the convexity cone of π_n, extensions of which appear in Part II of Appendix B. DeVore [1968] essentially generalized these results to T- and ET_2-systems.

Very little else was then done on the general theory except as it pertained to Gaussian-type quadrature formulae and f in the convexity cone of a WT-system (see Part II of Appendix B). In Pinkus [1976a] are to be found extensions of the first two above-mentioned results of Bojanic, DeVore [1966] to splines with simple fixed knots. It is only in this decade (and especially in the last three years) that a more systematic study of this theory has begun.

Theorem 5.2 is essentially to be found in Nürnberger [1985b], as is Theorem 5.9 and Exercise 4. Theorem 5.10 and Corollary 5.11 are from Strauss [1982]. Theorem 5.13 is contained in Pinkus, Totik [1986], while the related Theorem 5.14 is from Nürnberger [1985a]. The main idea used in the proof of Theorem 5.14 is to be found in Eggleston [1958, Theorem 18]. Proposition 5.15 is new. Theorem 5.17 is from Pinkus, Totik [1986]. Almost all the material of Sections 5 and 6 is from Pinkus, Strauss [1987]. However Theorem 5.19 for $K = [a, b]$ is essentially in Strauss [1982]. Property B for $K = [a, b]$ was also introduced in Strauss [1982] at the suggestion of R. DeVore. It was used to provide simpler proofs of the unicity property in $C^1[a, b]$ of the examples considered by DeVore [1968] (Proposition 5.24) and Pinkus [1976a] (Example 5.4). Example 5.3 is contained in Sommer, Strauss [1981] where there is also to be found a generalization of Example 5.4. Whitney's Extension Theorem used in Lemma 5.20 may be found in Hörmander [1983]. The proof of Theorem 5.22 uses various results from functional analysis. All these results are contained in Rudin [1973].

The best one-sided L^1-approximation problem may, via (5.2), also be viewed as a problem in semi-infinite optimization, or semi-infinite linear programming. This is an approach not taken here, but one considered in Chapter 7. Under fairly lenient assumptions, this problem can also be translated into a moment problem. This is done in Part II of Appendix B.

6
Discrete ℓ_1^m-Approximation

1. Introduction

In the previous chapters, we considered in detail both one- and two-sided L^1-approximation where the norm was given by

$$\|f\|_1 = \int_K |f| d\mu$$

with μ a non-atomic positive measure.

In this chapter we go to the other extreme and consider measures μ which are purely atomic. In fact we shall assume that μ has exactly m atoms. This corresponds to approximation in the vector space $\mathbb{R}^m$. The previous notation is now both cumbersome and inappropriate. We therefore redefine our problem in this new setting, introducing some new notation.

Let
$$W = \{\mathbf{w} : \mathbf{w} = (w_1, \ldots, w_m), \ w_i > 0, \ i = 1, \ldots, m\}.$$

We say that $\mathbf{w} \in W$ is a *weight* (corresponding to the previous measure μ). On $\mathbb{R}^m$, we define the $\ell_1^m(\mathbf{w})$-norm given by

$$\|\mathbf{x}\|_\mathbf{w} = \sum_{i=1}^m |x_i| w_i$$

where $\mathbf{x} = (x_1, \ldots, x_m) \in \mathbb{R}^m$.

In this short chapter we study the theoretical problem of both one- and two-sided $\ell_1^m(\mathbf{w})$-approximation from finite-dimensional subspaces. We defer to the next chapter the computational (algorithmic) aspects of this problem. In Section 2 we concern ourselves with the two-sided approximation problem. In Section 3 we study the analogous one-sided approximation problem. The theory obtained shall be essentially a mix of some of the results of Chapters 2–5. Certain properties are unfortunately lost, but some new results are also obtained.

2. Two-Sided ℓ_1^m-Approximation

Let U be an n-dimensional subspace of $\mathbb{R}^m$. Given $\mathbf{w} \in W$ and $\mathbf{b} \in \mathbb{R}^m$, we consider

$$(6.1) \qquad \min\{\|\mathbf{b} - \mathbf{u}\|_\mathbf{w} : \mathbf{u} \in U\}.$$

Since dim $U = n$, we have $U = \text{span}\{\mathbf{u}^1, \ldots, \mathbf{u}^n\}$ for some linearly independent $\mathbf{u}^i \in \mathbb{R}^m$, $i = 1, \ldots, n$. If we let A denote the $m \times n$ matrix whose jth column is the vector $\mathbf{u}^j$, then (6.1) may be (and often is) written in the form

$$(6.2) \qquad \min\{\|\mathbf{b} - A\mathbf{a}\|_{\mathbf{w}} : \mathbf{a} \in \mathbb{R}^n\}.$$

(We are here assuming that $\mathbf{a}$ and $\mathbf{b}$ are column vectors.) We shall, however, generally use the form (6.1).

The existence of a best approximant to each $\mathbf{b} \in \mathbb{R}^m$ is an immediate consequence of Corollary 1.3. The characterization of the best approximants follows from Theorem 2.1, which we now restate, translated to our present context.

Theorem 6.1. *Given $\mathbf{b} \in \mathbb{R}^m$, $\mathbf{w} \in W$, and a finite-dimensional subspace U of $\mathbb{R}^m$, then $\mathbf{u}^* \in U$ is a best approximant to $\mathbf{b}$ from U in the $\ell_1^m(\mathbf{w})$-norm if and only if*

$$\left| \sum_{i=1}^m \text{sgn}(b_i - u_i^*) u_i w_i \right| \leq \sum_{i \in Z(\mathbf{b} - \mathbf{u}^*)} |u_i| w_i$$

for all $\mathbf{u} = (u_1, \ldots, u_m) \in U$.

Alternatively, based on Theorem 1.9 (see also the proof of Theorem 2.1), we have

Proposition 6.2. *Let $\mathbf{b} \in \mathbb{R}^m$, $\mathbf{w} \in W$, and U be a finite-dimensional subspace of $\mathbb{R}^m$. Then $\mathbf{u}^* \in U$ is a best approximant to $\mathbf{b}$ from U in the $\ell_1^m(\mathbf{w})$-norm if and only if there exists a $\mathbf{y} \in \mathbb{R}^m$ satisfying*

1) $\|\mathbf{y}\|_\infty = 1$

2) $\displaystyle\sum_{i=1}^m y_i u_i w_i = 0$, *all $\mathbf{u} = (u_1, \ldots, u_m) \in U$*

3) $y_i = \text{sgn}(b_i - u_i^*)$ *for all $i \notin Z(\mathbf{b} - \mathbf{u}^*)$.*

It is to be understood that $\|\mathbf{y}\|_\infty = \max\{|y_i| : i = 1, \ldots, m\}$. In order to simplify our notation, we also write

$$(\mathbf{x}, \mathbf{y})_{\mathbf{w}} = \sum_{i=1}^m x_i y_i w_i$$

for every $\mathbf{x}, \mathbf{y} \in \mathbb{R}^m$.

As a special case of Corollary 2.5, we have the following sufficient criterion for determining when $\mathbf{b}$ has a unique best approximant from U.

Proposition 6.3. *Given $\mathbf{b} \in \mathbb{R}^m$, $\mathbf{w} \in W$, and $\mathbf{u}^* \in U$. If*

$$\left| \sum_{i=1}^m \text{sgn}(b_i - u_i^*) u_i w_i \right| < \sum_{i \in Z(\mathbf{b} - \mathbf{u}^*)} |u_i| w_i$$

for all $\mathbf{u} \in U$, $\mathbf{u} \neq \mathbf{0}$, then $\mathbf{u}^*$ is the unique best approximant to $\mathbf{b}$ from U in the $\ell_1^m(\mathbf{w})$-norm.

We shall show that the converse of the above proposition holds. We first prove a simple lemma. This lemma will also be later used in our study of $P_U(\mathbf{b})$.

Lemma 6.4. Let $\mathbf{b} \in \mathbb{R}^m$ and $\mathbf{w} \in W$. Let $\mathbf{u}^*$ be a best approximant to $\mathbf{b}$ from U in the $\ell_1^m(\mathbf{w})$-norm. Assume $\mathbf{u} \in U$, $\mathbf{u} \neq \mathbf{0}$, satisfies

$$(6.3) \qquad \left| \sum_{i=1}^{m} \operatorname{sgn}(b_i - u_i^*) u_i w_i \right| = \sum_{i \in Z(\mathbf{b} - \mathbf{u}^*)} |u_i| w_i.$$

a) If $\sum_{i=1}^{m} \operatorname{sgn}(b_i - u_i^*) u_i w_i = 0$, there exists an $\varepsilon > 0$ such that $\mathbf{u}^* + t\mathbf{u}$ is a best approximant to $\mathbf{b}$ for every t satisfying $|t| \leq \varepsilon$.

b) If $\delta(\sum_{i=1}^{m} \operatorname{sgn}(b_i - u_i^*) u_i w_i) > 0$, where $\delta \in \{-1, 1\}$, there exists an $\varepsilon > 0$ such that $\mathbf{u}^* + t\mathbf{u}$ is a best approximant to $\mathbf{b}$ for every t satisfying $0 \leq t\delta \leq \varepsilon$.

Proof. From (6.3), $\mathbf{b} \neq \mathbf{u}^*$. Furthermore, since we are in the finite-dimensional space $\mathbb{R}^m$,

$$\min\{|b_i - u_i^*| : i \notin Z(\mathbf{b} - \mathbf{u}^*)\} = c > 0.$$

Let $\mathbf{u} \in U$, $\mathbf{u} \neq \mathbf{0}$, satisfy (6.3). There then exists an $\varepsilon > 0$ such that, for all t satisfying $|t| \leq \varepsilon$,

$$\operatorname{sgn}(b_i - u_i^*) = \operatorname{sgn}(b_i - u_i^* - tu_i)$$

for every $i \notin Z(\mathbf{b} - \mathbf{u}^*)$. For every such t,

$$\|\mathbf{b} - \mathbf{u}^* - t\mathbf{u}\|_{\mathbf{w}} - \|\mathbf{b} - \mathbf{u}^*\|_{\mathbf{w}} = \sum_{i=1}^{m} \left[|b_i - u_i^* - tu_i| - |b_i - u_i^*| \right] w_i$$

$$= \sum_{i=1}^{m} \operatorname{sgn}(b_i - u_i^*)(b_i - u_i^* - tu_i - b_i + u_i^*) w_i$$

$$+ \sum_{i \in Z(\mathbf{b} - \mathbf{u}^*)} |tu_i| w_i$$

$$= -t \sum_{i=1}^{m} \operatorname{sgn}(b_i - u_i^*) u_i w_i + |t| \sum_{i \in Z(\mathbf{b} - \mathbf{u}^*)} |u_i| w_i.$$

Since (6.3) holds, we obtain

$$\|\mathbf{b} - \mathbf{u}^* - t\mathbf{u}\|_{\mathbf{w}} - \|\mathbf{b} - \mathbf{u}^*\|_{\mathbf{w}} = -t \sum_{i=1}^{m} \operatorname{sgn}(b_i - u_i^*) u_i w_i + |t| \left| \sum_{i=1}^{m} \operatorname{sgn}(b_i - u_i) u_i w_i \right|.$$

If $\sum_{i=1}^{m} \operatorname{sgn}(b_i - u_i^*) u_i w_i = 0$, then for all such t, i.e., $|t| \leq \varepsilon$, we have

$$\|\mathbf{b} - \mathbf{u}^* - t\mathbf{u}\|_{\mathbf{w}} = \|\mathbf{b} - \mathbf{u}^*\|_{\mathbf{w}}.$$

If $\delta(\sum_{i=1}^m \text{sgn}(b_i - u_i^*)u_i w_i) > 0$, $\delta \in \{-1, 1\}$, then

$$\|\mathbf{b} - \mathbf{u}^* - t\,\mathbf{u}\|_\mathbf{w} = \|\mathbf{b} - \mathbf{u}^*\|_\mathbf{w}$$

for all t satisfying $0 \le t\delta \le \varepsilon$. □

Putting together Proposition 6.3 and Lemma 6.4, we now have:

Theorem 6.5. *Let* $\mathbf{b} \in I\!\!R^m$, *and* $\mathbf{w} \in W$. *Then* $\mathbf{u}^*$ *is the unique best approximant to* $\mathbf{b}$ *from* U *in the* $\ell_1^m(\mathbf{w})$-*norm if and only if*

$$(6.4) \qquad \left| \sum_{i=1}^m \text{sgn}(b_i - u_i^*)u_i w_i \right| < \sum_{i \in Z(\mathbf{b}-\mathbf{u}^*)} |u_i| w_i$$

for all $\mathbf{u} \in U$, $\mathbf{u} \ne \mathbf{0}$.

The inequality (6.4), together with Corollary 2.6, implies that every unique best approximant is a strongly unique best approximant. From Corollary 2.6, we explicitly have:

Proposition 6.6. *Assume* $\mathbf{u}^* \in U$ *is the unique best approximant to* $\mathbf{b}$ *from* U *in the* $\ell_1^m(\mathbf{w})$-*norm. Set*

$$\gamma = \min_{\|\mathbf{u}\|_\mathbf{w}=1} \left[\sum_{i \in Z(\mathbf{b}-\mathbf{u}^*)} |u_i| w_i - \sum_{i=1}^m \text{sgn}(b_i - u_i^*)u_i w_i \right].$$

Then $\gamma > 0$ *and, for all* $\mathbf{u} \in U$,

$$\gamma\|\mathbf{u}^* - \mathbf{u}\|_\mathbf{w} \le \|\mathbf{b} - \mathbf{u}\|_\mathbf{w} - \|\mathbf{b} - \mathbf{u}^*\|_\mathbf{w}.$$

Furthermore, if $\gamma' > \gamma$, *there then exists a* $\tilde{\mathbf{u}} \in U$ *for which*

$$\gamma'\|\tilde{\mathbf{u}} - \mathbf{u}^*\|_\mathbf{w} > \|\mathbf{b} - \tilde{\mathbf{u}}\|_\mathbf{w} - \|\mathbf{b} - \mathbf{u}^*\|_\mathbf{w}.$$

Let us record one further property of the unique best approximant to $\mathbf{b}$ from U. In what follows, for any subset I of $\{1, \ldots, m\}$, $|I|$ denotes the number of elements of I.

Proposition 6.7. *Let* $\mathbf{b} \in I\!\!R^m$, *and* $\mathbf{w} \in W$. *Assume* $\mathbf{u}^*$ *is the unique best approximant to* $\mathbf{b}$ *from* U *in the* $\ell_1^m(\mathbf{w})$-*norm. If* $\mathbf{u} \in U$ *satisfies* $u_i = 0$ *for all* $i \in Z(\mathbf{b} - \mathbf{u}^*)$, *then* $\mathbf{u} = \mathbf{0}$. *In particular,* $|Z(\mathbf{b} - \mathbf{u}^*)| \ge \dim U$.

Proof. Assume there exists a $\mathbf{u} \in U$, $\mathbf{u} \ne \mathbf{0}$, such that $u_i = 0$ for all $i \in Z(\mathbf{b} - \mathbf{u}^*)$. Then

$$\sum_{i \in Z(\mathbf{b}-\mathbf{u}^*)} |u_i| w_i = 0.$$

A contradiction to the uniqueness of $\mathbf{u}^*$ now follows from Theorem 6.5. In particular $|Z(\mathbf{b} - \mathbf{u}^*)| \ge \dim U$, since otherwise there would exist a $\mathbf{u} \in U$, $\mathbf{u} \ne \mathbf{0}$, satisfying $u_i = 0$ for all $i \in Z(\mathbf{b} - \mathbf{u}^*)$. □

In other words, if dim $U = n$, and $\mathbf{u}^*$ is the unique best approximant to $\mathbf{b}$ from U, then $\mathbf{u}^*$ must interpolate to $\mathbf{b}$ on at least n indices over which U is linearly independent. This property does not necessarily hold in the case of continuous functions, nor does it hold in any of the ℓ_p^m-norms on $I\!\!R^m$ where $1 < p \leq \infty$.

If $P_U(\mathbf{b})$ is not a singleton, i.e., the best approximant to $\mathbf{b}$ from U is not unique, then this property need not hold with respect to every vector in $P_U(\mathbf{b})$. However it does hold for every extreme point of the convex set $P_U(\mathbf{b})$ (generalizing Proposition 6.7). This simple observation will help us understand the structure of $P_U(\mathbf{b})$.

Proposition 6.8. *Let* $\mathbf{b} \in I\!\!R^m$, *and* $\mathbf{w} \in W$. *Then* $\mathbf{u}^* \in P_U(\mathbf{b})$ *is an extreme point of* $P_U(\mathbf{b})$ *if and only if for any* $\mathbf{u} \in U$ *satisfying* $u_i = 0$ *for all* $i \in Z(\mathbf{b} - \mathbf{u}^*)$, *we have* $\mathbf{u} = \mathbf{0}$.

Proof. ($\Rightarrow$). This proof is essentially the same as the proof of Proposition 6.7. If there exists a $\mathbf{u} \in U$, $\mathbf{u} \neq \mathbf{0}$, for which $u_i = 0$ for all $i \in Z(\mathbf{b} - \mathbf{u}^*)$, then from Theorem 6.1 and Lemma 6.4, part (a), it follows that $\mathbf{u}^* + t\,\mathbf{u} \in P_U(\mathbf{b})$ for all t satisfying $|t| \leq \varepsilon$ for some $\varepsilon > 0$. But then $\mathbf{u}^*$ is not an extreme point of $P_U(\mathbf{b})$.

($\Leftarrow$). If $\mathbf{u}^* \in P_U(\mathbf{b})$ is not an extreme point of $P_U(\mathbf{b})$, there exists a $\mathbf{u} \in U$, $\mathbf{u} \neq \mathbf{0}$, and an $\varepsilon > 0$, such that $\mathbf{u}^* + t\,\mathbf{u} \in P_U(\mathbf{b})$ for all t satisfying $|t| \leq \varepsilon$. Let $0 < \delta \leq \varepsilon$ be such that

$$\operatorname{sgn}(b_i - u_i^*) = \operatorname{sgn}(b_i - u_i^* - tu_i)$$

for all $i \notin Z(\mathbf{b} - \mathbf{u}^*)$, and all $|t| \leq \delta$. We now reverse the argument in Lemma 6.4. Since

$$
\begin{aligned}
0 &= \|\mathbf{b} - \mathbf{u}^* - t\,\mathbf{u}\|_{\mathbf{w}} - \|\mathbf{b} - \mathbf{u}^*\|_{\mathbf{w}} \\
&= \sum_{i \notin Z(\mathbf{b}-\mathbf{u}^*)} \operatorname{sgn}(b_i - u_i^*)(b_i - u_i^* - tu_i - b_i + u_i^*)w_i + \sum_{i \in Z(\mathbf{b}-\mathbf{u}^*)} |tu_i|w_i \\
&= -t \sum_{i \notin Z(\mathbf{b}-\mathbf{u}^*)} \operatorname{sgn}(b_i - u_i^*)u_i w_i + |t| \sum_{i \in Z(\mathbf{b}-\mathbf{u}^*)} |u_i|w_i
\end{aligned}
$$

holds for $t = \pm\delta$, it therefore follows that

$$\sum_{i=1}^{m} \operatorname{sgn}(b_i - u_i^*)u_i w_i = \sum_{i \in Z(\mathbf{b}-\mathbf{u}^*)} |u_i|w_i = 0.$$

Thus $u_i = 0$ for all $i \in Z(\mathbf{b} - \mathbf{u}^*)$. $\square$

A closed bounded convex set in $I\!\!R^m$ may have an infinite number of extreme points (e.g. any disk in $I\!\!R^2$). The above proposition implies that $P_U(\mathbf{b})$ has at most $\binom{m}{n}$ extreme points where $n = \dim U$. This follows from the fact that every extreme point of $P_U(\mathbf{b})$ interpolates to $\mathbf{b}$ on at least n

indices on which U has full rank. We therefore have, in addition to Proposition 6.8,

Theorem 6.9. *For every* $\mathbf{b} \in I\!\!R^m$ *and* $\mathbf{w} \in W$, $P_U(\mathbf{b})$ *is a closed bounded convex set with a finite number of extreme points.*

In the next chapter we shall consider various algorithms for calculating best $\ell_1^m(\mathbf{w})$ approximants. In general there is no very simple way of calculating a best approximant to $\mathbf{b}$ from U. However one or two special cases do deserve mention.

Proposition 6.10. *Let* $\dim U = n$, *and* $\mathbf{w} \in W$. *Assume there exists a set of n distinct indices* $K = \{k_1, \ldots, k_n\}$ *in* $\{1, \ldots, m\}$ *for which*

$$(6.5) \qquad \sum_{i \notin K} |u_i| w_i \le \sum_{i \in K} |u_i| w_i$$

for all $\mathbf{u} \in U$. *Then given* $\mathbf{b} \in I\!\!R^m$, *there exists a unique* $\mathbf{u}^* \in U$ *for which* $b_{k_j} = u^*_{k_j}$, $j = 1, \ldots, n$. *Furthermore* $\mathbf{u}^* \in P_U(\mathbf{b})$. *If, in addition, strict inequality holds in (6.5) for all* $\mathbf{u} \in U$, $\mathbf{u} \ne \mathbf{0}$, *then* $P_U(\mathbf{b}) = \{\mathbf{u}^*\}$.

Proof. Since (6.5) holds, there cannot exist a $\mathbf{u} \in U$, $\mathbf{u} \ne \mathbf{0}$, satisfying $u_{k_j} = 0$, $j = 1, \ldots, n$. This implies, for each $\mathbf{b} \in I\!\!R^m$, the existence of a unique $\mathbf{u}^* \in U$ satisfying $b_{k_j} = u^*_{k_j}$, $j = 1, \ldots, n$. Now, for any $\mathbf{u} \in U$

$$\left| \sum_{i=1}^m \operatorname{sgn}(b_i - u^*_i) u_i w_i \right| \le \sum_{i \notin K} |u_i| w_i \le \sum_{i \in K} |u_i| w_i \le \sum_{i \in Z(\mathbf{b}-\mathbf{u}^*)} |u_i| w_i.$$

From Theorem 6.1 we obtain $\mathbf{u}^* \in P_U(\mathbf{b})$. The remaining claim of the proposition follows from Proposition 6.3. □

Condition (6.5) is a very stringent demand, since the interpolation indices $\{k_1, \ldots, k_n\}$ are then independent of $\mathbf{b}$. However, there is one special case of Proposition 6.10 well worth noting. For $m = n + 1$ one can always find $\{k_1, \ldots, k_n\}$ satisfying (6.5). In other words, for $m = n + 1$ the interpolation indices may be taken to be independent of $\mathbf{b}$.

Proposition 6.11. *Let* $\dim U = n$, $U \subset I\!\!R^{n+1}$, *and* $\mathbf{w} \in W$. *There exists a* $j \in \{1, \ldots, n + 1\}$ *for which*

$$|u_j| w_j \le \sum_{\substack{i=1 \\ i \ne j}}^{n+1} |u_i| w_i$$

for all $\mathbf{u} \in U$.

Proof. Since $\dim U = n$, there exists a (unique up to multiplication by a constant) non-trivial $\mathbf{y} \in I\!\!R^{n+1}$ satisfying

$$\sum_{i=1}^{n+1} y_i u_i w_i = 0$$

for all $\mathbf{u} \in U$. Assume $|y_j| = \|\mathbf{y}\|_\infty = 1$. Then

$$|u_j| w_j = |y_j u_j w_j| = \left| \sum_{\substack{i=1 \\ i \neq j}}^{n+1} y_i u_i w_i \right| \leq \sum_{\substack{i=1 \\ i \neq j}}^{n+1} |u_i| w_i. \qquad \square$$

Remark. There is yet another condition on U and $\mathbf{b}$ which allows us to construct, in a fairly easy manner, a best approximant to $\mathbf{b}$ from U in the $\ell_1^m(\mathbf{w})$-norm. This is when U is a T-system and $\mathbf{b}$ is in the convexity cone generated by U. There then exist indices $\{k_1, \ldots, k_n\}$ such that for all such $\mathbf{b}$ a best approximant is obtained by interpolating to $\mathbf{b}$ at the indices $\{k_1, \ldots, k_n\}$. The interested reader will find the analogue of this result in Part II of Appendix B.

If (6.5) holds, then the map which interpolates to $\mathbf{b}$ from U at the indices $\{k_1, \ldots, k_n\}$ is a *linear selection*. That is, it is a linear map which is also a metric selection. Other linear selections may exist if (6.5) holds and the $\{k_1, \ldots, k_n\}$ are not uniquely defined. For example, if $U = \mathrm{span}\{(1,1)\}$ and $\mathbf{w} = (1,1)$, then we could choose $k_1 = 1$ or $k_1 = 2$. The map $L(\mathbf{b}) = (\lambda b_1 + (1-\lambda)b_2, \lambda b_1 + (1-\lambda)b_2)$ is also a linear selection for every $\lambda \in [0,1]$. It is an interpolation map if $\lambda \in \{0,1\}$.

It is natural to ask for conditions on U implying the existence of a linear selection, since this is a most desirable property. A linear selection on U exists if and only if (6.5) holds for some $\{k_1, \ldots, k_n\}$, where $n = \dim U$. Proposition 6.10 is one half of this result. The other half is the content of this next theorem.

Theorem 6.12. *Let $U \subset \mathbb{R}^m$, $\dim U = n$, and $\mathbf{w} \in W$. If there exists a linear selection L onto U, then there exist indices $\{k_1, \ldots, k_n\}$ in $\{1, \ldots, m\}$ for which (6.5) holds for all $\mathbf{u} \in U$.*

Proof. Assume that L exists. Since L is a linear selection, the following two simple properties hold. Firstly, $L(\mathbf{b}) \in P_U(\mathbf{b})$ for every $\mathbf{b} \in \mathbb{R}^m$, and secondly $L(\mathbf{c}^1), \ldots, L(\mathbf{c}^m)$ span U for any basis $\mathbf{c}^1, \ldots, \mathbf{c}^m$ of $\mathbb{R}^m$.

We prove the theorem by an induction argument. We claim that for each r, $1 \leq r \leq n$, there exists a subspace V of U of dimension r, and indices $\{k_1, \ldots, k_r\} \subseteq \{1, \ldots, m\}$ such that

$$\sum_{\substack{i=1 \\ i \notin \{k_1, \ldots, k_r\}}}^{m} |v_i| w_i \leq \sum_{j=1}^{r} |v_{k_j}| w_{k_j}$$

for all $\mathbf{v} \in V$. For $r = n$, this is the statement of the theorem.

We first prove the case $r = 1$. Assume to the contrary that for every $\mathbf{u} \in U$, $\mathbf{u} \neq \mathbf{0}$, and each $k \in \{1, \ldots, m\}$

$$\sum_{\substack{i=1 \\ i \neq k}}^{m} |u_i| w_i > |u_k| w_k.$$

Let $\mathbf{e}^k$ denote the kth unit vector. Then for every $\mathbf{u} \in U$, $\mathbf{u} \neq \mathbf{0}$,

$$\left| \sum_{i=1}^m \mathrm{sgn}(e_i^k) u_i w_i \right| = |u_k| w_k < \sum_{\substack{i=1 \\ i \neq k}}^m |u_i| w_i = \sum_{i \in Z(\mathbf{e}^k)} |u_i| w_i.$$

From Proposition 6.3, we have that $P_U(\mathbf{e}^k) = \{\mathbf{0}\}$. Thus $L(\mathbf{e}^k) = \mathbf{0}$ for each $k = 1, \ldots, m$, contradicting the fact that $L(\mathbf{e}^1), \ldots, L(\mathbf{e}^m)$ span U. Thus there exists a $\mathbf{v} \in U$, $\mathbf{v} \neq \mathbf{0}$, and $k_1 \in \{1, \ldots, m\}$ for which

$$\sum_{\substack{i=1 \\ i \neq k_1}}^m |v_i| w_i \leq |v_{k_1}| w_{k_1}.$$

We now show how to advance the induction step. Assume that $1 \leq r < n$, and we are given $V \subset U$ of dimension r, and indices $\{k_1, \ldots, k_r\}$ such that

$$\sum_{\substack{i=1 \\ i \notin \{k_1, \ldots, k_r\}}}^m |v_i| w_i \leq \sum_{j=1}^r |v_{k_j}| w_{k_j}$$

for all $\mathbf{v} \in V$. Set

$$\widetilde{V} = \{\mathbf{u} : \mathbf{u} \in U, u_{k_j} = 0, j = 1, \ldots, r\}.$$

It is easily seen that $\dim V|_{\{k_1, \ldots, k_r\}} = r$. Thus $\dim \widetilde{V} = n - r$, and $U = V \oplus \widetilde{V}$.

We first claim that there exists a $k_{r+1} \in \{1, \ldots, m\}$ and $\widetilde{\mathbf{v}}^* \in \widetilde{V}$, $\widetilde{\mathbf{v}}^* \neq \mathbf{0}$, for which

$$\sum_{\substack{i=1 \\ i \neq k_{r+1}}}^m |\widetilde{v}_i^*| w_i \leq |\widetilde{v}_{k_{r+1}}^*| w_{k_{r+1}}.$$

(Note that $k_{r+1} \notin \{k_1, \ldots, k_r\}$ since $\widetilde{v}_{k_i}^* = 0$, $i = 1, \ldots, r$.) Assume not. That is

$$\sum_{\substack{i=1 \\ i \neq k}}^m |\widetilde{v}_i| w_i > |\widetilde{v}_k| w_k$$

for every $\widetilde{\mathbf{v}} \in \widetilde{V}$, $\widetilde{\mathbf{v}} \neq \mathbf{0}$, and $k \in \{1, \ldots, m\}$. For each $k \notin \{k_1, \ldots, k_r\}$, and $\mathbf{u} = \mathbf{v} + \widetilde{\mathbf{v}}$, $\mathbf{v} \in V$, $\widetilde{\mathbf{v}} \in \widetilde{V}$,

$$\left| \sum_{i=1}^m \mathrm{sgn}(e_i^k) u_i w_i \right| = |u_k| w_k = |v_k + \widetilde{v}_k| w_k \leq |v_k| w_k + |\widetilde{v}_k| w_k$$

$$\leq \sum_{j=1}^r |v_{k_j}| w_{k_j} - \sum_{\substack{i=1 \\ i \notin \{k_1, \ldots, k_r, k\}}}^m |v_i| w_i + \sum_{\substack{i=1 \\ i \neq k.}}^m |\widetilde{v}_i| w_i$$

$$\leq \sum_{j=1}^r |v_{k_j} + \widetilde{v}_{k_j}| w_{k_j} + \sum_{\substack{i=1 \\ i \notin \{k_1, \ldots, k_r, k\}}}^m |v_i + \widetilde{v}_i| w_i$$

$$\leq \sum_{\substack{i=1 \\ i \neq k}}^m |v_i + \widetilde{v}_i| w_i = \sum_{i \in Z(\mathbf{e}^k)} |u_i| w_i,$$

where we have used the induction hypothesis and the fact that $\tilde{v}_{k_j} = 0$, $j = 1, \ldots, r$.

Thus $\mathbf{0} \in P_U(\mathbf{e}^k)$. More importantly, from Proposition 2.4, see also Exercise 1, if $\mathbf{u} \in P_U(\mathbf{e}^k)$, then

$$\left| \sum_{i=1}^m \operatorname{sgn}(e_i^k) u_i w_i \right| = \sum_{i \in Z(\mathbf{e}^k)} |u_i| w_i \, .$$

Assume $\mathbf{u} = \mathbf{v} + \tilde{\mathbf{v}} \in P_U(\mathbf{e}^k)$, where $\mathbf{v} \in V$ and $\tilde{\mathbf{v}} \in \tilde{V}$. Then equality holds in the above series of inequalities and in particular

$$|\tilde{v}_k| w_k = \sum_{\substack{i=1 \\ i \neq k}}^m |\tilde{v}_i| w_i \, .$$

By hypothesis this implies that $\tilde{\mathbf{v}} = \mathbf{0}$. Thus $P_U(\mathbf{e}^k) \subseteq V$ for each $k \notin \{k_1, \ldots, k_r\}$, i.e., $L(\mathbf{e}^k) \in V$ for each such k. Let $\mathbf{v}^1, \ldots, \mathbf{v}^r$ be any basis for V. Then $\{\mathbf{v}^i\}_{i=1}^r \cup \{\mathbf{e}^k\}_{k \notin \{k_1, \ldots, k_r\}}$ is a basis for $\mathbb{R}^m$. However $\{L(\mathbf{v}^i)\}_{i=1}^r \cup \{L(\mathbf{e}^k)\}_{k \notin \{k_1, \ldots, k_r\}}$ span V and $\dim V = r < n$. This is a contradiction. There therefore exists a $k_{r+1} \in \{1, \ldots, m\} \backslash \{k_1, \ldots, k_r\}$ and $\tilde{\mathbf{v}}^* \in \tilde{V}$, $\tilde{\mathbf{v}}^* \neq \mathbf{0}$, such that

$$\sum_{\substack{i=1 \\ i \neq k_{r+1}}}^m |\tilde{v}_i^*| w_i \leq |\tilde{v}_{k_{r+1}}^*| w_{k_{r+1}} \, .$$

Set $V^* = V \oplus \{\tilde{\mathbf{v}}^*\}$. Then $\dim V^* = r + 1$. Furthermore, if $\mathbf{u} \in V^*$, then $\mathbf{u} = \mathbf{v} + \alpha \tilde{\mathbf{v}}^*$ for some $\alpha \in \mathbb{R}$ and

$$\sum_{\substack{i=1 \\ i \notin \{k_1, \ldots, k_{r+1}\}}}^m |u_i| w_i = \sum_{\substack{i=1 \\ i \notin \{k_1, \ldots, k_{r+1}\}}}^m |v_i + \alpha \tilde{v}_i^*| w_i \leq \sum_{\substack{i=1 \\ i \notin \{k_1, \ldots, k_{r+1}\}}}^m \left(|v_i| w_i + |\alpha \tilde{v}_i^*| w_i \right)$$

$$= \sum_{\substack{i=1 \\ i \notin \{k_1, \ldots, k_r\}}}^m |v_i| w_i - |v_{k_{r+1}}| w_{k_{r+1}} + |\alpha| \sum_{\substack{i=1 \\ i \neq k_{r+1}}}^m |\tilde{v}_i^*| w_i$$

$$\leq \sum_{j=1}^r |v_{k_j}| w_{k_j} - |v_{k_{r+1}}| w_{k_{r+1}} + |\alpha| |\tilde{v}_{k_{r+1}}^*| w_{k_{r+1}}$$

$$\leq \sum_{j=1}^r |v_{k_j}| w_{k_j} + |v_{k_{r+1}} + \alpha \tilde{v}_{k_{r+1}}^*| w_{k_{r+1}}$$

$$= \sum_{j=1}^{r+1} |v_{k_j} + \alpha \tilde{v}_{k_j}^*| w_{k_j} = \sum_{j=1}^{r+1} |u_{k_j}| w_{k_j} \, .$$

The induction step is advanced which proves the theorem. □

Every linear selection is a continuous selection, while the converse is but rarely true. In Chapters 2 and 3 we considered the problem of the existence of continuous selections onto finite-dimensional subspaces. Unlike in both the above-mentioned cases, we shall here prove that every finite-dimensional subspace possesses a continuous selection. In the course of the proof of this result, we shall use this next proposition.

Proposition 6.13. *Let U be a finite-dimensional subspace of $\mathbb{R}^m$, and $\mathbf{w} \in W$. Assume $\lim_{k \to \infty} \mathbf{b}^k = \mathbf{b}$, $\mathbf{u}^k \in P_U(\mathbf{b}^k)$ for each k, $\lim_{k \to \infty} \mathbf{u}^k = \mathbf{u}$, and $\mathbf{u}, \mathbf{v} \in P_U(\mathbf{b})$ where $\mathbf{u} \neq \mathbf{v}$. Then there exists a K and $\alpha > 0$ such that, for all $k \geq K$,*

$$\mathbf{u}^k + \alpha(\mathbf{v} - \mathbf{u}) \in P_U(\mathbf{b}^k).$$

Before proving this result, it is worth noting that one consequence thereof is that the set of $\mathbf{b} \in \mathbb{R}^m$ with more than one best $\ell_1^m(\mathbf{w})$ approximant is an open set.

Proof. Since $\mathbf{u}^k \in P_U(\mathbf{b}^k)$, there exists from Proposition 6.2 a $\mathbf{y}^k \in \mathbb{R}^m$ satisfying

1) $\|\mathbf{y}^k\|_\infty = 1$

2) $\displaystyle\sum_{i=1}^m y_i^k u_i w_i = 0$, all $\mathbf{u} = (u_1, \ldots, u_m) \in U$

3) $y_i^k = \operatorname{sgn}(b_i^k - u_i^k)$ for all $i \notin Z(\mathbf{b}^k - \mathbf{u}^k)$.

Equivalently to (3), we may also write

$$\sum_{i=1}^m y_i^k (b_i^k - u_i^k) w_i = \|\mathbf{b}^k - \mathbf{u}^k\|_{\mathbf{w}}.$$

Because $\lim_{k \to \infty} \mathbf{b}^k - \mathbf{u}^k = \mathbf{b} - \mathbf{u}$, there exists a K such that for all $k \geq K$, we have

$$(b_i^k - u_i^k) \operatorname{sgn}(b_i - u_i) \geq |b_i - u_i|/2$$

for all $i \notin Z(\mathbf{b} - \mathbf{u})$. Thus, for all $k \geq K$,

$3')\ y_i^k = \operatorname{sgn}(b_i - u_i)$ for all $i \notin Z(\mathbf{b} - \mathbf{u})$.

Since $\mathbf{v} \in P_U(\mathbf{b})$, it follows from (1), (2) and $(3')$ that, for $k \geq K$,

$$\|\mathbf{b} - \mathbf{u}\|_{\mathbf{w}} = \sum_{i=1}^m y_i^k (b_i - u_i) w_i = \sum_{i=1}^m y_i^k (b_i - v_i) w_i = \|\mathbf{b} - \mathbf{v}\|_{\mathbf{w}}.$$

This last equality further implies

$3'')\ y_i^k = \operatorname{sgn}(b_i - v_i)$ for all $i \notin Z(\mathbf{b} - \mathbf{v})$.

Choose $\alpha > 0$ so that

$$(b_i^k - u_i^k)(b_i^k - u_i^k - \alpha(v_i - u_i)) > 0$$

for all $i \notin Z(\mathbf{b} - \mathbf{u})$ and all $k \geq K$. The choice of K implies that such an α may be chosen independent of k.

The above choice of α implies that

$$y_i^k(b_i^k - u_i^k - \alpha(v_i - u_i))w_i = |b_i^k - u_i^k - \alpha(v_i - u_i)|w_i$$

for all $k \geq K$ and $i \notin Z(\mathbf{b} - \mathbf{u})$. It is easy to check that this equality actually holds for all i. For assume $i \in Z(\mathbf{b} - \mathbf{u})$. If in addition $i \in Z(\mathbf{b} - \mathbf{v})$, then the result follows from (3). If $i \notin Z(\mathbf{b} - \mathbf{v})$, then (3) together with (3″) prove the desired equality. Thus

$$y_i^k(b_i^k - u_i^k - \alpha(v_i - u_i))w_i = |b_i^k - u_i^k - \alpha(v_i - u_i)|w_i$$

for all $k \geq K$ and $i = 1, \ldots, m$.

Using (2) and (3) we finally obtain

$$\|\mathbf{b}^k - \mathbf{u}^k\|_\mathbf{w} = \sum_{i=1}^m y_i^k(b_i^k - u_i^k)w_i = \sum_{i=1}^m y_i^k(b_i^k - u_i^k - \alpha(v_i - u_i))w_i$$
$$= \sum_{i=1}^m |b_i^k - u_i^k - \alpha(v_i - u_i)|w_i = \|\mathbf{b}^k - \mathbf{u}^k - \alpha(\mathbf{v} - \mathbf{u})\|_\mathbf{w},$$

which proves that $\mathbf{u}^k + \alpha(\mathbf{v} - \mathbf{u}) \in P_U(\mathbf{b}^k)$ for all $k \geq K$.   □

We define a continuous selection $s(\cdot)$ onto U as follows. Let $\|\cdot\|_2$ be any strictly convex norm on $I\!\!R^m$, e.g., the Euclidean norm. For each $\mathbf{b} \in I\!\!R^m$, let s satisfy

> 1) $s(\mathbf{b}) \in P_U(\mathbf{b})$
>
> 2) $\|s(\mathbf{b})\|_2 \leq \|\mathbf{u}\|_2$ for all $\mathbf{u} \in P_U(\mathbf{b})$.

Since $P_U(\mathbf{b})$ is a convex, compact subset of U, the function s is uniquely defined.

Theorem 6.14. *Let U be a finite-dimensional subspace of $I\!\!R^m$, and $\mathbf{w} \in W$. The function s, as defined above, is a continuous selection onto U.*

Proof. Let $\lim_{k\to\infty} \mathbf{b}^k = \mathbf{b}$. Set $\mathbf{u}^k = s(\mathbf{b}^k)$. Assume there exists a subsequence of $\{\mathbf{u}^k\}$, again denoted $\{\mathbf{u}^k\}$, which converges to some $\mathbf{u} \in U$. From Theorem 1.17, $\mathbf{u} \in P_U(\mathbf{b})$. We must prove that $\mathbf{u} = s(\mathbf{b})$.

Assume that $\mathbf{v} \in P_U(\mathbf{b})$. From Proposition 6.13 there exists a K and $\alpha > 0$ such that for all $k \geq K$

$$\mathbf{u}^k + \alpha(\mathbf{v} - \mathbf{u}) \in P_U(\mathbf{b}^k).$$

Since $P_U(\mathbf{b}^k)$ is convex, we may assume that $0 < \alpha \leq 1$. From the definition of s,

$$\|\mathbf{u}^k\|_2 \leq \|\mathbf{u}^k + \alpha(\mathbf{v} - \mathbf{u})\|_2$$

for all $k \geq K$. Taking limits we obtain

$$\|\mathbf{u}\|_2 \leq \|\mathbf{u} + \alpha(\mathbf{v} - \mathbf{u})\|_2.$$

Moreover

$$\|\mathbf{u}\|_2 \le \|\mathbf{u} + \alpha(\mathbf{v} - \mathbf{u})\|_2 = \|(1-\alpha)\mathbf{u} + \alpha\mathbf{v}\|_2 \le (1-\alpha)\|\mathbf{u}\|_2 + \alpha\|\mathbf{v}\|_2 .$$

Thus $\|\mathbf{u}\|_2 \le \|\mathbf{v}\|_2$. Since this is true for any $\mathbf{v} \in P_U(\mathbf{b})$, we obtain $\mathbf{u} = s(\mathbf{b})$.
□

Let us now consider the question of unicity spaces. What are conditions on U such that to each $\mathbf{b} \in I\!\!R^m$ there exists a unique best approximant from U in the $\ell_1^m(\mathbf{w})$-norm? Unlike the situation in Chapter 2, i.e., ν a non-atomic positive measure and $U \subset L^1(B, \nu)$, unicity spaces may and do exist. Unlike the situation in Chapter 3 (and 4), the analogue of T-systems on $I\!\!R^m$ need not be unicity spaces. Our characterization of unicity spaces is a simpler form of Theorem 2.8 and should be considered as a special case thereof. The proof is the same, and we shall not repeat it. (See also Exercise 8 for an equivalent statement more analogous to Theorem 3.1.)

Theorem 6.15. *Let $U \subset I\!\!R^m$, dim $U = n$, and $\mathbf{w} \in W$. Then U is a unicity space for $\ell_1^m(\mathbf{w})$ if and only if for every $\mathbf{y} \in I\!\!R^m$ satisfying*

$$1)\ \|\mathbf{y}\|_\infty = 1$$
$$2)\ (\mathbf{y}, \mathbf{u})_{\mathbf{w}} = 0,\ \text{all } \mathbf{u} \in U$$

we have

$$\left|\{i : |y_i| < 1\}\right| \ge n.$$

The one-dimensional subspace spanned by the vector $\mathbf{u} = (u_1, \ldots, u_m)$ is a unicity space for $\ell_1^m(\mathbf{w})$ if and only if

$$\sum_{i=1}^m \varepsilon_i u_i w_i \ne 0$$

for every choice of $\varepsilon_i \in \{-1, 1\}$, $i = 1, \ldots, m$.

It is natural to ask, as we did in the previous chapters, about the size of the set

$$\{\mathbf{b} : P_U(\mathbf{b}) \text{ a singleton}\}.$$

In the previous chapters, we were always able to show that this set was large. (This was because we could alter functions almost arbitrarily on sets of small measure.) If U is a unicity space, then this set is, by definition, all of $I\!\!R^m$. However in our present setting this set may be very small. An extreme case is illustrated by $U = \text{span}\{(1,1)\}$ and $\mathbf{w} = (1,1)$. For *every* $\mathbf{b} \in I\!\!R^2$, $\mathbf{b} \notin U$, we have that $P_U(\mathbf{b})$ is not a singleton. In fact, if $\mathbf{b} = (b_1, b_2)$ where $b_1 > b_2$, then

$$P_U(\mathbf{b}) = \{(\alpha, \alpha) : b_1 \ge \alpha \ge b_2\}$$

with the analogous result for $b_1 < b_2$.

Are there spaces U which are unicity spaces in the $\ell_1^m(\mathbf{w})$-norm for all $\mathbf{w} \in W$? The answer is yes. They are easily delineated and are uninteresting.

If $\mathbf{e}^i$ denotes the ith unit vector in $I\!R^m$, and $U = \mathrm{span}\{\mathbf{e}^{i_1}, \ldots, \mathbf{e}^{i_n}\}$ for some n distinct indices in $\{1, \ldots, m\}$, then it is easily seen that U is a unicity space in the $\ell_1^m(\mathbf{w})$-norm for every $\mathbf{w} \in W$. We can apply Theorem 6.15 noting that every $\mathbf{y}$ satisfying (1) and (2) thereof necessarily has $y_{i_j} = 0$, $j = 1, \ldots, n$. Much more simply and to the point, the unique best approximant to each $\mathbf{b}$ from U in the $\ell_1^m(\mathbf{w})$-norm, for any $\mathbf{w} \in W$, is $\sum_{j=1}^n b_{i_j} \mathbf{e}^{i_j}$.

We claim that these are the only subspaces U of $I\!R^m$ which are unicity spaces in the $\ell_1^m(\mathbf{w})$-norm for all $\mathbf{w} \in W$. To this end, we define

$$\mathrm{supp}\, U = \{i : u_i \neq 0, \text{ some } \mathbf{u} \in U\}.$$

For every U satisfying $\dim U = n$, we have $|\mathrm{supp}\, U| \geq n$. Furthermore, as is easily seen, $\mathrm{supp}\, U = \{i_1, \ldots, i_n\}$ (i.e., $|\mathrm{supp}\, U| = n$) if and only if $U = \mathrm{span}\{\mathbf{e}^{i_1}, \ldots, \mathbf{e}^{i_n}\}$. We can now state:

Theorem 6.16. *Let $U \subset I\!R^m$, $\dim U = n$. Then U is a unicity space in the $\ell_1^m(\mathbf{w})$-norm for all $\mathbf{w} \in W$ if and only if $|\mathrm{supp}\, U| = n$.*

In the proof of Theorem 6.16, we shall use this next result.

Lemma 6.17. *Assume $U \subset I\!R^m$, $\dim U = n$, and $|\mathrm{supp}\, U| \geq n + 1$. There then exists a $\mathbf{y} \in I\!R^m$, $\mathbf{y} \neq \mathbf{0}$, satisfying $(\mathbf{y}, \mathbf{u})_e = 0$ for all $\mathbf{u} \in U$, where $\mathbf{e} = (1, 1, \ldots, 1)$, and such that*

$$\big|\{i : y_i = 0\}\big| \leq n - 1.$$

Proof. Let

$$Y (= U^\perp) = \{\mathbf{y} : (\mathbf{y}, \mathbf{u})_e = 0, \text{ all } \mathbf{u} \in U\}.$$

Since $\dim U = n$, we have $\dim Y = m - n$. For each $\mathbf{y} \in Y$, $\mathbf{y} \neq \mathbf{0}$, set

$$Z(\mathbf{y}) = \{i : y_i = 0\}.$$

Let $\mathbf{y}^* \in Y$, $\mathbf{y}^* \neq \mathbf{0}$, satisfy

$$|Z(\mathbf{y}^*)| = \min\{|Z(\mathbf{y})| : \mathbf{y} \in Y, \mathbf{y} \neq \mathbf{0}\}.$$

Assume $Z(\mathbf{y}^*) = \{i_1, \ldots, i_k\}$, and $k \geq n$. If there exists a $\mathbf{y} \in Y$ for which $y_{i_j} \neq 0$ for some $j \in \{1, \ldots, k\}$, then a contradiction to the minimality of $|Z(\mathbf{y}^*)|$ ensues by taking $\tilde{\mathbf{y}} = \mathbf{y}^* + \varepsilon \mathbf{y}$ for all but a finite number of values for ε. Thus $y_{i_j} = 0$, $j = 1, \ldots, k$, for all $\mathbf{y} \in Y$. Since $k \geq n$, and $\dim Y = m - n$, (implying $|\mathrm{supp}\, Y| \geq m - n$), it easily follows that $k = n$ and $|\mathrm{supp}\, Y| = m - n$. Thus $Y = \mathrm{span}\{\mathbf{e}^{i'_1}, \ldots, \mathbf{e}^{i'_{m-n}}\}$, where $\{i'_1, \ldots, i'_{m-n}\} \cup \{i_1, \ldots, i_n\} = \{1, \ldots, m\}$. Since $Y = U^\perp$, we have that $U = \mathrm{span}\{\mathbf{e}^{i_1}, \ldots, \mathbf{e}^{i_n}\}$, and $|\mathrm{supp}\, U| = n$. This is a contradiction.　□

Proof of Theorem 6.16. ($\Leftarrow$). Prior to the statement of Theorem 6.16 we showed that, if $|\text{supp}\, U| = n$, then U is a unicity space in the $\ell_1^m(\mathbf{w})$-norm for all $\mathbf{w} \in W$.

($\Rightarrow$). Assume $|\text{supp}\, U| \geq n + 1$. From Lemma 6.17, there exists a $\mathbf{y} \in \mathbb{R}^m$ satisfying

$$1)\ \|\mathbf{y}\|_\infty = 1$$

$$2)\ \sum_{i=1}^{m} y_i u_i = 0, \text{ all } \mathbf{u} \in U$$

$$3)\ |\{i : y_i = 0\}| \leq n - 1.$$

For $y_i \neq 0$, set $w_i^* = |y_i|$ and $y_i^* = y_i/w_i^*$. For $y_i = 0$, set $w_i^* = 1$ and $y_i^* = y_i$. Then

$$1')\ \|\mathbf{y}^*\|_\infty = 1$$

$$2')\ \sum_{i=1}^{m} y_i^* u_i w_i^* = 0, \text{ all } \mathbf{u} \in U$$

$$3')\ |\{i : |y_i^*| < 1\}| \leq n - 1.$$

From Theorem 6.15, U is not a unicity space in the $\ell_1^m(\mathbf{w}^*)$-norm. □

We close this section by considering a question, no analogue of which was asked in the previous chapters. Namely, given $U \subset \mathbb{R}^m$, what is the size of the set of $\mathbf{w} \in W$ for which U is a unicity space for $\ell_1^m(\mathbf{w})$? This same question could well have been asked in Chapter 3. The reason it was not is because we do not know the answer. However here we can prove:

Theorem 6.18. *Let U be a finite-dimensional subspace of $\mathbb{R}^m$. Then, for almost all $\mathbf{w} \in W$, U is a unicity space for $\ell_1^m(\mathbf{w})$.*

In proving Theorem 6.18, we shall use this next simple lemma.

Lemma 6.19. *Let $\mathbf{v}^1, \ldots, \mathbf{v}^k \in \mathbb{R}^m$, $\mathbf{v}^i \neq 0$, all i, and k finite. Then, for almost all $\mathbf{w} \in W$,*

$$\sum_{i=1}^{m} v_i^j w_i \neq 0$$

for $j = 1, \ldots, k$.

Proof. For each $j \in \{1, \ldots, k\}$, the $\mathbf{w} \in \mathbb{R}^m$ satisfying $\sum_{i=1}^{m} v_i^j w_i = 0$ form a hyperplane L_j in $\mathbb{R}^m$ (going through the origin). Almost all $\mathbf{w} \in W$ do not lie on any of the hyperplanes $\{L_j\}_{j=1}^{k}$. □

Proof of Theorem 6.18. Let $\dim U = n$, $1 \leq n \leq m - 1$. Our proof is based on Theorem 6.15. We shall show that for almost all $\mathbf{w} \in W$ there does not exist a $\mathbf{y} \in \mathbb{R}^m$ satisfying

$$1)\ \|\mathbf{y}\|_\infty = 1$$

$$2)\ (\mathbf{y}, \mathbf{u})_{\mathbf{w}} = 0, \text{ all } \mathbf{u} \in U$$

$$3)\ |\{i : |y_i| < 1\}| \leq n - 1.$$

We first rewrite these conditions in a more amenable form.

Set
$$Y \left(= U^{\perp}\right) = \left\{ \mathbf{y} : \sum_{i=1}^{m} y_i u_i = 0, \text{ all } \mathbf{u} \in U \right\}.$$

Obviously dim $Y = m - n$. For each $\mathbf{y} \in Y$, and $\mathbf{w} \in W$
$$0 = \sum_{i=1}^{m} y_i u_i = \sum_{i=1}^{m} (y_i/w_i) u_i w_i.$$

Thus $\mathbf{y} = (y_1, \ldots, y_m)$ satisfies condition (2) as above if and only if $\mathbf{y}(\mathbf{w}) = (y_1 w_1, \ldots, y_m w_m) \in Y$. Since the above condition (1) is simply a normalization, it follows that we must prove that for almost all $\mathbf{w} \in W$ there does not exist a $\mathbf{y} \in Y$, $\mathbf{y} \neq \mathbf{0}$, satisfying

(6.6)
$$\left| \left\{ i : \frac{|y_i|}{w_i} < \max_{j=1,\ldots,m} \frac{|y_j|}{w_j} \right\} \right| \leq n - 1.$$

For each $\mathbf{s} = (s_1, \ldots, s_{n-1})$, $1 \leq s_1 < \cdots < s_{n-1} \leq m$, choose $\mathbf{u}^{\mathbf{s}} \in U$, $\mathbf{u}^{\mathbf{s}} \neq \mathbf{0}$, satisfying $u_{s_j}^{\mathbf{s}} = 0$, $j = 1, \ldots, n-1$. Such a $\mathbf{u}^{\mathbf{s}}$ exists since dim $U = n$. The number of such $\mathbf{u}^{\mathbf{s}}$ is $r = \binom{m}{n-1}$. For each $\mathbf{u}^{\mathbf{s}}$ we construct the 2^m vectors of the form
$$\mathbf{v} = (\varepsilon_1 u_1^{\mathbf{s}}, \ldots, \varepsilon_m u_m^{\mathbf{s}})$$

where $\varepsilon_i \in \{-1, 1\}$, $i = 1, \ldots, m$. Many of these vectors are the same, but this is unimportant. Let $\mathbf{v}^1, \ldots, \mathbf{v}^k$, $k = 2^m r$ be the collection of all such vectors.

From Lemma 6.19, for almost all $\mathbf{w} \in W$, we have
$$\sum_{i=1}^{m} v_i^j w_i \neq 0$$

for $j = 1, \ldots, k$. In other words, for almost all $\mathbf{w} \in W$ we have

(6.7)
$$\sum_{i=1}^{m} \varepsilon_i u_i^{\mathbf{s}} w_i \neq 0$$

for all $\mathbf{s}$ as above, and all choices of $\varepsilon_i \in \{-1, 1\}$. We claim that for each $\mathbf{w} \in W$ satisfying (6.7), there does not exist a $\mathbf{y} \in Y$, $\mathbf{y} \neq \mathbf{0}$, satisfying (6.6).

Assume $\mathbf{w} \in W$ satisfies (6.7) and $\mathbf{y} \in Y$, $\mathbf{y} \neq \mathbf{0}$, satisfies (6.6). Choose $\mathbf{s} = (s_1, \ldots, s_{n-1})$, $1 \leq s_1 < \cdots < s_{n-1} \leq m$, such that
$$\left\{ i : \frac{|y_i|}{w_i} < \max_{j=1,\ldots,m} \frac{|y_j|}{w_j} \right\} \subseteq \{s_1, \ldots, s_{n-1}\}.$$

Let $\mathbf{u}^{\mathbf{s}}$ be as above. Thus if $u_i^{\mathbf{s}} \neq 0$, then $|y_i|/w_i = \sigma > 0$, independent of i. In other words, if $u_i^{\mathbf{s}} \neq 0$, then $y_i/w_i = \varepsilon_i \sigma$, where $\varepsilon_i \in \{-1, 1\}$. Since $\mathbf{y} \in Y$,
$$0 = \sum_{i=1}^{m} y_i u_i^{\mathbf{s}} = \sum_{i=1}^{m} \frac{y_i}{w_i} u_i^{\mathbf{s}} w_i = \sigma \sum_{i=1}^{m} \varepsilon_i u_i^{\mathbf{s}} w_i.$$

Since $\sigma > 0$, this contradicts (6.7). □

3. One-Sided ℓ_1^m-Approximation

As previously, U denotes an n-dimensional subspace of $\mathbb{R}^m$. For each $\mathbf{b} \in \mathbb{R}^m$, we set

$$\mathcal{U}(\mathbf{b}) = \{\mathbf{u} : \mathbf{u} \in U, \mathbf{u} \le \mathbf{b}\}$$

where by $\mathbf{u} \le \mathbf{b}$ we mean that $u_i \le b_i$ for each $i = 1, \ldots, m$. Assuming $\mathcal{U}(\mathbf{b}) \ne \emptyset$, we consider the problem

(6.8) $$\min\{\|\mathbf{b} - \mathbf{u}\|_{\mathbf{w}} : \mathbf{u} \in \mathcal{U}(\mathbf{b})\}$$

where $\mathbf{w} \in W$. Following our previous notation, $P_{\mathcal{U}(\mathbf{b})}(\mathbf{b})$ will denote the set of best one-sided $\ell_1^m(\mathbf{w})$ approximants to $\mathbf{b}$ from U. That is, $\mathbf{u}^* \in P_{\mathcal{U}(\mathbf{b})}(\mathbf{b})$ if $\mathbf{u}^* \in \mathcal{U}(\mathbf{b})$ and

$$\|\mathbf{b} - \mathbf{u}^*\|_{\mathbf{w}} \le \|\mathbf{b} - \mathbf{u}\|_{\mathbf{w}}$$

for all $\mathbf{u} \in \mathcal{U}(\mathbf{b})$. Equivalently $\mathbf{u}^* \in P_{\mathcal{U}(\mathbf{b})}(\mathbf{b})$ if and only if

(6.9) $$\sum_{i=1}^{m} u_i^* w_i \ge \sum_{i=1}^{m} u_i w_i$$

for all $\mathbf{u} = (u_1, \ldots, u_m) \in \mathcal{U}(\mathbf{b})$. It is this form (6.9), rather than (6.8), which will generally prove more useful.

The results of this section are very similar to the results of Sections 2, 3 and 4 of Chapter 5. Certain analogues, most notably Theorems 5.13 and 5.14, are not valid in this setting. As such we shall not even state the analogue of Theorem 5.2, nor its consequences. This material is deferred to the exercises. On the other hand certain results do hold which have no analogue in Chapter 5.

We start with a characterization result analogous to Theorem 5.1. Note that the restrictions of Theorem 5.1 have been dropped.

Theorem 6.20. *Assume that U is an n-dimensional subspace of $\mathbb{R}^m$, $\mathbf{w} \in W$, $\mathbf{b} \in \mathbb{R}^m$, and $\mathcal{U}(\mathbf{b}) \ne \emptyset$. Then $\mathbf{u}^* \in P_{\mathcal{U}(\mathbf{b})}(\mathbf{b})$ if and only if $\mathbf{u}^* \in \mathcal{U}(\mathbf{b})$ and if $\mathbf{u} \in U$ satisfies $u_j \le 0$ for all $j \in Z(\mathbf{b} - \mathbf{u}^*)$, then $\sum_{i=1}^{m} u_i w_i \le 0$.*

Proof. If $\sum_{i=1}^{m} u_i w_i = 0$ for all $\mathbf{u} \in U$, then $P_{\mathcal{U}(\mathbf{b})}(\mathbf{b}) = \mathcal{U}(\mathbf{b})$ and the theorem, while correct, is rather meaningless. We therefore assume that $\sum_{i=1}^{m} u_i w_i \ne 0$ for some $\mathbf{u} \in U$.

($\Leftarrow$). Let $\mathbf{u} \in \mathcal{U}(\mathbf{b})$. Then $u_j - u_j^* \le 0$ for all $j \in Z(\mathbf{b} - \mathbf{u}^*)$. Therefore $\sum_{i=1}^{m} (u_i - u_i^*) w_i \le 0$, i.e., $\sum_{i=1}^{m} u_i w_i \le \sum_{i=1}^{m} u_i^* w_i$. From (6.9), $\mathbf{u}^* \in P_{\mathcal{U}(\mathbf{b})}(\mathbf{b})$.

($\Rightarrow$). Assume that $\mathbf{u}^* \in P_{\mathcal{U}(\mathbf{b})}(\mathbf{b})$. Since there exists a $\mathbf{u} \in U$ for which $\sum_{i=1}^{m} u_i w_i \ne 0$, it easily follows that $Z(\mathbf{b} - \mathbf{u}^*) \ne \emptyset$. Let $\mathbf{v} \in U$ satisfy $v_j \le 0$ for all $j \in Z(\mathbf{b} - \mathbf{u}^*)$, and $\sum_{i=1}^{m} v_i w_i > 0$. Then for some $\varepsilon > 0$, ε small,

$\mathbf{u}^* + \varepsilon\mathbf{v} \in \mathcal{U}(\mathbf{b})$ and $\sum_{i=1}^m (u_i^* + \varepsilon v_i)w_i > \sum_{i=1}^m u_i^* w_i$. This contradicts the fact that $\mathbf{u}^* \in P_{\mathcal{U}(\mathbf{b})}(\mathbf{b})$. □

A condition implying the uniqueness of the best one-sided $\ell_1^m(\mathbf{w})$ approximant to a fixed $\mathbf{b}$ is obtained from an extension of the above method of proof.

Proposition 6.21. *Assume U is an n-dimensional subspace of $\mathbb{R}^m$, $\mathbf{w} \in W$, $\mathbf{b} \in \mathbb{R}^m$, and $\mathcal{U}(\mathbf{b}) \neq \emptyset$. Then $P_{\mathcal{U}(\mathbf{b})}(\mathbf{b}) = \{\mathbf{u}^*\}$ if and only if $\mathbf{u}^* \in \mathcal{U}(\mathbf{b})$, and if $\mathbf{u} \in U$ satisfies $u_j \leq 0$ for all $j \in Z(\mathbf{b} - \mathbf{u}^*)$ and $\sum_{i=1}^m u_i w_i \geq 0$, then $\mathbf{u} = \mathbf{0}$.*

Proof. ($\Leftarrow$). If $\mathbf{u} \in U$ satisfies $u_j \leq 0$ for all $j \in Z(\mathbf{b}-\mathbf{u}^*)$, then $\sum_{i=1}^m u_i w_i \leq 0$. Thus from Theorem 6.20, we obtain $\mathbf{u}^* \in P_{\mathcal{U}(\mathbf{b})}(\mathbf{b})$. Assume $\tilde{\mathbf{u}} \in P_{\mathcal{U}(\mathbf{b})}(\mathbf{b})$. Set $\mathbf{u} = \tilde{\mathbf{u}} - \mathbf{u}^*$. Then $u_j \leq 0$ for all $j \in Z(\mathbf{b} - \mathbf{u}^*)$, and $\sum_{i=1}^m u_i w_i = 0$. Thus by assumption, $\mathbf{u} = \mathbf{0}$. Therefore $\mathbf{u}^* = \tilde{\mathbf{u}}$ and $P_{\mathcal{U}(\mathbf{b})}(\mathbf{b}) = \{\mathbf{u}^*\}$.
($\Rightarrow$). Assume $P_{\mathcal{U}(\mathbf{b})}(\mathbf{b}) = \{\mathbf{u}^*\}$. From Theorem 6.20, if $\mathbf{u} \in U$ satisfies $u_j \leq 0$ for all $j \in Z(\mathbf{b} - \mathbf{u}^*)$ then $\sum_{i=1}^m u_i w_i \leq 0$. If $\sum_{i=1}^m u_i w_i = 0$, then $\mathbf{u}^* + \varepsilon\mathbf{u} \in P_{\mathcal{U}(\mathbf{b})}(\mathbf{b})$ for some $\varepsilon > 0$, ε small. Since $P_{\mathcal{U}(\mathbf{b})}(\mathbf{b}) = \{\mathbf{u}^*\}$, we therefore have $\mathbf{u} = \mathbf{0}$. □

Even if the best one-sided $\ell_1^m(\mathbf{w})$ approximant to $\mathbf{b}$ from U is not unique, we can still say something about $P_{\mathcal{U}(\mathbf{b})}(\mathbf{b})$ and its extreme points (see Proposition 6.8 and Theorem 6.9).

Proposition 6.22. *Assume U is an n-dimensional subspace of $\mathbb{R}^m$, $\mathbf{w} \in W$, $\mathbf{b} \in \mathbb{R}^m$, and $\mathcal{U}(\mathbf{b}) \neq \emptyset$. Then $\mathbf{u}^*$ is an extreme point of $P_{\mathcal{U}(\mathbf{b})}(\mathbf{b})$ if and only if for any $\mathbf{u} \in U$ satisfying $u_j = 0$ for all $j \in Z(\mathbf{b} - \mathbf{u}^*)$, we have $\mathbf{u} = \mathbf{0}$.*

Proof. ($\Rightarrow$). Assume $\mathbf{u}^* \in P_{\mathcal{U}(\mathbf{b})}(\mathbf{b})$ and there exists a $\mathbf{u} \in U$, $\mathbf{u} \neq \mathbf{0}$, satisfying $u_j = 0$ for all $j \in Z(\mathbf{b} - \mathbf{u}^*)$. If $\sum_{i=1}^m u_i w_i \neq 0$, then we contradict Theorem 6.20 by taking $\mathbf{u}$ or $-\mathbf{u}$. Thus $\sum_{i=1}^m u_i w_i = 0$ and $\mathbf{u}^* \pm \varepsilon\mathbf{u} \in P_{\mathcal{U}(\mathbf{b})}(\mathbf{b})$ for all $\varepsilon > 0$ sufficiently small. Thus $\mathbf{u}^*$ is not an extreme point of $P_{\mathcal{U}(\mathbf{b})}(\mathbf{b})$.
($\Leftarrow$). Assume $\mathbf{u}^* \in P_{\mathcal{U}(\mathbf{b})}(\mathbf{b})$, but $\mathbf{u}^*$ is not an extreme point of $P_{\mathcal{U}(\mathbf{b})}(\mathbf{b})$. There then exists a $\mathbf{u} \in U$, $\mathbf{u} \neq \mathbf{0}$, and an $\varepsilon > 0$ such that $\mathbf{u}^* + \delta\mathbf{u} \in P_{\mathcal{U}(\mathbf{b})}(\mathbf{b})$ for all δ satisfying $|\delta| \leq \varepsilon$. If $u_j \neq 0$ for some $j \in Z(\mathbf{b} - \mathbf{u}^*)$, then either $\mathbf{u}^* + \varepsilon\mathbf{u}$ or $\mathbf{u}^* - \varepsilon\mathbf{u}$ is not in $\mathcal{U}(\mathbf{b})$. Thus $u_j = 0$ for all $j \in Z(\mathbf{b} - \mathbf{u}^*)$. □

As a consequence of this result we see that $|Z(\mathbf{b} - \mathbf{u}^*)| \geq n$ for every extreme point $\mathbf{u}^*$ of $P_{\mathcal{U}(\mathbf{b})}(\mathbf{b})$. We also have:

Proposition 6.23. *If $\mathcal{U}(\mathbf{b}) \neq \emptyset$, then $P_{\mathcal{U}(\mathbf{b})}(\mathbf{b})$ is a closed bounded convex set with a finite number of extreme points.*

Let us now turn to the question of conditions on U implying that U is a unicity space in the problem of best one-sided $\ell_1^m(\mathbf{w})$-approximation. In other words, conditions on U such that for every $\mathbf{b} \in \mathbb{R}^m$ with $\mathcal{U}(\mathbf{b}) \neq \emptyset$, $P_{\mathcal{U}(\mathbf{b})}(\mathbf{b})$ is a singleton.

Analogous to Theorem 5.10 we have.

Theorem 6.24. *Let U be an n-dimensional subspace of $\mathbb{R}^m$ and $\mathbf{w} \in W$. Then U is a unicity space for one-sided $\ell_1^m(\mathbf{w})$-approximation if and only if for each $\mathbf{u} \in U$, $\mathbf{u} \neq \mathbf{0}$, the zero vector is not in $P_{\mathcal{U}(|\mathbf{u}|)}(|\mathbf{u}|)$.*

The proof of Theorem 6.24 is exactly the same as the proof of Theorem 5.10. We shall therefore not repeat it here.

Using the definition of one-sided approximants, we rewrite Theorem 6.24 as follows.

Corollary 6.25. *Let U be an n-dimensional subspace of $\mathbb{R}^m$ and $\mathbf{w} \in W$. Then U is a unicity space for one-sided $\ell_1^m(\mathbf{w})$-approximation if and only if for each $\mathbf{u} \in U$, $\mathbf{u} \neq \mathbf{0}$, there exists a $\mathbf{v} \in U$ satisfying $v_j \leq 0$ for all $j \in Z(\mathbf{u})$ and $\sum_{i=1}^{m} v_i w_i > 0$.*

The verification of such a condition is generally very difficult. One simpler, but special, condition which implies its veracity is the following.

Proposition 6.26. *Assume U is an n-dimensional subspace of $\mathbb{R}^m$ and $\mathbf{w} \in W$. If there exists a $\widetilde{\mathbf{u}} \in U$ satisfying $\sum_{i=1}^{m} \widetilde{u}_i w_i \neq 0$, and if*

$$V = \{\mathbf{v} : \mathbf{v} \in U, \sum_{i=1}^{m} v_i w_i = 0\}$$

has the property that $|Z(\mathbf{v})| \leq n - 2$ for every $\mathbf{v} \in V$, $\mathbf{v} \neq \mathbf{0}$, then U is a unicity space for one-sided $\ell_1^m(\mathbf{w})$-approximation.

Proof. The first condition on U is certainly necessary. Otherwise $\sum_{i=1}^{m} u_i w_i = 0$ for all $\mathbf{u} \in U$, and U cannot possibly be a unicity space.

Now dim $V = n-1$. The condition $|Z(\mathbf{v})| \leq n-2$ for every $\mathbf{v} \in V$, $\mathbf{v} \neq \mathbf{0}$, is equivalent to the demand that given any $n-1$ distinct indices $\{i_j\}_{j=1}^{n-1}$ in $\{1, \ldots, m\}$ and $n-1$ arbitrary data $\{\alpha_{i_j}\}_{j=1}^{n-1}$, there exists a unique $\mathbf{v} \in V$ satisfying $v_{i_j} = \alpha_{i_j}$, $j = 1, \ldots, n-1$.

Let $\mathbf{u} \in U$. If $\sum_{i=1}^{m} u_i w_i \neq 0$, then by taking $\mathbf{u}$ or $-\mathbf{u}$ the conditions of Corollary 6.25 are immediately verified. It remains to prove that the conditions of Corollary 6.25 hold for $\mathbf{v} \in V$, $\mathbf{v} \neq \mathbf{0}$. To this end, let $\widetilde{\mathbf{u}} \in U$ satisfy $\sum_{i=1}^{m} \widetilde{u}_i w_i > 0$. For given $\mathbf{v} \in V$, $\mathbf{v} \neq \mathbf{0}$, we have $|Z(\mathbf{v})| \leq n-2$. There exists, by the above, a $\widetilde{\mathbf{v}} \in V$ satisfying $(\widetilde{u} + \widetilde{v})_j = 0$ for each $j \in Z(\mathbf{v})$. Furthermore $\sum_{i=1}^{m} (\widetilde{u}_i + \widetilde{v}_i) w_i = \sum_{i=1}^{m} \widetilde{u}_i w_i > 0$. This proves the result. □

The exact analogue of Theorem 5.17 gives us explicit conditions on U implying that U is a unicity space for one-sided $\ell_1^m(\mathbf{w})$-approximation for all $\mathbf{w} \in W$.

Theorem 6.27. *Assume U is an n-dimensional subspace of $\mathbb{R}^m$. U is a unicity space for one-sided $\ell_1^m(\mathbf{w})$-approximation for all $\mathbf{w} \in W$ if and only if*

U has a basis of vectors $\mathbf{u}^1, \ldots, \mathbf{u}^n$ *satisfying*

$$a) \, \mathbf{u}^i \geq \mathbf{0}, \, i = 1, \ldots, n$$

$$b) \, \mathrm{supp} \, \mathbf{u}^i \cap \mathrm{supp} \, \mathbf{u}^j = \emptyset, \, \text{all} \, i \neq j.$$

The proof of Theorem 6.27 is an exact parallel of the proof of Theorem 5.17. We deduce from Theorem 6.27 that very few subspaces U are unicity spaces in this one-sided problem for all $\mathbf{w} \in W$. We do nevertheless have the following analogue of Theorem 6.18.

Theorem 6.28. *Let U be an n-dimensional subspace of $\mathbb{R}^m$. Then, for almost all $\mathbf{w} \in W$, U is a unicity space in the problem of one-sided $\ell_1^m(\mathbf{w})$-approximation.*

Proof. We prove this result by showing that for almost all $\mathbf{w} \in W$ the conditions of Corollary 6.25 hold. That is, for almost all $\mathbf{w} \in W$, U has the property that given any $\mathbf{u} \in U$, $\mathbf{u} \neq \mathbf{0}$, there exists a $\mathbf{v} \in U$ satisfying $v_j \leq 0$, all $j \in Z(\mathbf{u})$, and $\sum_{i=1}^m v_i w_i > 0$. Let S be the set of all $\mathbf{s} = (s_1, \ldots, s_k)$, where the $\{s_i\}_{i=1}^k$ are distinct integers in $\{1, \ldots, m\}$, with the property that $\mathbf{s} \in S$ if and only if

$$U_{\mathbf{s}} = \{\mathbf{u} : \mathbf{u} \in U, \, u_{s_i} = 0, \, i = 1, \ldots, k\}$$

is one-dimensional. (k is not fixed.) In other words, to each $\mathbf{s} \in S$ there exists a $\mathbf{u}^{\mathbf{s}} \in U$, $\mathbf{u}^{\mathbf{s}} \neq \mathbf{0}$, which is unique up to multiplication by a non-zero constant, satisfying $u_{s_i}^{\mathbf{s}} = 0$, $i = 1, \ldots, k$. Thus for each $\mathbf{u} \in U$, $\mathbf{u} \neq \mathbf{0}$, we have $Z(\mathbf{u}) \subseteq Z(\mathbf{u}^{\mathbf{s}})$ for some $\mathbf{s} \in S$. The number of $\mathbf{u}^{\mathbf{s}}$, $\mathbf{s} \in S$ is finite. From Lemma 6.19 we have that for almost all $\mathbf{w} \in W$

$$(6.10) \qquad \sum_{i=1}^m u_i^{\mathbf{s}} w_i \neq 0$$

for all $\mathbf{s} \in S$. Let $\mathbf{w}$ satisfy (6.10) for all $\mathbf{s} \in S$. For $\mathbf{u} \in U$, $\mathbf{u} \neq \mathbf{0}$, let $\mathbf{s} \in S$ satisfy $Z(\mathbf{u}) \subseteq Z(\mathbf{u}^{\mathbf{s}})$. Thus $u_j^{\mathbf{s}} = 0$, all $j \in Z(\mathbf{u})$, and therefore $\sum_{i=1}^m u_i^{\mathbf{s}} w_i \neq 0$. Multiplying $\mathbf{u}^{\mathbf{s}}$ by -1, if necessary, we verify the conditions of Corollary 6.25. This proves our theorem.   □

Exercises

1. Let U be a finite-dimensional subspace of $\mathbb{R}^m$, $\mathbf{w} \in W$ and $\mathbf{b} \in \mathbb{R}^m$. Assume that $\mathbf{u}^1, \mathbf{u}^2 \in P_U(\mathbf{b})$. Prove that

$$a) \, (b_i - u_i^1)(b_i - u_i^2) \geq 0, \, i = 1, \ldots, m$$

$$b) \, \sum_{i=1}^m \mathrm{sgn}(b_i - u_i^1)(u_i^2 - u_i^1) w_i = \sum_{i \in Z(\mathbf{b} - \mathbf{u}^1)} |u_i^2 - u_i^1| w_i.$$

2. Let $U = \mathrm{span}\{(1, 2, 3)\}$ and $\mathbf{w} = (1, 1, 1)$. Prove that, for any $\mathbf{b} \in \mathbb{R}^3$ a best $\ell_1^3(\mathbf{w})$ approximant to $\mathbf{b}$ from U is given by $\mathbf{u}^* = b_3(1/3, 2/3, 1)$. Prove

that $\mathbf{u}^*$ is not the unique best $\ell_1^3(\mathbf{w})$ approximant to $\mathbf{b}$ from U if and only if $(3b_1 - b_3)(3b_2 - 2b_3) > 0$.

3. Let $\dim U = n$, $U \in \mathbb{R}^{n+1}$, and $\mathbf{w} \in W$. Prove that either U is a unicity space for $\ell_1^{n+1}(\mathbf{w})$, or for every $\mathbf{b} \in \mathbb{R}^{n+1}\backslash U$ the set $P_U(\mathbf{b})$ is not a singleton.

4. Construct a two-dimensional subspace $U = \text{span}\{\mathbf{u}^1, \mathbf{u}^2\}$ of $\mathbb{R}^m$ (some m), a $\mathbf{b} \in \mathbb{R}^m$ and $\mathbf{u}^* \in U$ such that

$$\left| \sum_{i=1}^m \text{sgn}(b_i - u_i^*)u_i^j \right| \leq \sum_{i \in Z(\mathbf{b}-\mathbf{u}^*)} |u_i^j|$$

for $j = 1, 2$, and yet $\mathbf{u}^*$ is not a best $\ell_1^m(\mathbf{e})$ approximant to $\mathbf{b}$ from U.

5. Let U be an n-dimensional subspace of $\mathbb{R}^m$ and $\mathbf{w} \in W$. Let $\mathbf{b} \in \mathbb{R}^m$ and $\mathbf{u}^* \in U$ be such that $Z(\mathbf{b} - \mathbf{u}^*) = \{1,\ldots,n\}$. Assume that if $\mathbf{u} \in U$ satisfies $u_i = 0$, $i = 1,\ldots,n$, then $\mathbf{u} = \mathbf{0}$. Let $\mathbf{u}^j \in U$, $j = 1,\ldots,n$ be defined by $u_i^j = \delta_{ij}$, $i,j = 1,\ldots,n$. Prove that $\mathbf{u}^* \in P_U(\mathbf{b})$ if and only if

$$\left| \sum_{i=n+1}^m \text{sgn}(b_i - u_i^*)u_i^j w_i \right| \leq 1$$

for each $j = 1,\ldots,n$.

6. Let U be an n-dimensional subspace of $\mathbb{R}^m$ and $\mathbf{w} \in W$. Let $\mathbf{b} \in \mathbb{R}^m$ and $\mathbf{u}^* \in U$ be such that $Z(\mathbf{b} - \mathbf{u}^*) = \{1,\ldots,s\}$ where $s \geq n$. Assume that if $\mathbf{u} \in U$ satisfies $u_i = 0$, $i = 1,\ldots,n$, then $\mathbf{u} = \mathbf{0}$. Let $\mathbf{u}^j \in U$, $j = 1,\ldots,n$ be defined by $u_i^j = \delta_{ij}$, $i,j = 1,\ldots,n$. Prove that $\mathbf{u}^* \in P_U(\mathbf{b})$ if and only if there exist $\{y_i\}_{i=n+1}^s$ for which $|y_i| \leq 1$, $i = n+1,\ldots,s$, and

$$\left| \sum_{i=s+1}^m \text{sgn}(b_i - u_i^*)u_i^j w_i + \sum_{i=n+1}^s y_i u_i^j w_i \right| \leq 1,$$

for each $j = 1,\ldots,n$. (Hint: Use Proposition 6.2.)

7. Prove Proposition 6.6.

8. Let U be an n-dimensional subspace of $\mathbb{R}^m$ and $\mathbf{w} \in W$. Prove that U is a unicity space for $\ell_1^m(\mathbf{w})$ if and only if there does not exist a $\mathbf{y} \in \mathbb{R}^m$ and $\mathbf{u}^* \in U$, $\mathbf{u}^* \neq \mathbf{0}$, satisfying

$$1)\ \|\mathbf{y}\|_\infty = 1$$
$$2)\ (\mathbf{y}, \mathbf{u})_\mathbf{w} = 0, \quad \text{all } \mathbf{u} \in U$$
$$3)\ |y_i| = 1 \text{ if } u_i^* \neq 0.$$

9. Let U be an n-dimensional subspace of $\mathbb{R}^m$. Let $\mathbf{b} \in \mathbb{R}^m$ and $\mathbf{w} \in W$. Prove that $\mathbf{u}^*$ is a best one-sided $\ell_1^m(\mathbf{w})$ approximant to $\mathbf{b}$ from U if and only if $\mathbf{u}^* \in \mathcal{U}(\mathbf{b})$ and there exist indices $\{j_r\}_{r=1}^k$ in $\{1,\ldots,m\}$, $0 \leq k \leq n$, and positive numbers $\{\lambda_r\}_{r=1}^k$ for which

$$a)\ b_{j_r} - u_{j_r}^* = 0, \quad r = 1,\ldots,k,$$
$$b)\ \sum_{i=1}^m u_i w_i = \sum_{r=1}^k \lambda_r u_{j_r}, \quad \text{all } \mathbf{u} \in U.$$

(Remark: This is the finite-dimensional analogue of Theorem 5.2. Note that the assumptions on U in Theorem 5.2 have been dropped. Also, if $k = 0$, we understand the above to simply mean that $\sum_{i=1}^{m} u_i w_i = 0$ for all $\mathbf{u} \in U$.)

10. Let U be an n-dimensional subspace of $\mathbb{R}^m$. Given $\mathbf{b} \in \mathbb{R}^m$, assume that $\mathcal{U}(\mathbf{b}) \neq \emptyset$. For $\mathbf{w} \in W$, prove that

$$\max \left\{ \sum_{i=1}^{m} u_i w_i \; : \; \mathbf{u} \in \mathcal{U}(\mathbf{b}) \right\} =$$

$$\min \left\{ \sum_{i=1}^{m} b_i c_i w_i \; : \; c_i \geq 0, \, i = 1, \ldots, n, \; \sum_{i=1}^{m} u_i w_i = \sum_{i=1}^{m} u_i c_i w_i, \text{ all } \mathbf{u} \in U \right\}.$$

(Hint: Use Exercise 9.)

11. Prove that the analogue of Theorem 5.13 is not valid here.

12. Prove Theorem 6.24.

13. Prove Theorem 6.27.

Notes and References

As noted in the introduction, much of the material of this chapter is a generalization or particularization of results contained in Chapters 2–5. Lemma 6.4 (and Theorem 6.5) is in Rivlin [1969, Theorem 3.6]. Although it is only stated for algebraic polynomials, the proof is exactly the same. The equivalence of uniqueness and strong uniqueness as contained in Proposition 6.6 was known, (see Watson [1980, p.122] or Angelos, Schmidt [1983]). Propositions 6.7 and 6.8, and Theorem 6.9 are contained in various forms in Rice [1964], Rivlin [1969], and Watson [1980]. Proposition 6.10 is new, while Proposition 6.11 is to be found in Duris, Sreedharan [1968], see also Rivlin [1969, p.85]. Theorem 6.12 is a special case of a theorem in Lin [1985]. The proof therein was somewhat lacking. Both Proposition 6.13 and Theorem 6.14 are particular instances of a general result of Brown [1964]. Although we have tailored the proofs to the ℓ_1^m-norm, the more general case in Brown [1964] is both elegant and simple. Theorem 6.15 is a variant of Theorem 2.8 and is due to Phelps [1966]. Both Theorems 6.16 and 6.18 are new. Thanks are due to N. Alon for help in proving Theorem 6.18. The results of Section 3 are variants of results of Chapter 5, except for Proposition 6.26 and Theorem 6.28 which are new. Exercises 5 and 6 are the contents of Powell, Roberts [1980].

7
Algorithms

---

1. Introduction

In this final chapter we present various ideas for calculating best two-sided and one-sided L^1 approximants. The term algorithm is perhaps overly employed and too ambiguous. What we shall describe are ideas and procedures. There are no flow charts or computer programs. In other words, we consider the algorithms from a mathematical, rather than computational, point of view. We shall not even discuss stability or convergence rates. Some of the algorithms described herein may be very difficult to implement for various and sundry reasons. It is more our intention to illustrate various possible paths. The reader who is seriously interested in their implementation should consult the references at the end of the chapter.

The organization of this chapter is as follows. Section 2 contains a short discussion on gradients and subgradients. This material is used at various subsequent stages, and we felt that it was worth both separating out and highlighting. In Section 3 we very briefly discuss the problem of determining best one-sided ℓ_1^m approximants, while in Section 4 we present algorithms for best two-sided ℓ_1^m approximants. In Chapter 6 we consistently used arbitrary weight functions $\mathbf{w} \in W$. This cumbersome notation was employed because of the nature of some of the problems discussed. In Sections 3 and 4 we always set $\mathbf{w} = (1, \ldots, 1)$ since the weight functions have no bearing on the algorithms. Sections 5 and 6 are concerned with algorithms for calculating best one-sided and two-sided L^1 approximants, respectively, in the case of continuous functions. Various different approaches to the problem are discussed in each of these sections.

2. Gradients and Subgradients

In this section we introduce certain concepts which we hope will provide us with a perspective and insight into the problem of calculating best approximants. The material of this section will be mainly, but not only, used in developing algorithms for the two-sided approximation problem.

Let X be a normed linear space over $I\!R$, and $U = \operatorname{span}\{u_1, \ldots, u_n\}$ an n-dimensional subspace of X. For each $f \in X$, set

$$H(\mathbf{a}) = \left\| f - \sum_{i=1}^{n} a_i u_i \right\|$$

where $\mathbf{a} = (a_1, \ldots, a_n) \in \mathbb{R}^n$. In Exercise 12 of Chapter 1, we noted that H is continuous, convex, and $\lim_{|||\mathbf{a}||| \to \infty} H(\mathbf{a}) = \infty$, where $||| \cdot |||$ is any norm on $\mathbb{R}^n$. Thus the unconstrained (two-sided) problem of determining a best approximant to f from U is equivalent to that of finding the minimum of a given convex function H. (The one-sided approximation problem is an example of a problem of finding the minimum of H, restricted to a closed convex domain of definition.) This seemingly innocuous problem of constructively determining the minimum of a convex function is a major hurdle and should in no way be underestimated (except in the very much simpler case where $n = 1$). The study of this problem leads us to the important concepts of *gradients* and *subgradients*.

Let us suppose, more generally, that H is any convex function from $\mathbb{R}^n$ to $\mathbb{R}$, with the property that $\lim_{|||\mathbf{a}||| \to \infty} H(\mathbf{a}) = \infty$. From general considerations it follows that H is continuous and also that H is almost everywhere differentiable. The subgradients of H at $\mathbf{a}$ are defined as follows:

Definition 7.1. Let H be as above and $\mathbf{a} \in \mathbb{R}^n$. A vector $\mathbf{g} \in \mathbb{R}^n$ is said to be a *subgradient* to H at $\mathbf{a}$ if

$$H(\mathbf{b}) - H(\mathbf{a}) \geq (\mathbf{g}, \mathbf{b} - \mathbf{a})$$

for all $\mathbf{b} \in \mathbb{R}^n$ (where $(\cdot, \cdot)$ is the usual inner product of vectors in $\mathbb{R}^n$). We let $G(\mathbf{a})$ denote the set of subgradients to H at $\mathbf{a}$.

Each element of $G(\mathbf{a})$ corresponds to a supporting hyperplane to H at $\mathbf{a}$. Since H is convex, $G(\mathbf{a})$ is non-empty. Furthermore, as is readily shown, the set $G(\mathbf{a})$ is bounded, closed, and convex for each $\mathbf{a} \in \mathbb{R}^n$.

A gradient is simply defined.

Definition 7.2. Let H be as above and $\mathbf{a} \in \mathbb{R}^n$. If $G(\mathbf{a})$ is a singleton, then this singleton is called the *gradient* to H at $\mathbf{a}$.

Thus a gradient to H exists at $\mathbf{a}$ if and only if there is a unique supporting hyperplane to H at $\mathbf{a}$. As a consequence of Proposition 7.2, the above definition of a gradient implies the existence of the partial derivatives to H at $\mathbf{a}$.

Let us now deduce the usual simple criterion for determining when $\mathbf{a}^*$ is a minimum point of H. Such a minimum point exists.

Proposition 7.1. *Let H be as above and $\mathbf{a}^* \in \mathbb{R}^n$. Then $\mathbf{a}^*$ is a minimum point of H if and only if $\mathbf{0} \in G(\mathbf{a}^*)$.*

Proof. ($\Rightarrow$). If $\mathbf{a}^*$ is a minimum of H, then

$$H(\mathbf{b}) - H(\mathbf{a}^*) \geq 0$$

for all $\mathbf{b} \in \mathbb{R}^n$. Thus

$$H(\mathbf{b}) - H(\mathbf{a}^*) \geq (\mathbf{0}, \mathbf{b} - \mathbf{a}^*)$$

for all $\mathbf{b} \in \mathbb{R}^n$, and therefore $\mathbf{0} \in G(\mathbf{a}^*)$.

($\Leftarrow$). If $\mathbf{0} \in G(\mathbf{a}^*)$, then

$$H(\mathbf{b}) - H(\mathbf{a}^*) \geq (\mathbf{0}, \mathbf{b} - \mathbf{a}^*) = 0$$

for all $\mathbf{b} \in \mathbb{R}^n$. Thus $H(\mathbf{b}) \geq H(\mathbf{a}^*)$ for all $\mathbf{b} \in \mathbb{R}^n$. □

Since $G(\mathbf{a})$ is a compact convex set, it is uniquely determined by its extreme points. These extreme points are related to one-sided directional derivatives as follows.

Proposition 7.2. *Let H be as above and $\mathbf{a} \in \mathbb{R}^n$. For each $\mathbf{d} \in \mathbb{R}^n$*

$$\lim_{t \to 0^+} \frac{H(\mathbf{a} + t\mathbf{d}) - H(\mathbf{a})}{t} = H'_{\mathbf{d}}(\mathbf{a})$$

exists. Furthermore,

$$H'_{\mathbf{d}}(\mathbf{a}) = \max\{(\mathbf{g}, \mathbf{d}) : \mathbf{g} \in G(\mathbf{a})\}.$$

Proof. The first part of this result is a generalization of Proposition 1.4. Set

$$r(t) = \frac{H(\mathbf{a} + t\mathbf{d}) - H(\mathbf{a})}{t}.$$

We verify that the above limit exists by proving that $r(t)$ is non-decreasing and bounded below on $(0, \infty)$.

Let $0 < s < t < \infty$. Since $s/t \in (0,1)$ and H is convex,

$$H(\mathbf{a} + s\mathbf{d}) = H\left(\frac{s}{t}(\mathbf{a} + t\mathbf{d}) + \left(1 - \frac{s}{t}\right)\mathbf{a}\right) \leq \frac{s}{t}H(\mathbf{a} + t\mathbf{d}) + \left(1 - \frac{s}{t}\right)H(\mathbf{a}).$$

Thus

$$t[H(\mathbf{a} + s\mathbf{d}) - H(\mathbf{a})] \leq s[H(\mathbf{a} + t\mathbf{d}) - H(\mathbf{a})],$$

whence $r(s) \leq r(t)$. Now, take any $\mathbf{g} \in G(\mathbf{a})$. By definition,

$$H(\mathbf{a} + t\mathbf{d}) - H(\mathbf{a}) \geq (\mathbf{g}, t\mathbf{d}) = t(\mathbf{g}, \mathbf{d}).$$

Thus $r(t) \geq (\mathbf{g}, \mathbf{d})$ for all $t \in (0, \infty)$. Since $r(t)$ is bounded below, the desired limit exists.

The above also confirms that $H'_{\mathbf{d}}(\mathbf{a}) \geq (\mathbf{g}, \mathbf{d})$ for every $\mathbf{g} \in G(\mathbf{a})$. Since $G(\mathbf{a})$ is compact, we therefore have

$$H'_{\mathbf{d}}(\mathbf{a}) \geq \max\{(\mathbf{g}, \mathbf{d}) : \mathbf{g} \in G(\mathbf{a})\}.$$

It remains to prove that equality holds.

Let $W_1, W_2 \subset \mathbb{R}^{n+1}$ be defined by

$$W_1 = \{(\mathbf{b}, y) : \mathbf{b} \in \mathbb{R}^n, y \geq H(\mathbf{b})\}$$
$$W_2 = \{(\mathbf{a} + t\mathbf{d}, H(\mathbf{a}) + tH'_{\mathbf{d}}(\mathbf{a})) : t \geq 0\}.$$

Since H is convex, it is easily seen that W_1 is convex. (W_1 is called the epigraph of H.) W_2 is a ray. Furthermore, since $r(t) \geq H'_{\mathbf{d}}(\mathbf{a})$ for all $t > 0$, it follows that W_2 contains no point in the interior of W_1. However $(\mathbf{a}, H(\mathbf{a})) \in W_1 \cap W_2$. There therefore exists a $\widetilde{\mathbf{g}} = (\mathbf{g}, g_{n+1}) \in \mathbb{R}^{n+1} \backslash \{0\}$ such that

$$(\mathbf{g}, \mathbf{b} - \mathbf{a}) + g_{n+1}(y - H(\mathbf{a})) \geq 0 \geq (\mathbf{g}, t\mathbf{d}) + g_{n+1}tH'_{\mathbf{d}}(\mathbf{a})$$

for all $\mathbf{b} \in \mathbb{R}^n$, $y \geq H(\mathbf{b})$, and $t \geq 0$. Set $\mathbf{b} = \mathbf{a}$. From the left-hand-side of the inequality, $g_{n+1}(y - H(\mathbf{a})) \geq 0$ for all $y \geq H(\mathbf{a})$ which implies that $g_{n+1} \geq 0$. If $g_{n+1} = 0$, then $(\mathbf{g}, \mathbf{b} - \mathbf{a}) \geq 0$ for all $\mathbf{b} \in \mathbb{R}^n$, and therefore $\mathbf{g} = \mathbf{0}$. This contradicts our assumption that $\widetilde{\mathbf{g}} \neq \mathbf{0}$. Thus $g_{n+1} > 0$. Set $\mathbf{g}^* = -\mathbf{g}/g_{n+1}$. Then

$$H(\mathbf{b}) - H(\mathbf{a}) \geq (\mathbf{g}^*, \mathbf{b} - \mathbf{a})$$

for all $\mathbf{b} \in \mathbb{R}^n$, and

$$(\mathbf{g}^*, \mathbf{d}) \geq H'_{\mathbf{d}}(\mathbf{a}).$$

From the first inequality, $\mathbf{g}^* \in G(\mathbf{a})$. The second inequality implies that

$$\max\{(\mathbf{g}, \mathbf{d}) : \mathbf{g} \in G(\mathbf{a})\} = H'_{\mathbf{d}}(\mathbf{a}). \qquad \square$$

As a consequence of Proposition 7.2, we have:

Proposition 7.3. *Let H be as above and $\mathbf{a} \in \mathbb{R}^n$. Assume there exists a $\mathbf{d} \in \mathbb{R}^n$ for which*

$$\max\{(\mathbf{g}, \mathbf{d}) : \mathbf{g} \in G(\mathbf{a})\} < 0.$$

Then for all $t > 0$, t sufficiently small,

$$H(\mathbf{a} + t\mathbf{d}) < H(\mathbf{a}).$$

Conversely, if $H(\mathbf{a} + t_0\mathbf{d}) < H(\mathbf{a})$ for some $\mathbf{d} \in \mathbb{R}^n$ and $t_0 > 0$, then

$$\max\{(\mathbf{g}, \mathbf{d}) : \mathbf{g} \in G(\mathbf{a})\} < 0.$$

Proof. From Proposition 7.2,

$$\max\{(\mathbf{g}, \mathbf{d}) : \mathbf{g} \in G(\mathbf{a})\} = H'_{\mathbf{d}}(\mathbf{a}).$$

If $H'_{\mathbf{d}}(\mathbf{a}) < 0$, then

$$H(\mathbf{a} + t\mathbf{d}) < H(\mathbf{a})$$

for all $t > 0$, t sufficiently small.

Conversely, assume

$$H(\mathbf{a} + t_0\mathbf{d}) < H(\mathbf{a})$$

for some $\mathbf{d} \in \mathbb{R}^n$ and $t_0 > 0$. Then $r(t_0) < 0$. From the proof of Proposition 7.2, $r(t)$ is a non-decreasing function. Thus $H'_{\mathbf{d}}(\mathbf{a}) < 0$. $\qquad \square$

Based on the above, we call **d** a *descent direction* if $H'_\mathbf{d}(\mathbf{a}) < 0$. We can now discern the germ of an idea behind the construction of algorithms for this problem. The general plan is to find a good descent direction, and to employ this information in an efficient manner. Let us assume that we can find $G(\mathbf{a})$ or at least a descent direction. There then exist algorithms for finding the minimum of H (see Notes and References). Unfortunately these algorithms tend to be very poorly behaved, and one should certainly use any and all additional information that one has concerning H. If H is twice differentiable, then there are many good known algorithms, most based on generalizations of Newton's method (using the Hessian matrix) which are quite satisfactory in general. Newton's method essentially gives us an 'optimal' local descent direction. If H is only continuously differentiable, the situation is somewhat less satisfactory. If H is only convex and $G(\mathbf{a})$ is not a singleton for all $\mathbf{a} \in \mathbb{R}^n$, then we may expect serious difficulties. This last case is all too frequent when dealing with L^1-approximation.

Let us now assume that $f \in L^1(B, \nu)$ where ν is a positive measure (see Chapter 2), and $U = \text{span}\{u_1, \ldots, u_n\}$ is an n-dimensional subspace of $L^1(B, \nu)$. As above, set

$$H(\mathbf{a}) = \left\| f - \sum_{i=1}^{n} a_i u_i \right\|_1 ,$$

where $\mathbf{a} = (a_1, \ldots, a_n)$, and

$$\|f\|_1 = \int_B |f| d\nu .$$

We end this section by identifying $G(\mathbf{a})$.

Proposition 7.4. *Let* $\mathbf{a} \in \mathbb{R}^n$. *Then* $G(\mathbf{a})$ *is the set of all vectors* $\mathbf{g} = (g_1, \ldots, g_n)$, *where*

$$g_j = \int_{Z(f - \Sigma a_i u_i)} h u_j d\nu - \int_B \text{sgn}\left(f - \sum_{i=1}^{n} a_i u_i\right) u_j d\nu , \quad j = 1, \ldots, n,$$

and h *is any* $L^\infty(B, \nu)$ *function satisfying* $|h| \le 1$ ν *a.e. on* $Z(f - \sum_{i=1}^{n} a_i u_i)$.

Proof. Let $\mathbf{g} = (g_1, \ldots, g_n)$ be as in the statement of the proposition. For every $\mathbf{b} \in \mathbb{R}^n$,

$$(\mathbf{g}, \mathbf{b} - \mathbf{a}) = \sum_{j=1}^{n} g_j(b_j - a_j)$$

$$= \int_{Z(f - \Sigma a_i u_i)} h \left(\sum_{j=1}^{n} b_j u_j - \sum_{j=1}^{n} a_j u_j \right) d\nu$$

$$- \int_B \text{sgn}\left(f - \sum_{i=1}^{n} a_i u_i\right) \left(\sum_{j=1}^{n} b_j u_j - \sum_{j=1}^{n} a_j u_j \right) d\nu$$

$$= \int_{Z(f-\Sigma a_i u_i)} h\Big(\sum_{j=1}^{n} b_j u_j - f\Big)d\nu + \int_{B} \operatorname{sgn}(f - \sum_{i=1}^{n} a_i u_i)(f - \sum_{j=1}^{n} b_j u_j)d\nu$$

$$- \int_{B} \operatorname{sgn}(f - \sum_{i=1}^{n} a_i u_i)(f - \sum_{j=1}^{n} a_j u_j)d\nu$$

$$\le \Big\| f - \sum_{j=1}^{n} b_j u_j \Big\|_1 - \Big\| f - \sum_{j=1}^{n} a_j u_j \Big\|_1$$

$$= H(\mathbf{b}) - H(\mathbf{a}).$$

Each $\mathbf{g}$, as above, is a subgradient to H at $\mathbf{a}$.

Let $\widetilde{G}$ denote the set of all such $\mathbf{g}$. Then $\widetilde{G} \subseteq G(\mathbf{a})$, and $\widetilde{G}$ is both convex and compact. If $\widetilde{G} \ne G(\mathbf{a})$, there exists a $\mathbf{g}^* \in G(\mathbf{a})$ and a $\mathbf{d} \in \mathbb{R}^n$ for which

$$(\mathbf{g}, \mathbf{d}) < (\mathbf{g}^*, \mathbf{d})$$

for all $\mathbf{g} \in \widetilde{G}$. Thus

$$\max\{(\mathbf{g}, \mathbf{d}) : \mathbf{g} \in \widetilde{G}\} < \max\{(\mathbf{g}, \mathbf{d}) : \mathbf{g} \in G(\mathbf{a})\} = H'_{\mathbf{d}}(\mathbf{a}).$$

From the first method of proof of Theorem 2.1, we have that

$$H'_{\mathbf{d}}(\mathbf{a}) = \int_{Z(f-\Sigma a_i u_i)} \Big| \sum_{j=1}^{n} d_j u_j \Big| d\nu - \int_{B} \operatorname{sgn}(f - \sum_{i=1}^{n} a_i u_i) \sum_{j=1}^{n} d_j u_j d\nu .$$

Let $\mathbf{g} \in \widetilde{G}$ be as in the statement of the proposition with $h = \operatorname{sgn}(\sum_{j=1}^{n} d_j u_j)$ on $Z(f - \sum_{i=1}^{n} a_i u_i)$. Then $H'_{\mathbf{d}}(\mathbf{a}) = (\mathbf{g}, \mathbf{d})$. This contradicts the above strict inequality and therefore $\widetilde{G} = G(\mathbf{a})$. □

Note that $\mathbf{d}$ is a descent direction if and only if

$$\int_{B} \operatorname{sgn}(f - \sum_{i=1}^{n} a_i u_i) \sum_{j=1}^{n} d_j u_j d\nu > \int_{Z(f-\Sigma a_i u_i)} \Big| \sum_{j=1}^{n} d_j u_j \Big| d\nu .$$

This is yet another explanation of Theorem 2.1. Also note that H has a gradient at $\mathbf{a}$ if and only if $\nu(Z(f - \sum_{i=1}^{n} a_i u_i)) = 0$. It is then given by $\mathbf{g} = (g_1, \ldots, g_n)$ where

$$g_j = -\int_{B} \operatorname{sgn}(f - \sum_{i=1}^{n} a_i u_i) u_j d\nu ,$$

$j = 1, \ldots, n$.

In the above, we chose a particular basis for U. If we choose a different basis, then nothing is changed. However here we must introduce a caveat. Certain authors also define a *direction of steepest descent* to H at $\mathbf{a}$ to be the directions $\mathbf{d}^*$ ($\|\mathbf{d}^*\|_2 = 1$) for which

$$H'_{\mathbf{d}^*}(\mathbf{a}) = \min\{H'_{\mathbf{d}}(\mathbf{a}) : \|\mathbf{d}\|_2 = 1\}.$$

Thus if H has a gradient $\mathbf{g}$ at $\mathbf{a}$, then $\mathbf{d}^* = -\alpha\mathbf{g}$, for some $\alpha > 0$. It is not difficult to show that the directions of steepest descent are basis dependent (even in the case of the existence of a gradient). The reason for this is essentially that the above definition uses the ℓ_2^n-norm and is not norm independent. By changing our basis we have changed our norm on the coefficient vectors.

3. One-Sided ℓ_1^m-Approximation

The problem of best ℓ_1^m-approximation (both one- and two-sided) is a *linear programming* problem. No more and no less. This is especially fruitful when looking at the one-sided ℓ_1^m-approximation problem where the correspondance is rather simple. To explain, let $U = \text{span}\{\mathbf{u}^1, \ldots, \mathbf{u}^n\}$ be an n-dimensional subspace of $\mathbb{R}^m$ and $\mathbf{b} \in \mathbb{R}^m$. Assume $\mathcal{U}(\mathbf{b}) \neq \emptyset$, i.e., there exists a $\mathbf{u} \in U$ for which $\mathbf{u} \leq \mathbf{b}$. Our one-sided ℓ_1^m-approximation problem is then

$$\min\Big\{\sum_{i=1}^m |b_i - u_i| : \mathbf{u} = (u_1, \ldots, u_m) \in \mathcal{U}(\mathbf{b})\Big\}.$$

Since $\mathbf{b}$ is fixed and $b_i - u_i \geq 0$ for all $i = 1, \ldots, m$ and all $\mathbf{u} \in \mathcal{U}(\mathbf{b})$, $\mathbf{u}^*$ attains the above minimum if and only if it attains the maximum in

$$\max\Big\{\sum_{i=1}^m u_i : \mathbf{u} \in \mathcal{U}(\mathbf{b})\Big\}.$$

Because $\mathbf{u}^1, \ldots, \mathbf{u}^n$ is a basis for U, we can rewrite this in the form

$$\max \sum_{i=1}^m \sum_{j=1}^n a_j u_i^j$$

subject to : $\quad \sum_{j=1}^n a_j u_i^j \leq b_i, \; i = 1, \ldots, m.$

To put this in a somewhat more standard and convenient form, set $d_j = \sum_{i=1}^m u_i^j$, $j = 1, \ldots, n$ and $a_j = a_j^1 - a_j^2$, $j = 1, \ldots, n$. Then the above problem is equivalent to

$$\max \sum_{j=1}^n (a_j^1 - a_j^2)d_j$$

subject to:

$$\sum_{j=1}^n (a_j^1 - a_j^2)u_i^j \leq b_i, \quad i = 1, \ldots, m$$

$$a_j^1, a_j^2 \geq 0, \quad j = 1, \ldots, n.$$

This is a form which may be found in any book on linear programming. There is also a dual form to this problem and it is given by

$$\min \sum_{i=1}^m b_i c_i$$

subject to:

$$\sum_{i=1}^{m} c_i u_i^j = d_j, \qquad j = 1, \ldots, n$$

$$c_i \geq 0, \qquad i = 1, \ldots, m.$$

The equivalence of these two forms is a standard result in linear programming. We have moreover stated it in Exercise 10 of Chapter 6. It should be noted that this duality does not always hold for the continuous version of this problem. This is discussed in Part II of Appendix B.

We claim that the one-sided ℓ_1^m-approximation problem is an almost totally general form of a linear programming problem. To demonstrate this fact, consider any linear programming problem of the form

$$\max \sum_{j=1}^{n} a_j p_j$$

subject to :

$$\sum_{j=1}^{n} a_j u_i^j \leq b_i, \quad i = 1, \ldots, m.$$

The equivalence holds under certain minor restrictions. These restrictions are:
  1) There exist $\{a_j\}_{j=1}^{n}$ satisfying $\sum_{j=1}^{n} a_j u_i^j \leq b_i$, $i = 1, \ldots, m$.
  2) The maximum is in fact attained (the solution value is not ∞).
  3) The solution set is bounded.
   To verify this equivalence, note that if there exists a $\mathbf{d} = (d_1, \ldots, d_n) \neq \mathbf{0}$ satisfying

(7.1)

$$a) \sum_{j=1}^{n} d_j u_i^j \leq 0, \quad i = 1, \ldots, m$$

$$b) \sum_{j=1}^{n} d_j p_j \geq 0,$$

then either condition (2) or (3) is violated. Thus there exists no $\mathbf{d} \neq \mathbf{0}$ satisfying (7.1). Let $\widetilde{U}$ be the $(m+1) \times n$ matrix whose first m rows are simply the vectors $(u_i^1, \ldots, u_i^n)$, $i = 1, \ldots, m$, and whose last row is the vector $(-p_1, \ldots, -p_n)$. Thus the closed convex sets

$$A = \{\widetilde{U}\mathbf{d} : \mathbf{d} \in I\!\!R^n\}$$

and

$$B = \{\mathbf{x} : x_i \leq 0, \ i = 1, \ldots, m\}$$

satisfy $B \cap A = \{\mathbf{0}\}$. There therefore exists a $\mathbf{w} \in I\!\!R^{m+1}$ for which

$$(\mathbf{w}, \mathbf{x}) < 0 \leq (\mathbf{w}, \widetilde{U}\mathbf{d})$$

for all $\mathbf{x} \in B\backslash\{0\}$ and $\mathbf{d} \in I\!\!R^n$. This implies that $w_i > 0$, $i = 1, \ldots, m+1$ and $\mathbf{w}\tilde{U} = \mathbf{0}$. We may assume that $w_{m+1} = 1$. Rewriting $\mathbf{w}\tilde{U} = \mathbf{0}$, we have

$$\sum_{i=1}^{m} w_i u_i^j = p_j, \quad j = 1, \ldots, n.$$

Substituting in our original problem, we obtain its equivalence to:

$$\max \sum_{i=1}^{m} u_i w_i$$

subject to:

$$u_i \leq b_i, \quad i = 1, \ldots, m$$

$$\mathbf{u} = (u_1, \ldots, u_m) \in U,$$

where $U = \text{span}\{\mathbf{u}^1, \ldots, \mathbf{u}^n\}$. Since $w_i > 0$, $i = 1, \ldots, m$, this is just the one-sided ℓ_1^m-approximation problem with weight $\mathbf{w} = (w_1, \ldots, w_m)$.

Since our approximation problem is an almost totally general form of a linear programming problem, we refer the reader to any standard text on linear programming. Many different algorithms exist for solving linear programming problems. We shall not discuss the various algorithms here as they are beyond the scope of this work and it would be presumptuous of us to try to repeat in a few pages the contents of these myriad books.

4. Two-Sided ℓ_1^m-Approximation

The problem of two-sided ℓ_1^m-approximation can also be shown to be a linear programming problem. This fact is almost as straightforward as in the previous section. Let $U = \text{span}\{\mathbf{u}^1, \ldots, \mathbf{u}^n\}$ be an n-dimensional subspace of $I\!\!R^m$ and $\mathbf{b} \in I\!\!R^m$. The two-sided ℓ_1^m-approximation problem is:

$$\min \Big\{ \sum_{i=1}^{m} |b_i - u_i| : \mathbf{u} = (u_1, \ldots, u_m) \in U \Big\}$$

or

$$\min \Big\{ \sum_{i=1}^{m} \Big| b_i - \sum_{j=1}^{n} a_j u_i^j \Big| : a_1, \ldots, a_n \in I\!\!R \Big\}.$$

Let $c_i = b_i - \sum_{j=1}^{n} a_j u_i^j$, $i = 1, \ldots, m$, and write $c_i = c_i^1 - c_i^2$ where $c_i^1, c_i^2 \geq 0$, $i = 1, \ldots, m$. We also write $a_j = a_j^1 - a_j^2$, $j = 1, \ldots, n$. The above problem can then be rewritten as:

$$\min \sum_{i=1}^{m} c_i^1 + c_i^2$$

subject to:

$$c_i^1 - c_i^2 + \sum_{j=1}^{n} (a_j^1 - a_j^2) u_i^j = b_i, \quad i = 1, \ldots, m$$

$$c_i^1, c_i^2 \geq 0, \quad i = 1, \ldots, m$$

$$a_j^1, a_j^2 \geq 0, \quad j = 1, \ldots, n.$$

These two problems are equivalent since it is easily seen that any solution to the above must have $c_i^1 c_i^2 = 0$, $i = 1, \ldots, m$. Thus $c_i^1 + c_i^2 = |c_i^1 - c_i^2| = |c_i|$, $i = 1, \ldots, m$.

The dual version of this problem (see Exercise 8 of Chapter 1) is given by:

$$\max \sum_{i=1}^{m} b_i y_i$$

subject to:

$$|y_i| \leq 1, \quad i = 1, \ldots, m$$
$$\sum_{i=1}^{m} y_i u_i^j = 0, \quad j = 1, \ldots, n.$$

This dual version can also be brought into more standard form.

Usual linear programming techniques can be applied to solve either of the above problems. We refer the reader to any standard text on linear programming. However the special nature of the problem suggests that it be taken into account. We shall describe two algorithms, based on simplex method ideas, which work directly with the approximation problem. Before describing these two algorithms, let us detail certain subprocesses.

Assume that we are given $\mathbf{b} \in \mathbb{R}^m$, and $\tilde{\mathbf{u}}, \mathbf{v} \in U$, where $\mathbf{v}$ is a descent direction for $\mathbf{b} - \tilde{\mathbf{u}}$ (see Section 2). That is, we assume that

$$c = \sum_{i=1}^{m} \operatorname{sgn}(b_i - \tilde{u}_i) v_i - \sum_{i \in Z(\mathbf{b} - \tilde{\mathbf{u}})} |v_i| > 0.$$

Consider the problem

$$\min_t \| \mathbf{b} - \tilde{\mathbf{u}} - t\mathbf{v} \|_1$$

(this is called a *line search*). Set $F(t) = \| \mathbf{b} - \tilde{\mathbf{u}} - t\mathbf{v} \|_1$. It is readily seen that $F(t)$ is convex, and in fact piecewise linear. Thus the above minimum is uniquely attained on some interval $[t_1, t_2]$, where $t_1 \leq t_2$. From the fact that $c > 0$, it follows that $t_1 > 0$. We shall characterize $[t_1, t_2]$ and give a simple procedure for finding it.

To this end, set

$$A = \{ j : (b_j - \tilde{u}_j) v_j > 0 \}.$$

Since $c > 0$, we have that $A \neq \emptyset$. Let

$$B = \{ (b_j - \tilde{u}_j) / v_j : j \in A \},$$

and let $0 < s_1 < \cdots < s_p$ denote the distinct values of B. Set

$$A_i = \{ j : b_j - \tilde{u}_j = s_i v_j \}, \quad i = 1, \ldots, p.$$

Thus $A = \bigcup_{i=1}^{p} A_i$.

Proposition 7.5. *Let $F(t)$ be as previously defined and $c > 0$. Then*

$$\min_t F(t) = \min_{i=1,\dots,p} F(s_i)\,.$$

Furthermore, $\min\{F(t) : t \in \mathbb{R}\} = F(s_k)$ *where k is the unique index in* $\{1,\dots,p\}$ *for which*

$$2 \sum_{i \in A_1 \cup \cdots \cup A_{k-1}} \operatorname{sgn}(b_i - \widetilde{u}_i) v_i \ - c < 0$$

and

$$2 \sum_{i \in A_1 \cup \cdots \cup A_k} \operatorname{sgn}(b_i - \widetilde{u}_i) v_i \ - c \geq 0\,.$$

This minimum is uniquely attained at s_k if strict inequality holds in the latter inequality. If equality holds, then the minimum is attained exactly on $[s_k, s_{k+1}]$ *(and $k \leq p - 1$).*

Proof. For convenience, set $s_0 = 0$ and $s_{p+1} = \infty$. For $j \in \{1,\dots,p+1\}$, and $t \in (s_{j-1}, s_j)$, we explicitly calculate $F(t)$.

If $i \notin A$, $i \notin Z(\mathbf{b} - \widetilde{\mathbf{u}})$, then

$$(b_i - \widetilde{u}_i) v_i \leq 0$$

and therefore $\operatorname{sgn}(b_i - \widetilde{u}_i - t v_i) = \operatorname{sgn}(b_i - \widetilde{u}_i)$. For $i \in A_1 \cup \cdots \cup A_{j-1}$, we have

$$(b_i - \widetilde{u}_i)/v_i < t\,,$$

and therefore $\operatorname{sgn}(b_i - \widetilde{u}_i - t v_i) = -\operatorname{sgn}(b_i - \widetilde{u}_i)$. If $i \in A_j \cup \cdots \cup A_p$, then

$$(b_i - \widetilde{u}_i)/v_i > t\,,$$

and $\operatorname{sgn}(b_i - \widetilde{u}_i - t v_i) = \operatorname{sgn}(b_i - \widetilde{u}_i)$.

A simple calculation now shows that

$$F(t) = \|\mathbf{b} - \widetilde{\mathbf{u}} - t\mathbf{v}\|_1 = \|\mathbf{b} - \widetilde{\mathbf{u}}\|_1$$
$$- 2 \sum_{i \in A_1 \cup \cdots \cup A_{j-1}} |b_i - \widetilde{u}_i| + t\Big(2 \sum_{i \in A_1 \cup \cdots \cup A_{j-1}} \operatorname{sgn}(b_i - \widetilde{u}_i) v_i \ - c\Big)\,.$$

Thus F is linear on (s_{j-1}, s_j) for each $j = 1,\dots,p+1$ (a similar statement holds on $(-\infty, 0)$) and the slope of F on (s_{j-1}, s_j) is

$$2 \sum_{i \in A_1 \cup \cdots \cup A_{j-1}} \operatorname{sgn}(b_i - \widetilde{u}_i) v_i \ - c\,.$$

For $j = 1$ this value is negative, and for $j = p + 1$ it is positive. Since F is convex, the proposition follows. $\qquad\square$

If we are given a descent direction $\mathbf{v}$ for $\mathbf{b} - \widetilde{\mathbf{u}}$, we now have a simple straightforward procedure for computing

$$\min_t \|\mathbf{b} - \widetilde{\mathbf{u}} - t\mathbf{v}\|_1\,.$$

The next problem we address is that of constructively determining whether $\widetilde{\mathbf{u}}$ is a best approximant to $\mathbf{b}$ from U and, if not, then of finding a descent direction. Here we shall consider various cases. We start with the simplest.

Case 1. Assume $|Z(\mathbf{b}-\widetilde{\mathbf{u}})| = n$, and U is linearly independent over $Z(\mathbf{b}-\widetilde{\mathbf{u}})$, i.e., $\dim U|_{Z(\mathbf{b}-\widetilde{\mathbf{u}})} = n$.

With no loss of generality, let us assume that $Z(\mathbf{b}-\widetilde{\mathbf{u}}) = \{1,\ldots,n\}$. We present two methods (which are totally equivalent).

Method 1a. From Exercise 5 of Chapter 6, we have the following:

Let $\mathbf{v}^j \in U$, $j = 1,\ldots,n$, satisfy $v_i^j = \delta_{ij}$, $i,j = 1,\ldots,n$. (By our assumption, such $\mathbf{v}^j$ exist and are unique.) The vector $\widetilde{\mathbf{u}}$ is a best approximant to $\mathbf{b}$ if and only if

$$\left| \sum_{i=1}^m \text{sgn}(b_i - \widetilde{u}_i)v_i^j \right| \leq 1, \quad j = 1,\ldots,n.$$

If the above does not hold for some j_0, then

$$\left| \sum_{i=1}^m \text{sgn}(b_i - \widetilde{u}_i)v_i^{j_0} \right| > 1 = \sum_{i \in Z(\mathbf{b}-\widetilde{\mathbf{u}})} |v_i^{j_0}|.$$

Thus either $\mathbf{v}^{j_0}$ or $-\mathbf{v}^{j_0}$ is a descent direction.

Method 1b. In Method 1a we first calculated $\mathbf{v}^1,\ldots,\mathbf{v}^n$. We may get around this as follows. Let $\mathbf{u}^1,\ldots,\mathbf{u}^n$ be any basis for U. Set

$$h_j = \sum_{i=1}^m \text{sgn}(b_i - \widetilde{u}_i)u_i^j, \quad j = 1,\ldots,n.$$

Since U has dimension n over the indices $\{1,\ldots,n\}$, there exist unique coefficients $\{\lambda_k\}_{k=1}^n$ satisfying

$$h_j = \sum_{k=1}^n \lambda_k u_k^j, \quad j = 1,\ldots,n.$$

The vector $\widetilde{\mathbf{u}}$ is a best approximant to $\mathbf{b}$ from U if and only if $|\lambda_k| \leq 1$, $k = 1,\ldots,n$. This fact is easily proved. We prove one direction by noting that if $|\lambda_{k_0}| > 1$ for some k_0, then we can construct a unique $\mathbf{v} \in U$ satisfying

$$v_i = \begin{cases} 0, & i = 1,\ldots,n,\ i \neq k_0 \\ \text{sgn}\,\lambda_{k_0}, & i = k_0. \end{cases}$$

A simple calculation gives

$$\sum_{i=1}^m \text{sgn}(b_i - \widetilde{u}_i)v_i > 1 = \sum_{i \in Z(\mathbf{b}-\widetilde{\mathbf{u}})} |v_i|.$$

Therefore $\mathbf{v}$ is a descent direction. (This $\mathbf{v}$ is an appropriate $\pm\mathbf{v}^{j_0}$ from Method 1a.)

The more difficult problem encountered is contained in this next Case 2. Note that this is a generalization of Case 1.

Case 2. Assume $|Z(\mathbf{b} - \widetilde{\mathbf{u}})| = s \geq n$, and U is linearly independent over $Z(\mathbf{b} - \widetilde{\mathbf{u}})$.

We again assume, without loss of generality, that $Z(\mathbf{b} - \widetilde{\mathbf{u}}) = \{1, \ldots, s\}$. For convenience we shall also assume that $s > n$ in Method 2a.

Method 2a. Assume U is linearly independent over $\{1, \ldots, n\}$. Let $\mathbf{v}^j \in U$, $j = 1, \ldots, n$, satisfy $v_i^j = \delta_{ij}$, $i, j = 1, \ldots, n$. From Exercise 6 of Chapter 6, $\widetilde{\mathbf{u}}$ is a best approximant to $\mathbf{b}$ from U if and only if there exist $\{y_i\}_{i=n+1}^{s}$ satisfying $|y_i| \leq 1$, $i = n+1, \ldots, s$, for which

$$\left| \sum_{i=1}^{m} \operatorname{sgn}(b_i - \widetilde{u}_i)v_i^j + \sum_{i=n+1}^{s} y_i v_i^j \right| \leq 1,$$

for $j = 1, \ldots, n$. Unfortunately, this fact is difficult to verify and, if not true, does not easily give us a descent direction.

Method 2b. We first develop a theoretical tool. This fact is well known in linear programming. We shall prove it using approximation theory techniques.

Let $\mathbf{r} = (r_1, \ldots, r_{n-1})$ be any set of $n-1$ indices in $\{1, \ldots, s\}$ on which U has dimension $n - 1$. Equivalently, for each such $\mathbf{r}$ the vector $\mathbf{v}^{\mathbf{r}} \in U$, $\mathbf{v}^{\mathbf{r}} \neq \mathbf{0}$, which satisfies $v_{r_i}^{\mathbf{r}} = 0$, $i = 1, \ldots, n - 1$, is unique up to multiplication by a non-zero constant.

Proposition 7.6. *Under the above assumptions, $\widetilde{\mathbf{u}}$ is a best approximant to $\mathbf{b}$ from U if and only if*

$$\left| \sum_{i=1}^{m} \operatorname{sgn}(b_i - \widetilde{u}_i)v_i^{\mathbf{r}} \right| \leq \sum_{i=1}^{s} |v_i^{\mathbf{r}}|$$

for all $\mathbf{v}^{\mathbf{r}}$ as above.

Proof. If $\widetilde{\mathbf{u}}$ is a best approximant, then the above inequality holds for every $\mathbf{u} \in U$, and in particular for the $\mathbf{v}^{\mathbf{r}}$. The content of the proposition is the converse direction, i.e., it suffices to check only this smaller finite set of vectors.

Assume $\widetilde{\mathbf{u}}$ is not a best approximant to $\mathbf{b}$ from U. Then for any $\mathbf{y} = (y_1, \ldots, y_s)$ satisfying

$$\sum_{i=1}^{m} \operatorname{sgn}(b_i - \widetilde{u}_i)u_i + \sum_{i=1}^{s} y_i u_i = 0$$

for all $\mathbf{u} \in U$, it follows from Proposition 6.2 that $\|\mathbf{y}\|_\infty > 1$.

Let

$$W = \left\{ \mathbf{w} : \mathbf{w} = (w_1, \ldots, w_s), \sum_{i=1}^{s} w_i u_i = 0, \text{ for all } \mathbf{u} \in U \right\}.$$

Since $\dim U|_{\{1,\ldots,s\}} = n$, we have that $\dim W = s - n$. Let $\mathbf{y}^* = (y_1^*, \ldots, y_s^*)$ satisfy

$$\sum_{i=1}^{m} \operatorname{sgn}(b_i - \tilde{u}_i)u_i + \sum_{i=1}^{s} y_i^* u_i = 0$$

for all $\mathbf{u} \in U$. Such a $\mathbf{y}^*$ exists. Then $\mathbf{y} = (y_1, \ldots, y_s)$ also satisfies the above equality if and only if $\mathbf{y} = \mathbf{y}^* - \mathbf{w}$ for some $\mathbf{w} \in W$.

Since $\tilde{\mathbf{u}}$ is not a best approximant to $\mathbf{b}$ from U,

$$\min\{\|\mathbf{y}^* - \mathbf{w}\|_\infty : \mathbf{w} \in W\} = c > 1.$$

Because W is a subspace, there exists a $\mathbf{w}^* \in W$ satisfying

$$\|\mathbf{y}^* - \mathbf{w}^*\|_\infty = c.$$

Since $\dim W = s - n$, we have from the general characterization theorem for best approximation in ℓ_∞^s, the existence of indices $1 \le m_1 < \cdots < m_k \le s$, $1 \le k \le s - n + 1$, and non-zero constants $\{\lambda_{m_j}\}_{j=1}^{k}$, for which

  a) $(y_{m_j}^* - w_{m_j}^*) = (\operatorname{sgn} \lambda_{m_j})c$, $j = 1, \ldots, k$
  b) $\sum_{j=1}^{k} \lambda_{m_j} w_{m_j} = 0$, all $\mathbf{w} \in W$
  c) W is of dimension $k - 1$ on $\{m_1, \ldots, m_k\}$.

Let $\boldsymbol{\lambda}^* = (\lambda_1^*, \ldots, \lambda_s^*)$, where

$$\lambda_i^* = \begin{cases} 0, & i \notin \{m_1, \ldots, m_k\} \\ \lambda_{m_j}, & i = m_j, \ j = 1, \ldots, k. \end{cases}$$

From (b),

$$\sum_{i=1}^{s} \lambda_i^* w_i = 0$$

for all $\mathbf{w} \in W$. Based on our definition of W, there exists a $\mathbf{v} \in U$ satisfying $v_i = \lambda_i^*$, $i = 1, \ldots, s$. Thus $v_i = 0$ on some $s - k$ indices in $\{1, \ldots, s\}$. Since $k \le s - n + 1$, this implies that $v_i = 0$ on at least $n - 1$ indices in $\{1, \ldots, s\}$. From (c), it follows that $\mathbf{v} = \mathbf{v}^{\mathbf{r}}$ as previously defined. That is, there exist $\{r_1, \ldots, r_{n-1}\}$ in $\{1, \ldots, s\}$ for which $v_{r_i} = 0$, $i = 1, \ldots, n-1$, and $\dim U|_{\{r_1, \ldots, r_{n-1}\}} = n - 1$.

Now,

$$\left| \sum_{i=1}^{m} \operatorname{sgn}(b_i - \tilde{u}_i)v_i^{\mathbf{r}} \right| = \left| \sum_{i=1}^{s} y_i^* v_i^{\mathbf{r}} \right| = \left| \sum_{i=1}^{s} (y_i^* - w_i^*)\lambda_i^* \right|$$

$$= \left| \sum_{j=1}^{k} (y_{m_j}^* - w_{m_j}^*)\lambda_{m_j} \right| = c\sum_{j=1}^{k} |\lambda_{m_j}| = c\sum_{i=1}^{s} |\lambda_i^*| = c\sum_{i=1}^{s} |v_i^{\mathbf{r}}| > \sum_{i \in Z(\mathbf{b} - \tilde{\mathbf{u}})} |v_i^{\mathbf{r}}|,$$

since $c > 1$ and $Z(\mathbf{b} - \tilde{\mathbf{u}}) = \{1, \ldots, s\}$. This proves the proposition.   □

Method 2b therefore involves checking the appropriate inequalities over all possible $\mathbf{v^r}$. Unfortunately there may be $\binom{s}{n-1}$ such vectors. For $s > n$ this case corresponds to a degenerate vertex. If $s = n$ this is just Method 1a.

Note that if there exists a $\mathbf{v^{r^*}}$ which does not satisfy the inequalities of Proposition 7.6, then either $\mathbf{v^{r^*}}$ or $-\mathbf{v^{r^*}}$ is a descent direction.

Before continuing to Case 3, we record two additional facts which will prove useful.

I. For any $\tilde{\mathbf{u}} \in U$, the quantity $\|\mathbf{b} - \tilde{\mathbf{u}}\|_1$ is by definition an upper bound for

$$\min\{\|\mathbf{b} - \mathbf{u}\|_1 : \mathbf{u} \in U\}.$$

From the above analysis, we can also obtain a lower bound for this desired quantity. Recall (Exercise 8 of Chapter 1) that if $\mathbf{y} \in I\!\!R^m$ satisfies $\|\mathbf{y}\|_\infty \le 1$ and $(\mathbf{y}, \mathbf{u}) = 0$ for all $\mathbf{u} \in U$, then the value $|(\mathbf{y}, \mathbf{b})|$ is a lower bound for our 'error'. Thus if $Z(\mathbf{b} - \tilde{\mathbf{u}}) = \{1, \ldots, s\}$, and $\mathbf{y}^* = (y_1^*, \ldots, y_s^*)$ satisfies

$$\sum_{i=1}^{m} \mathrm{sgn}(b_i - \tilde{u}_i)u_i + \sum_{i=1}^{s} y_i^* u_i = 0$$

for all $\mathbf{u} \in U$, then

$$\|\mathbf{b} - \tilde{\mathbf{u}}\|_1 / \max\{1, \|\mathbf{y}^*\|_\infty\}$$

is a lower bound for

$$\min\{\|\mathbf{b} - \mathbf{u}\|_1 : \mathbf{u} \in U\}.$$

II. Let $\mathbf{v^{r^*}}$ be as in Method 2b (or Method 1) contradicting the inequality of Proposition 7.6. For convenience assume that $\mathbf{v^{r^*}}$ is a descent direction for $\mathbf{b} - \tilde{\mathbf{u}}$, and $\min\{\|\mathbf{b} - \tilde{\mathbf{u}} - t\mathbf{v^{r^*}}\|_1 : t \in I\!\!R\}$ is uniquely attained on $[t_1, t_2]$. Set $\mathbf{r}^* = (r_1^*, \ldots, r_{n-1}^*)$. Then $\{r_1^*, \ldots, r_{n-1}^*\} \subseteq Z(\mathbf{b} - \tilde{\mathbf{u}} - t\mathbf{v^{r^*}})$ for every t. From Proposition 7.5 it follows that, for both $i = 1$ and $i = 2$, the set $Z(\mathbf{b} - \tilde{\mathbf{u}} - t_i\mathbf{v^{r^*}})$ contains at least one additional index k_i such that $v_{k_i}^{r^*} \ne 0$. Since $\mathbf{v^{r^*}}$ was the unique (up to multiplication by a constant) element in U vanishing on $\{r_1^*, \ldots, r_{n-1}^*\}$, this implies that $\dim U|_{Z(\mathbf{b}-\tilde{\mathbf{u}}-t_i\mathbf{v^{r^*}})} = n$. In other words, if we start in either Case 1 or Case 2, use descent direction $\mathbf{v^{r^*}}$ (if it exists) and choose the appropriate t_1 or t_2, then we remain within Case 1 or 2.

We know consider the complement to Cases 1 and 2.

Case 3. Assume $|Z(\mathbf{b} - \tilde{\mathbf{u}})| = k$, and $\dim U|_{Z(\mathbf{b}-\tilde{\mathbf{u}})} = r < n$.

In the above we allow for $k = 0$, i.e., $Z(\mathbf{b} - \tilde{\mathbf{u}}) = \emptyset$. Assume that $Z(\mathbf{b} - \tilde{\mathbf{u}}) = \{1, \ldots, k\}$. Set

$$V = \{\mathbf{v} : \mathbf{v} \in U, v_i = 0, i = 1, \ldots, k\}.$$

By assumption, V is a subspace of U of dimension $n - r$. If

$$\sum_{i=1}^{m} \mathrm{sgn}(b_i - \tilde{u}_i)v_i > 0$$

for some $\mathbf{v} \in V$, then $\widetilde{\mathbf{u}}$ is not a best approximant to $\mathbf{b}$ from U, and $\mathbf{v}$ is a descent direction. Note that if

$$\min_t \|\mathbf{b} - \widetilde{\mathbf{u}} - t\mathbf{v}\|_1$$

is uniquely attained on $[t_1, t_2]$, then $\dim U|_{Z(\mathbf{b}-\widetilde{\mathbf{u}})} < \dim U|_{Z(\mathbf{b}-\widetilde{\mathbf{u}}-t_i\mathbf{v})}$, $i = 1, 2$.

Now assume that

$$\sum_{i=1}^m \operatorname{sgn}(b_i - \widetilde{u}_i)v_i = 0$$

for all $\mathbf{v} \in V$. Here we can use one of two methods.

Method 3a. Let $\mathbf{v} \in V$. Then there exist $t_1 < 0 < t_2$ given by

$$t_1 = \max\{(b_i - \widetilde{u}_i)/v_i : (b_i - \widetilde{u}_i)v_i < 0\}$$
$$t_2 = \min\{(b_i - \widetilde{u}_i)/v_i : (b_i - \widetilde{u}_i)v_i > 0\}$$

with the property that

$$\min_t \|\mathbf{b} - \widetilde{\mathbf{u}} - t\mathbf{v}\|_1$$

is uniquely attained on $[t_1, t_2]$. Thus $Z(\mathbf{b} - \widetilde{\mathbf{u}}) \subset Z(\mathbf{b} - \widetilde{\mathbf{u}} - t_i\mathbf{v})$ for $i = 1, 2$, and

$$\dim U|_{Z(\mathbf{b}-\widetilde{\mathbf{u}})} < \dim U|_{Z(\mathbf{b}-\widetilde{\mathbf{u}}-t_i\mathbf{v})},$$

$i = 1, 2$. We have essentially moved to the edge of a flat, and therefore after a finite number of steps, we shall revert to one of the previous cases.

Method 3b. Let W be any r-dimensional subspace of U which satisfies $\dim W|_{\{1,\dots,k\}} = r$. Thus

$$U = W \oplus V.$$

Now $\widetilde{\mathbf{u}}$ is a best approximant to $\mathbf{b}$ from U, i.e., $\mathbf{0}$ is a best approximant to $\mathbf{b} - \widetilde{\mathbf{u}}$ from U, if and only if $\mathbf{0}$ is a best approximant to $\mathbf{b} - \widetilde{\mathbf{u}}$ from W. Since W is a subspace of U, one direction is immediate. Assume that $\mathbf{0}$ is a best approximant to $\mathbf{b} - \widetilde{\mathbf{u}}$ from W. Then

$$\left|\sum_{i=1}^m \operatorname{sgn}(b_i - \widetilde{u}_i)w_i\right| \le \sum_{i \in Z(\mathbf{b}-\widetilde{\mathbf{u}})} |w_i|$$

for all $\mathbf{w} \in W$. Let $\mathbf{u} \in U$. Then $\mathbf{u} = \mathbf{w} + \mathbf{v}$ for some $\mathbf{w} \in W$ and $\mathbf{v} \in V$. Now,

$$\sum_{i=1}^m \operatorname{sgn}(b_i - \widetilde{u}_i)v_i = \sum_{i \in Z(\mathbf{b}-\widetilde{\mathbf{u}})} |v_i| = 0.$$

Thus

$$\left|\sum_{i=1}^m \operatorname{sgn}(b_i - \widetilde{u}_i)u_i\right| \le \sum_{i \in Z(\mathbf{b}-\widetilde{\mathbf{u}})} |u_i|$$

and $\widetilde{\mathbf{u}}$ is a best approximant to $\mathbf{b}$ from U.

Since dim $W = r$ and dim $W|_{\{1,\ldots,k\}} = r$, we have reverted to either Case 1 or Case 2 (with respect to W) and we can apply the methods thereof.

We now describe two algorithms, both of which are variants of the simplex method.

Algorithm 1. This algorithm is a direct application of our line search analysis (Proposition 7.5) and Cases 1 and 2. Let $\mathbf{b} \in I\!\!R^m$. We start with a $\mathbf{u}^0 \in U$ which, for convenience, we take so that dim $U|_{Z(\mathbf{b}-\mathbf{u}^0)} = n$, i.e., find n indices over which U is linearly independent and let $\mathbf{u}^0$ interpolate to $\mathbf{b}$ on these n indices.

Given $\mathbf{u}^k \in U$ with dim $U|_{Z(\mathbf{b}-\mathbf{u}^k)} = n$, we apply the methods of Case 1 or 2 to determine whether $\mathbf{u}^k$ is a best approximant to $\mathbf{b}$ from U. If $\mathbf{u}^k$ is not a best approximant to $\mathbf{b}$, we apply Method 2b to obtain a descent direction $\mathbf{v}^r$ of the form given in Proposition 7.6. We then determine the exact interval $[t_1, t_2]$ on which

$$\min_t \|\mathbf{b} - \mathbf{u}^k - t\mathbf{v}^r\|_1$$

is attained, and set $\mathbf{u}^{k+1} = \mathbf{u}^k + t_i \mathbf{v}^r$ for $i = 1$ or 2. This is the algorithm.

Note that dim $U|_{Z(\mathbf{b}-\mathbf{u}^{k+1})} = n$ and we may therefore continue this process, if necessary. Furthermore $\|\mathbf{b} - \mathbf{u}^{k+1}\|_1 < \|\mathbf{b} - \mathbf{u}^k\|_1$. This algorithm will give us a best approximant after a finite number of steps. This follows from the fact (Proposition 6.8) that every extreme point $\mathbf{u}^*$ of the set of best approximants satisfies dim $U|_{Z(\mathbf{b}-\mathbf{u}^*)} = n$, and from the fact that the number of $\mathbf{u} \in U$ for which dim $U|_{Z(\mathbf{b}-\mathbf{u})} = n$ is finite (at most $\binom{m}{n}$).

Some additional comments are in order. Using Method 2b with $s = |Z(\mathbf{b} - \mathbf{u}^k)|$ much larger than n is a serious drawback. However we shall in general have $|Z(\mathbf{b} - \mathbf{u}^k)| = n$ (i.e., be in Case 1) since if $\mathbf{u}^k$ interpolates to $\mathbf{b}$ at n points over which U is linearly independent, then the probability of an arbitrary $\mathbf{b} \in I\!\!R^m$ agreeing with $\mathbf{u}^k$ on any other index is zero. We shall also generally have $t_1 = t_2$, which somewhat eases the computation. In addition, if we can more easily find a descent direction $\mathbf{v}$ for $\mathbf{b} - \mathbf{u}^k$, which is not of the above form, then $\mathbf{u}^{k+1}$ may not satisfy the hypothesis dim $U|_{Z(\mathbf{b}-\mathbf{u}^{k+1})} = n$. However, we can then apply the ideas in Case 3 and return to our desired state after at most n steps. In this way we can also start with an arbitrary $\mathbf{u}^0$. It is also suggested that we simultaneously compute (at relatively little extra cost) the lower bound as given after the discussion of Method 2b.

Algorithm 2. This algorithm is mathematically less elegant than the previous algorithm. None the less it is totally equivalent to Algorithm 1, and the reader is urged to check this for himself. It however looks somewhat different and is included for this reason.

Let $\mathbf{b} \in I\!\!R^m$ and $\mathbf{u}^0$ be as described in Algorithm 1. Choose $\{k_1, \ldots, k_n\} \subseteq Z(\mathbf{b} - \mathbf{u}^0)$ where dim $U|_{\{k_1,\ldots,k_n\}} = n$, and let $\mathbf{v}^j \in U$, $j = 1, \ldots, n$, satisfy $v_{k_i}^j = \delta_{ij}$, $i, j = 1, \ldots, n$. Then $\mathbf{v}^1, \ldots, \mathbf{v}^n$ is a basis for U.

The idea of this algorithm is to cycle through $\mathbf{v}^1, \ldots, \mathbf{v}^n$ upgrading them and the $\{k_1, \ldots, k_n\}$ at each step until a level set is reached. At this point a determination is made as to whether a best approximant has been reached, and if not, a new basis is defined and the process is repeated.

Given $\mathbf{v}^1, \ldots, \mathbf{v}^n$ as above, determine $[t_1, t_2]$ for which

$$\min_t \|\mathbf{b} - \mathbf{u}^0 - t\mathbf{v}^1\|_1$$

is attained exactly on $[t_1, t_2]$. Set $\mathbf{u}^1 = \mathbf{u}^0 - t_i\mathbf{v}^1$ with either $i = 1$ or $i = 2$. Now $\{k_2, \ldots, k_n\} \subseteq Z(\mathbf{b} - \mathbf{u}^1)$ and there exists a $k_1' \in Z(\mathbf{b} - \mathbf{u}^1)$, $k_1' \notin \{k_2, \ldots, k_n\}$ such that $\dim U|_{\{k_1', k_2, \ldots, k_n\}} = n$. We upgrade $\mathbf{v}^1, \ldots, \mathbf{v}^n$ in the sense that we define

$$\mathbf{w}^j = \mathbf{v}^j - \frac{v_{k_1'}^j}{v_{k_1'}^1}\mathbf{v}^1, \quad j = 2, \ldots, n$$

$$\mathbf{w}^1 = \frac{1}{v_{k_1'}^1}\mathbf{v}^1.$$

and redefine the $\mathbf{w}^j$ as $\mathbf{v}^j$, $j = 1, \ldots, n$, and k_1 as k_1'. We then again apply the above method, where we replace $\mathbf{v}^1$ by $\mathbf{v}^2$. We continue in this manner with cycling. That is, after applying this process to $\mathbf{v}^n$ we return to $\mathbf{v}^1$.

Since $\dim U|_{Z(\mathbf{b} - \mathbf{u}^k)} = n$ and $\|\mathbf{b} - \mathbf{u}^{k+1}\|_1 \leq \|\mathbf{b} - \mathbf{u}^k\|_1$ for each k, we must reach a stage where $\|\mathbf{b} - \mathbf{u}^{k+n}\|_1 = \|\mathbf{b} - \mathbf{u}^k\|_1$ for some k. It is here that we should apply Method 2b to $\mathbf{b} - \mathbf{u}^k$ (or $\mathbf{b} - \mathbf{u}^j$ for any $k \leq j \leq k+n$). If we can find a descent direction $\mathbf{v}^\mathbf{r}$ with $\mathbf{r} = (r_1, \ldots, r_{n-1})$ as in Proposition 7.6, we then define a new basis $\mathbf{v}^1, \ldots, \mathbf{v}^n$, as previously, based on $\{k_1, \ldots, k_n\} \subseteq Z(\mathbf{b} - \mathbf{u}^k)$, where $r_i = k_{i+1}$, $i = 1, \ldots, n-1$, and $\dim U|_{\{k_1, \ldots, k_n\}} = n$. (Thus $\mathbf{v}^1$ or $-\mathbf{v}^1$ is a descent direction.) We now return to the previous cycling procedure. For the very same reasons as given in the discussion of Algorithm 1, we shall, after a finite number of steps, necessarily reach a best approximant.

5. One-Sided L^1-Approximation

We assume that K is a compact subset of $\mathbb{R}^m$ satisfying $K = \overline{\text{int } K}$, U is an n-dimensional subspace of $C(K)$, and μ is any 'admissible' measure on K, i.e., μ is non-atomic, positive and finite, and $\mu(A) > 0$ for every open set A.

We assume that we are given an $f \in C(K)$ with $\mathcal{U}(f) \neq \emptyset$, i.e., there exists a $u \in U$ satisfying $u \leq f$. The problem we shall discuss is:

$$(7.2) \qquad \max\left\{ \int_K u \, d\mu : u \in \mathcal{U}(f) \right\}.$$

Finding a best one-sided (from below) $L^1(K, \mu)$ approximant to f from U is equivalent to finding a $u \in \mathcal{U}(f)$ satisfying (7.2). We assume that $u^0 \in \mathcal{U}(f)$ is a solution to (7.2), and

$$\sigma_0 = \int_K u^0 d\mu.$$

We shall describe four algorithms for calculating (7.2). Of these four algorithms, the first is a total discretization, the second is a partial discretization, the third is a partial discretization with optimization, and the fourth involves no discretization.

Algorithm 1. This algorithm is based on a total discretization of the problem. We presume that the discrete problem (see Section 3) can be solved.

For each natural number m, we are given points $x_1^m, \ldots, x_m^m \in K$ and positive numbers $\delta_1^m, \ldots, \delta_m^m$, satisfying

$$(7.3) \qquad \lim_{m \to \infty} \sum_{i=1}^{m} \delta_i^m f(x_i^m) = \int_K f \, d\mu$$

for all $f \in C(K)$. For convenience only, we assume that $\sum_{i=1}^{m} \delta_i^m = \mu(K)$ for all m.

One simple example of such a discretization is when $K = [a, b]$, $a = x_0^m < x_1^m < \cdots < x_m^m = b$, and $\delta_i^m = (x_i^m - x_{i-1}^m)$, $i = 1, \ldots, m$, where $\lim_{m \to \infty} \max\{|x_i^m - x_{i-1}^m| : i = 1, \ldots, m\} = 0$. Many other examples may be constructed.

Note that equation (7.3) implies that the points $\{x_i^m\}_{i=1}^{m}$ become dense in K, i.e., for each open set A of K, there exists an M such that $\{x_1^m, \ldots, x_m^m\} \cap A \neq \emptyset$ for all $m \geq M$.

For each m, we set

$$(7.4) \quad \sigma_m = \max\left\{ \sum_{i=1}^{m} \delta_i^m u(x_i^m) : u \in U, \ u(x_i^m) \leq f(x_i^m), \ i = 1, \ldots, m \right\}.$$

(This is an abuse of notation in that it may be that $\sigma_m = \infty$.) The algorithm here is simple to state. Solve (7.4) and let $m \uparrow \infty$. Before proving the convergence of the algorithm, we need to pursue some technical facts.

Lemma 7.7. *There exists an M such that the $\{\sigma_m\}_{m \geq M}$ of (7.4) is a bounded sequence, and if u^m is any solution to (7.4), then the set of functions $\{u^m\}_{m \geq M}$ is also uniformly bounded.*

Proof. Since $\mathcal{U}(f) \neq \emptyset$, it immediately follows from (7.3) that the σ_m are uniformly bounded from below. We first prove that the σ_m are uniformly bounded from above for all m sufficiently large.

Assume that this is not the case. There therefore exists a subsequence $\{m_k\}_{k=1}^{\infty}$ and $v^{m_k} \in U$ satisfying $v^{m_k}(x_i^{m_k}) \leq f(x_i^{m_k})$, $i = 1, \ldots, m_k$, and $\sum_{i=1}^{m_k} \delta_i^{m_k} v^{m_k}(x_i^{m_k}) \geq k$, all k. Let $u_1, \ldots, u_n$ be any basis for U, and $v^{m_k} = \sum_{j=1}^{n} a_j^{m_k} u_j$. On a subsequence, again denoted $\{m_k\}$, we have the existence of an $r \in \{1, \ldots, n\}$ for which

$$|a_r^{m_k}| = \max\{|a_j^{m_k}| : j = 1, \ldots, n\}$$

for all k, and $a_r^{m_k}$ is of one sign ε for all k. Set $b_j^{m_k} = a_j^{m_k}/a_r^{m_k}$, $j = 1, \ldots, n$. On another subsequence, again denoted $\{m_k\}$, we have the existence of $\{b_j\}_{j=1}^n$ where

$$\lim_{k \to \infty} b_j^{m_k} = b_j, \quad j = 1, \ldots, n.$$

Thus $|b_j| \leq 1$, $j = 1, \ldots, n$, and $b_r = 1$. Set $v = \sum_{j=1}^n b_j u_j$. Then $v \neq 0$.

Because U is finite-dimensional, $v^{m_k}/a_r^{m_k}$ converges uniformly to v as $k \uparrow \infty$. Furthermore

$$\varepsilon \sum_{i=1}^{m_k} \delta_i^{m_k} v^{m_k}(x_i^{m_k})/a_r^{m_k} \geq k/|a_r^{m_k}|.$$

From continuity considerations, it therefore follows that

$$\lim_{k \to \infty} \varepsilon \sum_{i=1}^{m_k} \delta_i^{m_k} v^{m_k}(x_i^{m_k})/a_r^{m_k} = \varepsilon \int_K v \, d\mu.$$

Since $\int_K v \, d\mu$ is finite, we necessarily have that $\lim_{k \to \infty} |a_r^{m_k}| = \infty$, and $\varepsilon \int_K v \, d\mu \geq 0$. Now,

$$v^{m_k}(x_i^{m_k})/|a_r^{m_k}| \leq f(x_i^{m_k})/|a_r^{m_k}|, \quad i = 1, \ldots, m_k,$$

for all k. Because the $\{x_i^{m_k}\}$ become dense in K, it follows that $\varepsilon v \leq 0$ on all of K. We have therefore constructed a $v \in U$, $v \neq 0$, satisfying $\int_K \varepsilon v \, d\mu \geq 0$ and $\varepsilon v \leq 0$ on K, where $\varepsilon \in \{-1, 1\}$. This is impossible and the σ_m are therefore uniformly bounded for all m sufficiently large.

Let u^m be any solution to (7.4). The above argument can be easily modified to prove that the $\{u^m\}$ are uniformly bounded for all m sufficiently large. The details are left to the reader.   □

We now prove the convergence result.

Theorem 7.8. *In the above algorithm, we have*

$$\lim_{m \to \infty} \sigma_m = \sigma_0.$$

Furthermore, every convergent subsequence of the $\{u^m\}$ converges to a solution of (7.2).

Proof. Let $\{u^{m_k}\}_{k=1}^\infty$ be a convergent subsequence of $\{u^m\}$. Such a subsequence exists from Lemma 7.7 since U is finite-dimensional. Let $u^* \in U$ be such that the u^{m_k} converge (uniformly) to u^* as $k \uparrow \infty$.

We first prove that $u^* \in \mathcal{U}(f)$. Assume not. Let $x^* \in K$ satisfy

$$f(x^*) - u^*(x^*) = -c < 0.$$

Since f and u^* are continuous on K, there exists an open neighborhood A of x^* such that

$$|f(x^*) - f(y)| < c/4$$

and

$$|u^*(x^*) - u^*(y)| < c/4$$

for all $y \in A$. Because u^{m_k} converges uniformly to u^* on K, there exists an M_1 such that, for all $m_k \geq M_1$,

$$\|u^{m_k} - u^*\|_\infty < c/4.$$

Finally, since (7.3) holds there exists an M_2 such that $\{x_1^m, \ldots, x_m^m\} \cap A \neq \emptyset$ for all $m \geq M_2$.

We put these facts together to obtain a contradiction. Assume $x_{j_k}^{m_k} \in A$, where $m_k \geq \max\{M_1, M_2\}$. Recall that $f(x_{j_k}^{m_k}) - u^{m_k}(x_{j_k}^{m_k}) \geq 0$. Thus

$$
\begin{aligned}
-c &= f(x^*) - u^*(x^*) \\
&= \left(f(x^*) - f(x_{j_k}^{m_k})\right) + \left(f(x_{j_k}^{m_k}) - u^{m_k}(x_{j_k}^{m_k})\right) + \\
&\quad \left(u^{m_k}(x_{j_k}^{m_k}) - u^*(x_{j_k}^{m_k})\right) + \left(u^*(x_{j_k}^{m_k}) - u^*(x^*)\right) \\
&\geq -|f(x^*) - f(x_{j_k}^{m_k})| - |u^{m_k}(x_{j_k}^{m_k}) - u^*(x_{j_k}^{m_k})| - |u^*(x_{j_k}^{m_k}) - u^*(x^*)| \\
&> -\frac{3c}{4}.
\end{aligned}
$$

This contradiction implies that $u^* \in \mathcal{U}(f)$.

Set

$$\sigma_* = \int_K u^* d\mu.$$

Since the u^{m_k} converge uniformly to u^*, it is easily seen that $\lim_{k \to \infty} \sigma_{m_k} = \sigma_*$. Because $u^* \in \mathcal{U}(f)$, it follows by definition that $\sigma_* \leq \sigma_0$. Recall that $u^0 \in P_{\mathcal{U}(f)}(f)$ satisfies

$$\sigma_0 = \int_K u^0 d\mu.$$

By definition,

$$\sigma_{m_k} \geq \sum_{i=1}^{m_k} \delta_i^{m_k} u^0(x_i^{m_k}),$$

for all k. From (7.3),

$$\sigma_* = \lim_{k \to \infty} \sigma_{m_k} \geq \int_K u^0 d\mu = \sigma_0.$$

Thus $\sigma_* = \sigma_0$, and u^* is a solution of (7.2).

Since $\lim_{k \to \infty} \sigma_{m_k} = \sigma_0$ for every subsequence $\{m_k\}$ on which $\{u^{m_k}\}$ converges, and the $\{u^m\}$ are uniformly bounded for m sufficiently large, we finally obtain

$$\lim_{m \to \infty} \sigma_m = \sigma_0.$$

This proves the theorem.   □

Note that if the solution to (7.2) (i.e., u^0) is unique, then

$$\lim_{m \to \infty} u^m = u^0$$

and the convergence is uniform.

Algorithm 2. For each $m \in \mathbb{N}$, let $x_1^m, \ldots, x_m^m \in K$. We assume that the sequence $\{x_i^m\}_{i=1}^m$ becomes dense in K. Let $u_1, \ldots, u_n$ be any given basis for U, and set

$$p_j = \int_K u_j d\mu, \quad j = 1, \ldots, n.$$

For each m, we set

(7.5) $\sigma_m = \max \left\{ \sum_{j=1}^n a_j p_j : \sum_{j=1}^n a_j u_j(x_i^m) \leq f(x_i^m), \ i = 1, \ldots, m \right\}.$

The algorithm is the following. Solve (7.5) and let $m \uparrow \infty$. Note that if for some m there exists a solution u^m of (7.5) satisfying $u^m \leq f$ on all of K, then we are finished. As in Algorithm 1, we first need a technical lemma.

Lemma 7.9. *There exists an M such that the $\{\sigma_m\}_{m \geq M}$ of (7.5) are uniformly bounded, and if u^m is any solution to (7.5), then the set of functions $\{u^m\}_{m \geq M}$ is uniformly bounded.*

We do not prove Lemma 7.9, since its proof is totally analogous to the proof of Lemma 7.7. In fact since the functional to be maximized in (7.5) is fixed independent of m, the proof is even easier. We do prove the convergence result.

Theorem 7.10. *In the above algorithm,*

$$\lim_{m \to \infty} \sigma_m = \sigma_0.$$

Furthermore, every convergent subsequence of the $\{u^m\}$ converges to a solution of (7.2).

Proof. Let $\{u^{m_k}\}$ be a subsequence of $\{u^m\}$ which converges to u^*. Since U is finite-dimensional, this convergence is uniform and $u^* \in U$. Set

$$\sigma_* = \int_K u^* d\mu.$$

Then $\lim_{k \to \infty} \sigma_{m_k} = \sigma_*$.

By definition $\sigma_m \geq \sigma_0$ for all m. Thus $\sigma_* \geq \sigma_0$. From the method of proof of Theorem 7.8 it follows that $u^* \in \mathcal{U}(f)$. Thus $\sigma_* \leq \sigma_0$. Therefore $\sigma_* = \sigma_0$ and u^* is a solution to (7.2). It now follows, as in the proof of Theorem 7.8, that

$$\lim_{m \to \infty} \sigma_m = \sigma_0. \qquad \qquad \square$$

Algorithm 3. This algorithm is a variant of Algorithm 2. It involves maximizing the very same functional, subject to a finite number of inequalities as in Algorithm 2. The difference is in the choice of the points $\{x_i\}$. After an initial choice, points are added in an 'optimal' manner.

Let $u_1, \ldots, u_n$ be any basis for U and set

$$p_j = \int_K u_j \, d\mu, \quad j = 1, \ldots, n.$$

For ease of exposition, we first fix some notation. Let

$$A = \{ \mathbf{a} : \mathbf{a} = (a_1, \ldots, a_n), \ \sum_{j=1}^{n} a_j u_j \leq f \}.$$

For any $\mathbf{a} \in \mathbb{R}^n$, we also set

$$g(\mathbf{a}; x) = f(x) - \sum_{j=1}^{n} a_j u_j(x)$$

and

$$G(\mathbf{a}) = \min_{x \in K} g(\mathbf{a}; x).$$

Thus $\mathbf{a} \in A$ if and only if $G(\mathbf{a}) \geq 0$.

In this algorithm, we start with a set $B_M = \{x_1, \ldots, x_M\}$ of points in K, where we assume that the points are chosen so that there exists no $\mathbf{d} = (d_1, \ldots, d_n) \neq \mathbf{0}$ satisfying

(7.6)
$$a) \ \sum_{j=1}^{n} d_j u_j(x_i) \leq 0, \quad i = 1, \ldots, M$$

$$b) \ \sum_{j=1}^{n} d_j p_j \geq 0.$$

Equivalently, there exists no $u \in U \backslash \{0\}$ satisfying $u(x_i) \leq 0$, $i = 1, \ldots, M$, and $\int_K u \, d\mu \geq 0$. Note that this is exactly (7.1). Thus the problem

$$\max \sum_{j=1}^{n} a_j p_j$$

subject to :
$$\sum_{j=1}^{n} a_j u_j(x_i) \leq f(x_i), \quad i = 1, \ldots, M$$

has a finite maximum and the solution set is bounded. We shall need somewhat more.

Lemma 7.11. *Assume the $\{x_i\}_{i=1}^{M}$ are given such that there exists no $\mathbf{d} \in \mathbb{R}^n \backslash \{0\}$ satisfying (7.6). Let $C_1 < C_2$ be any fixed constants. Then the set of $\mathbf{a} \in \mathbb{R}^n$ satisfying*

(7.7)
$$a) \ \sum_{j=1}^{n} a_j u_j(x_i) \leq f(x_i), \quad i = 1, \ldots, M$$

$$b) \ C_1 \leq \sum_{j=1}^{n} a_j p_j \leq C_2$$

is bounded.

Proof. Assume that this set is unbounded. Thus there exists a sequence of $\{\mathbf{a}^r\}_{r=1}^{\infty}$ in $I\!\!R^n$ satisfying (7.7), and an index $k \in \{1, \ldots, n\}$ such that

$$a) \; |a_k^r| = \max\{|a_j^r| : j = 1, \ldots, n\}$$

$$b) \; \lim_{r \to \infty} \varepsilon a_k^r = \infty, \quad \text{for some } \varepsilon \in \{-1, 1\}.$$

Let $d_j^r = a_j^r/a_k^r$, $j = 1, \ldots, n$. On a subsequence, again denoted by $\{r\}$, we have

$$\lim_{r \to \infty} d_j^r = d_j, \quad j = 1, \ldots, n,$$

i.e., the limits exist. Thus $|d_j| \le 1$, $j = 1, \ldots, n$, and $d_k = 1$. Since the $\mathbf{a}^r$ satisfy (7.7), it follows after dividing by a_k^r and letting $r \uparrow \infty$, that

$$\varepsilon \sum_{j=1}^{n} d_j u_j(x_i) \le 0, \quad i = 1, \ldots, M$$

$$\sum_{j=1}^{n} d_j p_j = 0.$$

However this contradicts our assumption with respect to (7.6). The lemma is proved.   □

We now describe the algorithm. Let $B_M = \{x_1, \ldots, x_M\}$ be as given above so that there exists no $\mathbf{d} \in I\!\!R^n \backslash \{0\}$ satisfying (7.6). Assume that we are given $B_m = \{x_1, \ldots, x_m\}$ for some $m \ge M$. Then B_{m+1} is obtained as follows.

We first solve the finite problem

$$\sigma_m = \max\left\{ \sum_{j=1}^{n} a_j p_j : \sum_{j=1}^{n} a_j u_j(x_i) \le f(x_i), \; i = 1, \ldots, m \right\}.$$

Since $\{x_1, \ldots, x_M\} \subseteq B_m$, this problem has a solution $\mathbf{a}^m = (a_1^m, \ldots, a_n^m)$. If $G(\mathbf{a}^m) \ge 0$, we are finished. (We have found a solution to our original problem.) We therefore assume that $G(\mathbf{a}^m) < 0$. Let $x_{m+1} \in K$ satisfy $G(\mathbf{a}^m) = g(\mathbf{a}^m; x_{m+1})$. Set $B_{m+1} = B_m \cup \{x_{m+1}\}$.

This is the algorithm. In what follows we assume that the algorithm does not terminate after a finite number of steps.

Theorem 7.12. *In the above algorithm*

$$\lim_{m \to \infty} \sigma_m = \sigma_0.$$

Furthermore $\{\mathbf{a}^m\}$ is a bounded sequence, and if $\mathbf{a}^*$ is any cluster point of this sequence, then $\sum_{i=1}^{n} a_i^* u_i$ solves (7.2).

Proof. By definition, $\{\sigma_m\}$ is a non-increasing sequence bounded below by σ_0. Thus, for each m,

$$\sum_{j=1}^{n} a_j^m u_j(x_i) \le f(x_i), \quad i = 1, \ldots, M$$

$$\sigma_0 \le \sum_{j=1}^{n} a_j^m p_j \le \sigma_M.$$

From Lemma 7.11, the $\{\mathbf{a}^m\}$ form a bounded sequence.

Let $\mathbf{a}^* = (a_1^*, \ldots, a_n^*)$ be any cluster point of $\{\mathbf{a}^m\}$, and $\sigma_* = \sum_{j=1}^n a_j^* p_j$. Then

$$\lim_{m \to \infty} \sigma_m = \sigma_* \geq \sigma_0 \,.$$

If $\mathbf{a}^* \in A$, i.e., $\sum_{j=1}^n a_j^* u_j \leq f$, then $\sigma_* \leq \sigma_0$ and the theorem is proved. We shall prove that $\mathbf{a}^* \in A$.

Assume $\mathbf{a}^* \notin A$, i.e., $G(\mathbf{a}^*) < 0$. Set

$$A_m = \{\mathbf{a} : \mathbf{a} = (a_1, \ldots, a_n), \sum_{j=1}^n a_j u_j(x_i) \leq f(x_i), \ i = 1, \ldots, m\} \,.$$

Then

$$A_M \supset A_{M+1} \supset \cdots \supset A$$

by definition. Furthermore $\mathbf{a}^* \in \bigcap_{m=M}^\infty A_m$. Since $\lim_{r \to \infty} \mathbf{a}^{m_r} = \mathbf{a}^*$ on some subsequence $\{m_r\}$, and U is finite-dimensional, the functions $\sum_{j=1}^n a_j^{m_r} u_j$ uniformly converge to $\sum_{j=1}^n a_j^* u_j$ on K. Thus there exists an M_1 such that, for all $m \geq M_1$,

$$\left\| \sum_{j=1}^n a_j^* u_j - \sum_{j=1}^n a_j^{m_r} u_j \right\|_\infty < -\frac{1}{2} G(\mathbf{a}^*) \,.$$

Let $m_r \geq M_1$. Then

$$G(\mathbf{a}^{m_r}) = g(\mathbf{a}^{m_r}; x_{m_r+1}) = f(x_{m_r+1}) - \sum_{j=1}^n a_j^{m_r} u_j(x_{m_r+1}) \,.$$

Since $\mathbf{a}^* \in \bigcap_{m=M}^\infty A_m$, we have $\mathbf{a}^* \in A_{m_r+1}$, and therefore

$$g(\mathbf{a}^*; x_{m_r+1}) = f(x_{m_r+1}) - \sum_{j=1}^n a_j^* u_j(x_{m_r+1}) \geq 0 \,.$$

Thus

$$G(\mathbf{a}^{m_r}) = g(\mathbf{a}^{m_r}; x_{m_r+1})$$

$$= g(\mathbf{a}^*; x_{m_r+1}) + \sum_{j=1}^n (a_j^* - a_j^{m_r}) u_j(x_{m_r+1})$$

$$\geq \sum_{j=1}^n (a_j^* - a_j^{m_r}) u_j(x_{m_r+1})$$

$$> \frac{1}{2} G(\mathbf{a}^*) \,.$$

In other words $G(\mathbf{a}^{m_r}) > (1/2) G(\mathbf{a}^*)$ for all $m_r \geq M_1$. But G is continuous on $\mathbb{R}^n$, and $\lim_{r \to \infty} \mathbf{a}^{m_r} = \mathbf{a}^*$. Thus $G(\mathbf{a}^*) \geq (1/2) G(\mathbf{a}^*)$. Since $G(\mathbf{a}^*) < 0$, this is a contradiction. Thus $\mathbf{a}^* \in A$.   □

Algorithm 4. This algorithm involves absolutely no discretization, and is based on ideas from Section 2. However we do assume that U contains a strictly positive function (and therefore $\mathcal{U}(f) \neq \emptyset$ for all $f \in C(K)$). We choose and fix a basis $u_1, \ldots, u_n$ for U where u_1 is strictly positive and

$$\int_K u_1 \, d\mu = 1 \,,$$

while

$$\int_K u_i \, d\mu = 0, \quad i = 2, \ldots, n \,.$$

Thus our problem (7.2) can be reformulated as

$$\max\{a_1 : \sum_{i=1}^{n} a_i u_i \leq f\} \,.$$

For each $\mathbf{a} = (a_2, \ldots, a_n) \in \mathbb{R}^{n-1}$, define

$$h(\mathbf{a}) = \min\left\{ \frac{f(x) - \sum_{i=2}^{n} a_i u_i(x)}{u_1(x)} : x \in K \right\} \,.$$

Since u_1 is strictly positive

$$\max\{a_1 : \sum_{i=1}^{n} a_i u_i \leq f\} = \max\{h(\mathbf{a}) : \mathbf{a} \in \mathbb{R}^{n-1}\} \,.$$

(In this algorithm we shall consider vectors in $\mathbb{R}^{n-1}$ as being indexed from 2 to n.)

The function h is a continuous concave function on $\mathbb{R}^{n-1}$. The continuity is obvious. The concavity may be proved as follows. Let $\mathbf{a}, \mathbf{b} \in \mathbb{R}^{n-1}$. By definition

$$f \geq h(\mathbf{a})u_1 + \sum_{i=2}^{n} a_i u_i$$

$$f \geq h(\mathbf{b})u_1 + \sum_{i=2}^{n} b_i u_i$$

on all of K. Thus, for any $\lambda \in [0, 1]$,

$$f \geq (\lambda h(\mathbf{a}) + (1 - \lambda)h(\mathbf{b}))u_1 + \sum_{i=2}^{n}(\lambda a_i + (1 - \lambda)b_i)u_i$$

on all of K, which implies that

$$h(\lambda \mathbf{a} + (1 - \lambda)\mathbf{b}) \geq \lambda h(\mathbf{a}) + (1 - \lambda)h(\mathbf{b}) \,.$$

Thus h is concave.

We also have by definition that h is finite on $\mathbb{R}^{n-1}$. We claim that

$$\lim_{|||\mathbf{a}||| \to \infty} h(\mathbf{a}) = -\infty$$

where $||| \cdot |||$ is any norm on $\mathbb{R}^{n-1}$. To see this, set $V = \text{span}\{u_2, \ldots, u_n\}$. Since V is a finite-dimensional subspace of $C(K)$, and $\int_K v \, d\mu = 0$ for all $v \in V$, we necessarily have

$$\lim_{|||\mathbf{a}||| \to \infty} \max \left\{ \sum_{i=2}^{n} a_i u_i(x) : x \in K \right\} = \infty.$$

Thus

$$\lim_{|||\mathbf{a}||| \to \infty} \min \left\{ \frac{f(x) - \sum_{i=2}^{n} a_i u_i(x)}{u_1(x)} : x \in K \right\} = -\infty,$$

i.e., $\lim_{|||\mathbf{a}||| \to \infty} h(\mathbf{a}) = -\infty$.

Maximizing h over $\mathbb{R}^{n-1}$ is therefore a problem of maximizing a concave function. We can apply the theory of Section 2 to this problem ($-h$ is convex). To this end we identify the subgradients of h at $\mathbf{a}$. Note that, for h concave on $\mathbb{R}^{n-1}$, a vector $\mathbf{g} \in \mathbb{R}^{n-1}$ is a subgradient to h at $\mathbf{a}$ if

$$h(\mathbf{b}) - h(\mathbf{a}) \le (\mathbf{g}, \mathbf{b} - \mathbf{a})$$

for all $\mathbf{b} \in \mathbb{R}^{n-1}$.

For given $\mathbf{a} \in \mathbb{R}^{n-1}$, set

$$Z(\mathbf{a}) = \left\{ x : \left(f - h(\mathbf{a})u_1 - \sum_{i=2}^{n} a_i u_i \right)(x) = 0 \right\}.$$

By definition $Z(\mathbf{a}) \ne \emptyset$ for each $\mathbf{a} \in \mathbb{R}^{n-1}$. For $x \in Z(\mathbf{a})$, set

$$\mathbf{g}^x = \left(-u_2(x)/u_1(x), \ldots, -u_n(x)/u_1(x) \right).$$

Let $G(\mathbf{a})$ denote the convex hull of the set of vectors $\{\mathbf{g}^x : x \in Z(\mathbf{a})\}$. $G(\mathbf{a})$ is closed since $Z(\mathbf{a})$ is closed.

Proposition 7.13. *$G(\mathbf{a})$ is the set of subgradients to h at $\mathbf{a}$.*

Proof. We first prove that, for each $x \in Z(\mathbf{a})$ the vector $\mathbf{g}^x$ is a subgradient to h at $\mathbf{a}$. Let $\mathbf{b} \in \mathbb{R}^{n-1}$. Then

$$f(x) = h(\mathbf{a})u_1(x) + \sum_{i=2}^{n} a_i u_i(x)$$

$$\ge h(\mathbf{b})u_1(x) + \sum_{i=2}^{n} b_i u_i(x).$$

Since $u_1(x) > 0$, we have

$$h(\mathbf{b}) - h(\mathbf{a}) \le \sum_{i=2}^{n} -\left(\frac{u_i(x)}{u_1(x)} \right)(b_i - a_i) = (\mathbf{g}^x, \mathbf{b} - \mathbf{a}).$$

It remains to prove that all subgradients to h at $\mathbf{a}$ are in $G(\mathbf{a})$. It follows, as in the proof of Proposition 7.4 (see also Proposition 7.2), that it suffices to prove that, for each $\mathbf{d} \in \mathbb{R}^{n-1}$, there exists a $\mathbf{g} \in G(\mathbf{a})$ for which

$$h'_\mathbf{d}(\mathbf{a}) = \lim_{t \to 0^+} \frac{h(\mathbf{a} + t\mathbf{d}) - h(\mathbf{a})}{t} = (\mathbf{g}, \mathbf{a}).$$

We therefore calculate $h'_{\mathbf{d}}(\mathbf{a})$.

Set

$$g(x) = \frac{f(x) - \sum_{i=2}^{n} a_i u_i(x)}{u_1(x)} - h(\mathbf{a}),$$

and $v_i(x) = u_i(x)/u_1(x)$, $i = 2, \ldots, n$. Note that $\min\{g(x) : x \in K\} = 0$.
Now

$$\frac{h(\mathbf{a} + t\mathbf{d}) - h(\mathbf{a})}{t} = \frac{1}{t}\left[\min_{x \in K} \left\{ \frac{f(x) - \sum_{i=2}^{n}(a_i + td_i)u_i(x)}{u_1(x)} \right\} \right.$$

$$\left. - \min_{x \in K} \left\{ \frac{f(x) - \sum_{i=2}^{n} a_i u_i(x)}{u_1(x)} \right\} \right]$$

$$= \frac{1}{t} \min_{x \in K} \left[g(x) - t \sum_{i=2}^{n} d_i v_i(x) \right].$$

For $x \in Z(g)$ $(= Z(\mathbf{a}))$,

$$\frac{1}{t}\left[g(x) - t \sum_{i=2}^{n} d_i v_i(x) \right] = - \sum_{i=2}^{n} d_i v_i(x).$$

Thus

$$h'_{\mathbf{d}}(\mathbf{a}) \leq \min_{x \in Z(\mathbf{a})} - \sum_{i=2}^{n} d_i v_i(x) = \min_{x \in Z(\mathbf{a})} (\mathbf{g}^x, \mathbf{d}).$$

If equality holds, we have proved our result. We therefore assume that equality
does not hold. Set

$$c^* = \min_{x \in Z(\mathbf{a})} - \sum_{i=2}^{n} d_i v_i(x) = - \max_{x \in Z(\mathbf{a})} \sum_{i=2}^{n} d_i v_i(x).$$

Assume there exists a $\delta > 0$ for which

$$h'_{\mathbf{d}}(\mathbf{a}) \leq c^* - \delta.$$

Let $t_k \downarrow 0$ and $x_k \in K$ satisfy

$$h'_{\mathbf{d}}(\mathbf{a}) = \lim_{k \to \infty} \frac{1}{t_k} \left[g(x_k) - t_k \sum_{i=2}^{n} d_i v_i(x_k) \right] \leq c^* - \delta.$$

Because the limit exists, it follows that

$$\lim_{k \to \infty} g(x_k) = 0.$$

Since K is compact, there exists a subsequence of the $\{x_k\}$, again denoted by
$\{x_k\}$, converging to some x^*. g is continuous and therefore $g(x^*) = 0$, i.e.,
$x^* \in Z(\mathbf{a})$.

The function $\sum_{i=2}^{n} d_i v_i$ is continuous and there exists a K_1 such that for
all $k \geq K_1$

$$\left| \sum_{i=2}^{n} d_i v_i(x_k) - \sum_{i=2}^{n} d_i v_i(x^*) \right| < \frac{\delta}{2}.$$

Thus for $k \geq K_1$

$$c^* + \sum_{i=2}^{n} d_i v_i(x_k) = - \max_{x \in Z(\mathbf{a})} \sum_{i=2}^{n} d_i v_i(x) + \sum_{i=2}^{n} d_i v_i(x_k)$$

$$\leq - \sum_{i=2}^{n} d_i v_i(x^*) + \sum_{i=2}^{n} d_i v_i(x_k) < \frac{\delta}{2}.$$

There also exists a K_2 such that for all $k \geq K_2$

$$g(x_k) \leq t_k [c^* - \frac{\delta}{2} + \sum_{i=2}^{n} d_i v_i(x_k)].$$

Thus, for $k \geq \max\{K_1, K_2\}$,

$$g(x_k) \leq t_k \left[c^* - \frac{\delta}{2} + \sum_{i=2}^{n} d_i v_i(x_k)\right] < 0.$$

However $g(x) \geq 0$ for all $x \in K$. This contradiction proves the proposition. □

As we have shown, the subgradients to h at $\mathbf{a}$ are easily determined theoretically. We can apply any subgradient algorithm. We shall not discuss the various subgradient algorithms which appear in the literature. This would take us too far afield. Suffice it to say that convergence is generally very slow, at least in theory.

6. Two-Sided L^1-Approximation

We again assume that K is a compact subset of $\mathbb{R}^m$ satisfying $K = \overline{\text{int } K}$, U is an n-dimensional subspace of $C(K)$, and μ is any 'admissible' measure. We illustrate four different approaches (algorithms) to the solution of our problem of finding a $u^* \in U$ such that

$$\|f - u^*\|_1 = \min\{\|f - u\|_1 : u \in U\}$$

for a given $f \in C(K)$.

Algorithm 1. This algorithm is based on a discretization of our problem.

For each $m \in \mathbb{N}$, let $x_1^m, \ldots, x_m^m \in K$ and let $\delta_1^m, \ldots, \delta_m^m$ be strictly positive numbers such that

$$(7.8) \qquad \lim_{m \to \infty} \sum_{i=1}^{m} \delta_i^m f(x_i^m) = \int_K f \, d\mu$$

for each $f \in C(K)$. For convenience only, we assume that $\sum_{i=1}^{m} \delta_i^m = \mu(K)$ for all m.

We shall only consider $m \geq M_1$ where $\{x_1^m, \ldots, x_m^m\}$ is such that for all $m \geq M_1$ we have $\dim U|_{\{x_1^m, \ldots, x_m^m\}} = n$. This is in no way a restriction since, from the above assumption of convergence to the integral, there must exist, for each open set A of K, an M (dependent on A) such that, for all $m \geq M$,

$\{x_1^m, \ldots, x_m^m\} \cap A \neq \emptyset$. Now take $y_1, \ldots, y_n$ for which $\dim U|_{\{y_1, \ldots, y_n\}} = n$. By continuity there exist disjoint open sets B_{y_i} containing y_i such that if $z_i \in B_{y_i}$, $i = 1, \ldots, n$, then $\dim U|_{\{z_1, \ldots, z_n\}} = n$. From the above, there exists an M_1 such that for all $m \geq M_1$ we have $\{x_1^m, \ldots, x_m^m\} \cap B_{y_i} \neq \emptyset$ for each $i = 1, \ldots, n$. Thus $\dim U|_{\{x_1^m, \ldots, x_m^m\}} = n$ for all $m \geq M_1$.

Fix $f \in C(K)$. Let $u^m \in U$ be a solution to :

$$\sigma_m = \min\left\{ \sum_{i=1}^m \delta_i^m |f(x_i^m) - u(x_i^m)| : u \in U \right\}.$$

We first prove the following.

Lemma 7.14. *There exists an M_2 such that the $\{u^m\}_{m \geq M_2}$ are uniformly bounded.*

Proof. Let M be such that, for all $m \geq M$, $\dim U|_{\{x_1^m, \ldots, x_m^m\}} = n$, and

$$\sum_{i=1}^m \delta_i^m |f(x_i^m)| \leq \|f\|_1 + 1.$$

Then, for all $m \geq M$,

$$\sum_{i=1}^m \delta_i^m |u^m(x_i^m)| \leq \sum_{i=1}^m \delta_i^m |f(x_i^m) - u^m(x_i^m)| + \sum_{i=1}^m \delta_i^m |f(x_i^m)| \leq 2(\|f\|_1 + 1).$$

That is, there exists a constant $C > 0$ such that

$$\sum_{i=1}^m \delta_i^m |u^m(x_i^m)| \leq C$$

for all $m \geq M$.

Let $u^m = \sum_{j=1}^n a_j^m u_j$, where $u_1, \ldots, u_n$ is a basis for U. If the $\{u^m\}$ are not uniformly bounded, there exists a subsequence $\{m_k\}$ on which

$$(a)\ |a_r^{m_k}| = \max\{|a_j^{m_k}| : j = 1, \ldots, n\}$$
$$(b)\ \lim_{k \to \infty} |a_r^{m_k}| = \infty.$$

Let $b_j^{m_k} = a_j^{m_k}/a_r^{m_k}$, $j = 1, \ldots, n$. Then $|b_j^{m_k}| \leq 1 = |b_r^{m_k}|$, $j = 1, \ldots, n$, and there exists a subsequence of $\{m_k\}$, again denoted by $\{m_k\}$, on which

$$\lim_{k \to \infty} b_j^{m_k} = b_j, \quad j = 1, \ldots, n.$$

Thus

$$\sum_{i=1}^{m_k} \delta_i^{m_k} \left| \sum_{j=1}^n b_j^{m_k} u_j(x_i^{m_k}) \right| \leq \frac{C}{|a_r^{m_k}|}.$$

Let $r \to \infty$. The right-hand-side of the inequality tends to zero, while it is easily seen that the left-hand-side of the inequality tends to $\int_K |\sum_{j=1}^n b_j u_j| d\mu$. Since $\sum_{j=1}^m b_j u_j \neq 0$, we have arrived at a contradiction.   □

The algorithm is of course given by finding σ_m and u^m at each step and letting m tend to infinity. We prove:

Theorem 7.15. *Let*

$$\sigma_0 = \min\{\|f - u\|_1 : u \in U\}.$$

Then $\lim_{m\to\infty} \sigma_m = \sigma_0$. *Furthermore, every convergent subsequence of the* $\{u^m\}$ *converges to a* $u^* \in U$ *satisfying* $\|f - u^*\|_1 = \sigma_0$.

Proof. Let $u^0 \in U$ satisfy $\|f - u^0\|_1 = \sigma_0$. Then

$$\sigma_m(u^0) = \sum_{i=1}^m \delta_i^m |f(x_i^m) - u^0(x_i^m)| \geq \sum_{i=1}^m \delta_i^m |f(x_i^m) - u^m(x_i^m)| = \sigma_m.$$

From (7.8), $\lim_{m\to\infty} \sigma_m(u^0) = \sigma_0$. Therefore $\overline{\lim}_{m\to\infty} \sigma_m \leq \sigma_0$. Assume $\lim_{k\to\infty} u^{m_k} = u^*$. It is a simple matter to prove that

$$\lim_{k\to\infty} \sigma_{m_k} = \|f - u^*\|_1.$$

Thus $\|f - u^*\|_1 \leq \sigma_0$. But $\|f - u^*\|_1 \geq \sigma_0$ by definition. Therefore u^* is a best approximant to f from U. Furthermore $\lim_{k\to\infty} \sigma_{m_k} = \sigma_0$ on every subsequence $\{m_k\}$ for which $\{u^{m_k}\}$ converges. Thus $\lim_{m\to\infty} \sigma_m = \sigma_0$. $\quad\square$

Algorithm 2. This and the next algorithm are gradient methods. We recall that $u^* \in U$ is a best approximant to $f \in C(K)$ if and only if

$$\left| \int_K \operatorname{sgn}(f - u^*)u \, d\mu \right| \leq \int_{Z(f-u^*)} |u| d\mu.$$

for all $u \in U$. We also have from Proposition 7.4 that, if $u_1, \ldots, u_n$ is any basis for U, then the subgradients at a given $\widetilde{u} \in U$, with respect to the above basis, are given by $\mathbf{g} = (g_1, \ldots, g_n)$ where

$$g_i = -\int_K \operatorname{sgn}(f - \widetilde{u})u_i d\mu + \int_{Z(f-\widetilde{u})} h u_i d\mu, \quad i = 1, \ldots, n$$

and h is any function in $L^\infty(Z(f - \widetilde{u}), \mu)$ satisfying $|h| \leq 1$ μ a.e. on $Z(f - \widetilde{u})$. Thus a gradient to $f - \widetilde{u}$ exists if and only if $\mu(Z(f - \widetilde{u})) = 0$. In both this and the next algorithm, we assume that $f \in C(K)$ is such that $\mu(Z(f - u)) = 0$ for all $u \in U$. This assumption, while restrictive, is not overly restrictive. It is also a necessary condition for the convergence of these algorithms to their correct values.

The algorithm to be presented here is simply a standard gradient algorithm. Given $\widetilde{u} \in U$, we have from Proposition 7.4 and our assumption that a descent direction is given by

$$v = \sum_{i=1}^n \left(\int_K \operatorname{sgn}(f - \widetilde{u})u_i d\mu \right) u_i.$$

(This descent direction does depend on the choice of the basis.) The algorithm is as follows. Let u^1 be any starting point in U. Assume that we are given u^k. If

$$\int_K \text{sgn}(f - u^k)u_i d\mu = 0, \quad i = 1, \dots, n,$$

then u^k is a best approximant to f from U. Assume that this is not the case. Set

$$v^k = \sum_{i=1}^n \left(\int_K \text{sgn}(f - u^k)u_i d\mu \right) u_i.$$

Find t^k such that

$$\min_t \| f - u^k - tv^k \|_1 = \| f - u^k - t^k v^k \|_1.$$

Such a t^k exists (and is positive). Set $u^{k+1} = u^k + t^k v^k$, and continue this process.

We shall prove the convergence of this algorithm. We first prove two ancillary lemmas which will also be used in the analysis of Algorithm 3.

Lemma 7.16. *Assume* $g_m, g \in C(K)$, $\lim_{m \to \infty} \| g_m - g \|_\infty = 0$, *and* $\mu(Z(g)) = 0$. *Then*

$$\lim_{m \to \infty} \int_K |\text{sgn } g_m - \text{sgn } g| d\mu = 0.$$

Proof. Set $G_m = \{ x : \text{sgn } g_m(x) \neq \text{sgn } g(x) \}$. Then

$$\int_K |\text{sgn } g_m - \text{sgn } g| d\mu \leq 2\mu(G_m).$$

It therefore suffices to prove that

$$\lim_{m \to \infty} \mu(G_m) = 0.$$

Assume that $\lim_{m \to \infty} \mu(G_m) \neq 0$. There then exists a $\delta > 0$ and a subsequence $\{ m_k \}_{k=1}^\infty$ for which $\mu(G_{m_k}) \geq \delta > 0$ for all k.

Set

$$A_n = \{ x : |g(x)| < \frac{1}{n} \}.$$

For each fixed n there exists an M_n such that, for all $m_k \geq M_n$,

$$\| g_{m_k} - g \|_\infty < \frac{1}{n}.$$

Now, if $x \notin A_n$ then $|g(x)| \geq 1/n$, and thus for all $m_k \geq M_n$

$$\text{sgn } g(x) = \text{sgn } g_{m_k}(x).$$

Thus, if $x \notin A_n$ then $x \notin G_{m_k}$ for all $m_k \geq M_n$. Therefore $\mu(A_n) \geq \mu(G_{m_k}) \geq \delta$ for each n.

By definition

$$A_1 \supseteq A_2 \supseteq A_3 \supseteq \cdots .$$

Thus

$$\delta \le \lim_{n \to \infty} \mu(A_n) = \mu(\bigcap_{n=1}^{\infty} A_n) = \mu(Z(g)).$$

But $\mu(Z(g)) = 0$, a contradiction. □

Lemma 7.17. *Assume $\mu(Z(f - u)) = 0$ for all $u \in U$. We are given $u^k, u^* \in U$, and a subsequence $\{k_m\}$ of $\{k\}$ such that:*

a) $\|f - u^{k+1}\|_1 \le \|f - u^k\|_1$, *all* k

b) $\lim\limits_{m \to \infty} u^{k_m} = u^*$

c) $\int_K \mathrm{sgn}(f - u^{k+1})u^{k+1}d\mu = \int_K \mathrm{sgn}(f - u^{k+1})u^k d\mu$, *all* k.

Set

$$C_k = \{x : \mathrm{sgn}(f - u^{k+1})(x) \ne \mathrm{sgn}(f - u^k)(x)\} .$$

Then

$$\lim_{m \to \infty} \mu(C_{k_m}) = 0 .$$

Proof. From (a), (c), and $\mu(Z(f - u^{k_m})) = \mu(Z(f - u^{k_m+1})) = 0$,

$$\begin{aligned}
0 \le & \|f - u^{k_m}\|_1 - \|f - u^{k_m+1}\|_1 \\
= & \int_K \mathrm{sgn}(f - u^{k_m})(f - u^{k_m})d\mu - \int_K \mathrm{sgn}(f - u^{k_m+1})(f - u^{k_m+1})d\mu \\
= & \int_K [\mathrm{sgn}(f - u^{k_m}) - \mathrm{sgn}(f - u^{k_m+1})](f - u^{k_m})d\mu \\
= & 2 \int_{C_{k_m}} \mathrm{sgn}(f - u^{k_m})(f - u^{k_m})d\mu \\
= & 2 \int_{C_{k_m}} |f - u^{k_m}|d\mu .
\end{aligned}$$

Since the $\|f - u^k\|_1$ are a non-increasing sequence of non-negative numbers,

$$\lim_{m \to \infty} \|f - u^{k_m}\|_1 - \|f - u^{k_m+1}\|_1 = 0 .$$

Thus

$$\lim_{m \to \infty} \int_{C_{k_m}} |f - u^{k_m}|d\mu = 0 .$$

We claim that

$$\lim_{m \to \infty} \int_{C_{k_m}} |f - u^*|d\mu = 0 .$$

This follows from (b) and the inequality

$$\int_{C_{k_m}} |f - u^*|d\mu \le \int_{C_{k_m}} |f - u^{k_m}|d\mu + \int_{C_{k_m}} |u^{k_m} - u^*|d\mu .$$

Since $f - u^*$ is a fixed function in $C(K)$, and $\mu(Z(f - u^*)) = 0$, it now easily follows that

$$\lim_{m \to \infty} \mu(C_{k_m}) = 0.$$ □

We now prove the convergence of the above algorithm. For $k = 1, 2, \ldots$, set $\sigma_k = \|f - u^k\|_1$ and,

$$\sigma_0 = \min\{\|f - u\|_1 : u \in U\}.$$

Theorem 7.18. *Let u^k, σ_k and σ_0 be as defined above. Then*

$$\lim_{k \to \infty} \sigma_k = \sigma_0.$$

Furthermore the $\{u^k\}$ are uniformly bounded and every convergent subsequence of the u^k converges to a best approximant to f from U.

Proof. By definition,

$$\|f - u^{k+1}\|_1 \le \|f - u^k\|_1, \quad k = 1, 2, \ldots$$

i.e., the σ_k are a non-increasing sequence of non-negative numbers. From this fact it easily follows that the $\{u^k\}$ are uniformly bounded and $\lim_{k \to \infty} \sigma_k$ exists.

Let u^* be a cluster point of the sequence $\{u^k\}$. Thus $\lim_{m \to \infty} u^{k_m} = u^*$ on some subsequence $\{k_m\}$. It suffices to prove that u^* is a best approximant to f from U.

Associated with u^{k_m}, we determined v^{k_m} and t^{k_m}, where

$$v^{k_m} = \sum_{i=1}^{n} \left(\int_K \operatorname{sgn}(f - u^{k_m}) u_i d\mu \right) u_i$$

and t^{k_m} is given by

$$\min_t \|f - u^{k_m} - t v^{k_m}\|_1 = \|f - u^{k_m} - t^{k_m} v^{k_m}\|_1.$$

From Lemma 7.16, we have that

$$\lim_{m \to \infty} \int_K \operatorname{sgn}(f - u^{k_m}) u_i d\mu = \int_K \operatorname{sgn}(f - u^*) u_i d\mu, \quad i = 1, \ldots, n.$$

We claim that in addition

$$\lim_{m \to \infty} \int_K \operatorname{sgn}(f - u^{k_m+1}) u_i d\mu = \int_K \operatorname{sgn}(f - u^*) u_i d\mu, \quad i = 1, \ldots, n.$$

This is a consequence of Lemmas 7.16 and 7.17. That is,

$$\left| \int_K [\operatorname{sgn}(f - u^{k_m+1}) - \operatorname{sgn}(f - u^*)] u_i d\mu \right|$$

$$\le \int_K |\operatorname{sgn}(f - u^{k_m+1}) - \operatorname{sgn}(f - u^{k_m})| u_i d\mu$$

$$+ \int_K |\operatorname{sgn}(f - u^{k_m}) - \operatorname{sgn}(f - u^*)| u_i d\mu$$

$$\le \|u_i\|_\infty \left[\int_K |\operatorname{sgn}(f - u^{k_m+1}) - \operatorname{sgn}(f - u^{k_m})| d\mu \right.$$

$$\left. + \int_K |\operatorname{sgn}(f - u^{k_m}) - \operatorname{sgn}(f - u^*)| d\mu \right].$$

As $m \uparrow \infty$, the first integral tends to zero from Lemma 7.17, while the second integral tends to zero from Lemma 7.16.

By construction,

$$
\begin{aligned}
0 &= \int_K \operatorname{sgn}(f - u^{k_m+1}) v^{k_m} d\mu \\
&= \sum_{i=1}^{n} \left(\int_K \operatorname{sgn}(f - u^{k_m+1}) u_i d\mu \right) \cdot \left(\int_K \operatorname{sgn}(f - u^{k_m}) u_i d\mu \right).
\end{aligned}
$$

Letting $m \uparrow \infty$, we obtain

$$
0 = \sum_{i=1}^{n} \left(\int_K \operatorname{sgn}(f - u^*) u_i d\mu \right)^2.
$$

Thus

$$
\int_K \operatorname{sgn}(f - u^*) u_i d\mu = 0, \quad i = 1, \ldots, n,
$$

which implies that u^* is a best approximant to f from U. □

In the above algorithm it was necessary to find, at each step, a best approximant from the one-dimensional subspace $\operatorname{span}\{v^k\}$, i.e., a line search. Let us consider this problem in somewhat more detail.

Assume that we are given $f - \tilde{u}$ where $\tilde{u} \in U$, and we wish to find a $\tilde{t}$ satisfying

$$
\min_{t} \|f - \tilde{u} - tv\|_1 = \|f - \tilde{u} - \tilde{t}v\|_1
$$

where $v \in U$, $v \neq 0$. Under our hypothesis $\mu(Z(f - u)) = 0$ for all $u \in U$, this is equivalent to finding a $\tilde{t}$ for which

$$
\int_K \operatorname{sgn}(f - \tilde{u} - \tilde{t}v) v \, d\mu = 0.
$$

For each $t \in \mathbb{R}$, let

$$
I(t) = \int_K \operatorname{sgn}(f - \tilde{u} - tv) v \, d\mu.
$$

It is easily seen that $\lim_{t \to \infty} I(t) = -\|v\|_1$ and $\lim_{t \to -\infty} I(t) = \|v\|_1$. Furthermore, since

$$
I(t_0) = -\frac{d}{dt} \|f - \tilde{u} - tv\|_1|_{t=t_0}
$$

(which exists since $\mu(Z(f - u)) = 0$ for all $u \in U$), and $\|f - \tilde{u} - tv\|_1$ is convex in t, we have that $I(t)$ is a continuous non-increasing function of t. (This can also be proved directly.) Thus there exists an interval $[t_1, t_2]$, $-\infty < t_1 \leq t_2 < \infty$, such that $I(t) = 0$ if and only if $t \in [t_1, t_2]$. What this analysis proves is that this line search is equivalent to the problem of finding a zero of a monotone continuous function defined on $\mathbb{R}$.

Algorithm 3. This algorithm is a variant of the previous algorithm. It is actually a weaker version of that algorithm. We include it more for theoretical interest. We again assume that $\mu(Z(f - u)) = 0$ for all $u \in U$.

The algorithm runs as follows. Let $u_1, \ldots, u_n$ be a fixed basis for U. Let u^0 be any starting point in U. Assume that we are given u^k. Let $k+1 = rn+i$, $i \in \{1, \ldots, n\}$, $r \geq 0$.

Set

$$u^{k+1} = u^k + t^k u_i$$

where t^k is chosen so that

$$\min_t \|f - u^k - tu_i\|_1 = \|f - u^k - t^k u_i\|_1$$

i.e., $\int_K \mathrm{sgn}(f - u^k - t^k u_i) u_i d\mu = 0$.

In other words, unlike in the previous algorithm where we always chose a descent direction dependent on where we were, here we search in n fixed linearly independent directions, cyclically.

Note that $\|f - u^{k+1}\|_1 \leq \|f - u^k\|_1$ for all k, and therefore the $\{u^k\}$ are uniformly bounded. In the proof of the convergence of this algorithm we use Lemmas 7.16 and 7.17. Note that the conditions (a) and (c) of Lemma 7.17 hold.

As usual, set

$$\sigma_k = \|f - u^k\|_1,$$

and

$$\sigma_0 = \min_{u \in U} \|f - u\|_1.$$

Theorem 7.19. *Under the above assumptions*

$$\lim_{k \to \infty} \sigma_k = \sigma_0.$$

Furthermore the $\{u^k\}$ are uniformly bounded and every convergent subsequence of the u^k converges to a best approximant to f from U.

Proof. Assume $\lim_{m \to \infty} u^{k_m} = u^*$. We must prove that

$$\int_K \mathrm{sgn}(f - u^*) u_i d\mu = 0, \quad i = 1, \ldots, n.$$

Without loss of generality, we may assume that $k_m = r_m n + i$ for some fixed $i \in \{1, \ldots, n\}$. For convenience, we set $i = 1$.

Since the $\{u^k\}$ are uniformly bounded, there exists a subsequence of $\{r_m\}$, which we again denote by $\{r_m\}$, on which $\{u^{r_m n+i}\}$ converges uniformly to some $\tilde{u}^i$, $i = 2, \ldots, n-1$. Set

$$C_{r_m n+i} = \{x : \mathrm{sgn}(f - u^{r_m n+i})(x) \neq \mathrm{sgn}(f - u^{r_m n+i+1})(x)\}$$

$i = 1, \ldots, n-1$.

We now prove that

$$\int_K \mathrm{sgn}(f - u^*) u_i d\mu = 0, \quad i = 1, \ldots, n.$$

Case 1. $i = 1$.

Since

$$\int_K \text{sgn}(f - u^{r_m n+1})u_1 d\mu = 0$$

(by definition), and $\lim_{m\to\infty} u^{r_m n+1} = u^*$, it follows from Lemma 7.16 that

$$\int_K \text{sgn}(f - u^*)u_1 d\mu = 0.$$

Case 2. $i \in \{2, \ldots, n\}$.

For each i,

$$\left|\int_K \text{sgn}(f - u^*)u_i d\mu\right| \leq \left|\int_K [\text{sgn}(f - u^*) - \text{sgn}(f - u^{r_m n+1})]u_i d\mu\right|$$
$$+ \left|\int_K \text{sgn}(f - u^{r_m n+1})u_i d\mu\right|.$$

From Lemma 7.16, we have

$$\lim_{m\to\infty} \int_K [\text{sgn}(f - u^*) - \text{sgn}(f - u^{r_m n+1})]u_i d\mu = 0.$$

It therefore suffices to prove that

$$\lim_{m\to\infty} \int_K \text{sgn}(f - u^{r_m n+1})u_i d\mu = 0.$$

Now

$$\int_K \text{sgn}(f - u^{r_m n+1})u_i d\mu = \sum_{j=1}^{i-1} \int_K [\text{sgn}(f - u^{r_m n+j}) - \text{sgn}(f - u^{r_m n+j+1})]u_i d\mu$$

since

$$\int_K \text{sgn}(f - u^{r_m n+i})u_i d\mu = 0,$$

by construction. Thus

$$\left|\int_K \text{sgn}(f - u^{r_m n+1})u_i d\mu\right|$$

$$\leq \sum_{j=1}^{i-1} \int_K |\text{sgn}(f - u^{r_m n+j}) - \text{sgn}(f - u^{r_m n+j+1})||u_i|d\mu$$

$$\leq \|u_i\|_\infty \sum_{j=1}^{i-1} \left(\int_K |\text{sgn}(f - u^{r_m n+j}) - \text{sgn}(f - u^{r_m n+j+1})|d\mu\right)$$

$$\leq 2\|u_i\|_\infty \sum_{j=1}^{i-1} \mu(C_{r_m n+j}).$$

From Lemma 7.17,

$$\lim_{m\to\infty} \mu(C_{r_m n+j}) = 0, \quad j = 1, \ldots, n-1$$

since the $\{u^{r_m n + j}\}$ converge uniformly to $\widetilde{u}^j$. Thus

$$\int_K \operatorname{sgn}(f - u^*) u_i \, d\mu = 0 \,, \quad i = 1, \dots, n \,.$$

This proves that u^* is a best approximant to f from U.   □

Remark. We again emphasize the fact that, without the assumption that $\mu(Z(f - u)) = 0$ for all $u \in U$, this and the previous algorithm need not converge to the optimal value.

Approach 4. What we shall present here is not an algorithm. It is rather an attempt to explain in part when and how a Newton type method can be applied. Any Newton method employs both the gradient and the Hessian matrix. If

$$H(\mathbf{a}) = \left\| f - \sum_{i=1}^n a_i u_i \right\|_1 ,$$

where $\mathbf{a} = (a_1, \dots, a_n)$, then the Hessian matrix is given by

$$\left(\frac{\partial^2 H}{\partial a_i \partial a_j} \right)_{i,j=1}^n$$

if it exists. Since H is convex, this matrix is positive semi-definite, and corresponds to the quadratic term in the Taylor expansion of H. The linear term in the Taylor expansion is given by the gradient. A Newton type method uses a descent direction based on the first two terms of this expansion, rather than on just the first. If H has a gradient $\mathbf{g}$ at $\mathbf{a}$, and a Hessian matrix F at $\mathbf{a}$ which is invertible, then Newton's method uses the descent direction $\mathbf{d} = F^{-1}\mathbf{g}$ (and involves no line search).

We shall formulate certain specific conditions under which the Hessian matrix exists, and calculate it. As was noted numerous times, the gradient of H at $\mathbf{a}$ exists if and only if $\mu(Z(f - \sum_{i=1}^n a_i u_i)) = 0$. The Hessian matrix will only exist under much more stringent conditions.

We here assume that $K = [a, b]$, f and U are $C^1[a, b]$ functions, and $f - \sum_{i=1}^n a_i^* u_i$ has only a finite number of zeros in $[a, b]$ given by

$$a < t_1 < \cdots < t_m < b$$

(i.e., none at the endpoints) and

$$(f - \sum_{i=1}^n a_i^* u_i)'(t_j) \neq 0, \quad j = 1, \dots, m \,.$$

Under these assumptions, we shall calculate the Hessian of

$$H(a_1, \dots, a_n) = \left\| f - \sum_{i=1}^n a_i u_i \right\|_1$$

at $a_i = a_i^*$, $i = 1, \ldots, n$.

Proposition 7.20. *Under the above assumptions*

$$\frac{\partial^2 H}{\partial a_k \partial a_j}\bigg|_{\mathbf{a}=\mathbf{a}^*} = 2 \sum_{\ell=1}^{m} \frac{u_j(t_\ell)u_k(t_\ell)}{|(f - \sum_{i=1}^{n} a_i^* u_i)'(t_\ell)|} .$$

Proof. Let us recall that

$$H_j(\mathbf{a}^*) = \frac{\partial H}{\partial a_j}\bigg|_{\mathbf{a}=\mathbf{a}^*} = -\int_a^b \mathrm{sgn}(f - \sum_{i=1}^{n} a_i^* u_i)u_j d\mu .$$

From our assumptions, it is easily seen that H_j exists and is continuous in a neighborhood of $\mathbf{a}^*$, for each $j = 1, \ldots, n$.

Let $\mathbf{e}^k$ denote the kth unit vector in $\mathbb{R}^n$. We shall prove that

$$\lim_{\varepsilon \to 0} \frac{H_j(\mathbf{a}^* + \varepsilon \mathbf{e}^k) - H_j(\mathbf{a}^*)}{\varepsilon} = 2 \sum_{\ell=1}^{m} \frac{u_j(t_\ell)u_k(t_\ell)}{|(f - \sum_{i=1}^{n} a_i^* u_i)'(t_\ell)|} .$$

Set $g = f - \sum_{i=1}^{n} a_i^* u_i$. Then

$$\frac{1}{\varepsilon}[H_j(\mathbf{a}^* + \varepsilon \mathbf{e}^k) - H_j(\mathbf{a}^*)] = \frac{1}{\varepsilon}\left(\int_a^b [\mathrm{sgn}\, g - \mathrm{sgn}(g - \varepsilon u_k)]u_j d\mu\right).$$

Since g has only the ordered zeros $\{t_\ell\}_{\ell=1}^{m}$ in $[a, b]$, and $g'(t_\ell) \neq 0$, $\ell = 1, \ldots, m$, it follows that $g'(t_\ell)g'(t_{\ell+1}) < 0$, $\ell = 1, \ldots, m-1$. Assume without loss of generality that $g'(t_1) > 0$. Set $t_0 = a$ and $t_{m+1} = b$. Then

$$(-1)^\ell g(t) > 0, \quad t \in (t_{\ell-1}, t_\ell),$$

$\ell = 1, \ldots, m+1$, and

$$\int_a^b \mathrm{sgn}(g)u_j d\mu = \sum_{\ell=1}^{m+1}(-1)^\ell \int_{t_{\ell-1}}^{t_\ell} u_j d\mu .$$

For ε small, $g - \varepsilon u_k$ also has exactly m zeros in $[a, b]$ at

$$a < t_1(\varepsilon) < \cdots < t_m(\varepsilon) < b,$$

$\lim_{\varepsilon \to 0} t_\ell(\varepsilon) = t_\ell$, $\ell = 1, \ldots, m$, and

$$(-1)^\ell (g - \varepsilon u_k)(t) > 0, \quad t \in (t_{\ell-1}(\varepsilon), t_\ell(\varepsilon)),$$

$\ell = 1, \ldots, m+1$, where $t_0(\varepsilon) = a$ and $t_{m+1}(\varepsilon) = b$. Thus

$$\begin{aligned}
\frac{1}{\varepsilon}\int_a^b [\mathrm{sgn}\, g - \mathrm{sgn}(g - \varepsilon u_k)]u_j d\mu &= \frac{1}{\varepsilon}\sum_{\ell=1}^{m+1}(-1)^\ell \left[\int_{t_{\ell-1}}^{t_\ell} u_j d\mu - \int_{t_{\ell-1}(\varepsilon)}^{t_\ell(\varepsilon)} u_j d\mu\right] \\
&= \frac{2}{\varepsilon}\sum_{\ell=1}^{m}(-1)^\ell \int_{t_\ell(\varepsilon)}^{t_\ell} u_j d\mu \\
&= 2 \sum_{\ell=1}^{m}(-1)^\ell \frac{(t_\ell - t_\ell(\varepsilon))}{\varepsilon} u_j(s_\ell(\varepsilon)),
\end{aligned}$$

where $\lim_{\varepsilon \to 0} s_\ell(\varepsilon) = t_\ell$, $\ell = 1, \ldots, m$. From Taylor's Theorem,

$$t_\ell(\varepsilon) - t_\ell = \frac{\varepsilon u_k(t_\ell(\varepsilon))}{(g - \varepsilon u_k)'(t_\ell)} + \text{lower order terms}.$$

These 'lower order terms' have no bearing here and we disregard them.
Thus

$$\lim_{\varepsilon \to 0} \frac{H_j(\mathbf{a}^* + \varepsilon \mathbf{e}^k) - H_j(\mathbf{a}^*)}{\varepsilon} = \lim_{\varepsilon \to 0} 2 \sum_{\ell=1}^{m} (-1)^\ell \frac{(t_\ell - t_\ell(\varepsilon))}{\varepsilon} u_j(s_\ell(\varepsilon))$$

$$= \lim_{\varepsilon \to 0} 2 \sum_{\ell=1}^{m} (-1)^{\ell+1} \frac{u_k(t_\ell(\varepsilon)) u_j(s_\ell(\varepsilon))}{(g - \varepsilon u_k)'(t_\ell)}$$

$$= 2 \sum_{\ell=1}^{m} (-1)^{\ell+1} \frac{u_k(t_\ell) u_j(t_\ell)}{g'(t_\ell)}.$$

Since $(-1)^{\ell+1} g'(t_\ell) > 0$, $\ell = 1, \ldots, m$, and $g = f - \sum_{i=1}^{n} a_i^* u_i$, we obtain the claim of the proposition. $\qquad\qquad\qquad\qquad\qquad\qquad\qquad\qquad\qquad\qquad$ □

Notes and References

The short treatment of gradients and subgradients in Section 2 (except for Proposition 7.4) may be found in Shor [1985]. Both Shor [1985] and Kiwiel [1985] are excellent sources for a thorough discussion of subgradient algorithms.

The observation that the two-sided ℓ_1^m-approximation problem could be formulated as a linear programming problem, as explained at the beginning of Section 4, was first made by Wagner [1959]. Barrodale and associates have studied this problem in detail using this linear programming problem formulation, but exploiting the special nature of the problem. This work, with generalizations to constrained problems, may be found in Barrodale, Young [1966], Barrodale [1970], and Barrodale, Roberts [1973], [1978]. Work on these same problems, using this approach, may also be found in Abdelmalek [1974], [1975], [1985], Armstrong, Hultz [1977], and Spyropoulos, Kiountouzis, Young [1973].

Both algorithms considered in Section 4 are gradient methods. The basic ideas behind Algorithm 1 may be found in Claerbout, Muir [1973]. With modifications, Algorithm 1, the line search analysis of Proposition 7.5, and many of the ideas in Cases 1 and 3 are in Bartels, Conn, Sinclair [1978]. Algorithm 2 is a variant of an algorithm of Usow [1967b]. His original algorithm was unnecessarily restricted to T-systems. T-systems have almost no relevance in ℓ_1^m-approximation. More importantly however, the proposed algorithm did not necessarily converge to the correct value. Originally, as proposed by Usow, this algorithm was missing the step of looking for a descent direction once a flat where $\|\mathbf{b} - \mathbf{u}^k\|_1 = \|\mathbf{b} - \mathbf{u}^{k+n}\|_1$ had been attained. A different algorithm, based on the dual problem, was suggested by Robers, Ben-Israel [1969]. An extensive study of algorithms for the two-sided ℓ_1^m-approximation problem may be found in Chapter 3 of Osborne [1985].

Algorithms for one-sided L^1-approximation (Section 5) have received relatively little consideration. A special case of Algorithm 1 is due to Lewis [1970]. He restricted his attention to T-systems and $K = [a, b]$. We have employed some of his ideas, but the proofs given here are simpler and more direct. The idea behind Algorithm 2 is similar to that found in Hettich, Zencke [1982, p.75]. Algorithm 3 is taken from Hettich, Zencke [1982, p.156]. Algorithm 4 was developed jointly with H. Strauss.

Algorithms for two-sided L^1-approximation (Section 6) have received somewhat more attention. Algorithm 1, i.e., Theorem 7.15, was originally proved for algebraic polynomials on $K = [a, b]$, with $a = x_0^m < \cdots < x_m^m = b$ and $\delta_i^m = x_i^m - x_{i-1}^m$, $i = 1, \ldots, m$, by Motzkin, Walsh [1956, Theorem 7.2]. Theorem 7.15 is a special case of a theorem of Kripke [1964]. Algorithm 2 we were unable to find in the literature in the form stated. However it is a standard gradient algorithm (cf. Zangwill [1969]). Algorithm 3 is a variant of an algorithm of Usow [1967a]. Usow proposed the algorithm for T-systems on $K = [a, b]$. The convergence proof is totally different. As mentioned in the text, both Algorithms 2 and 3 do not necessarily converge to the correct value σ_0 if $\mu(Z(f - u)) \neq 0$ for some $u \in U$. Usow incorrectly supposed that, in Algoritm 3, it was sufficient to assume that $\mu(Z(f - u^k)) \neq 0$ for all k. Marti [1975] proposed a method of circumventing this difficulty. However his method is restricted to a subclass of T-systems, and involves searches in an infinite number of directions. Algorithm 3 goes under various names. It is sometimes called a cyclic coordinate algorithm (see Zangwill [1969, p.111]). But it is also essentially an alternating algorithm (due to von Neumann), or a Diliberto, Straus algorithm (see Light, Cheney [1985]). Lemmas 7.16 and 7.17, which elegantly prove the convergence of Algorithms 2 and 3, are taken from Light, Holland [1984].

Watson [1981] used the Hessian matrix, as given in Approach 4, to construct an algorithm for L^1-approximation. This in turn is a generalization of a method proposed by Glashoff, Schultz [1979], where the Hessian matrix is also exploited but never explicitly named. The calculation of the Hessian, as given in Proposition 7.20, is taken from Wolfe [1976].

Further discussions of algorithms for L^1-approximation may be found in Rivlin [1969] and Watson [1980]. We purposely did not include, in this chapter, algorithms utilizing specialized structures of the underlying approximating subspaces. Obviously, however, one can and should tailor an algorithm to the subspace. In this connection, it is worth noting that, if the subspace is a T-system (or even a WT-system), then a trusted and true starting point u^0 in any of the algorithms in Section 6 is given by interpolating f at the 'canonical points' as in (B.1) of Appendix B.

Appendix A
T- and WT-Systems

As the title indicates, this appendix contains a collection of facts on T- and WT-systems. It is not our intention to provide a comprehensive treatise on this topic. A more all-inclusive study of T- and WT-systems may be found in Karlin, Studden [1966], Krein, Nudel'man [1977], and Zielke [1979]. Our purpose is to compile, in a cohesive fashion, some of the results used in the previous chapters. Part I contains a few facts concerning T-systems. In Part II we consider WT-systems. In the problem of L^1-approximation, WT-systems play a more fundamental role that T-systems (see Chapter IV). As such, the emphasis of this appendix is more on the subject of WT-systems. Appendix B also contains a great deal of material on T- and WT-systems.

Part I. T-Systems

Chebyshev systems (the name was given by Bernstein [1926]) are abbreviated T-systems since at one time the Cyrillic transliteration gave us the spelling Tchebycheff (and variants thereof). They are fundamental in the study of approximation theory. At times T-systems and T-spaces are referred to as Haar systems and Haar spaces, respectively. Formally a system in this context is meant to be a finite sequence of functions, and a space is their linear span. None the less, we shall use the term *system* for both the system and the space. We will however differentiate between T-systems and Haar systems.

Definition 1. Let B be a compact Hausdorff space and $C(B)$ the set of real-valued continuous functions on B. The n-dimensional subspace U of $C(B)$ is said to be a *Haar system* if no $u \in U$, $u \neq 0$, has more than $n - 1$ distinct zeros in B.

An equivalent formulation of this definition is the following: U is a Haar system on B if given any basis $u_1, \ldots, u_n$ for U, and any n distinct points $x_1, \ldots, x_n$ in B, then $\det(u_i(x_j))_{i,j=1}^n \neq 0$ (see the proof of Proposition 1). Alternatively, given any n distinct points $x_1, \ldots, x_n$ in B, and any arbitrary n data (values) $\alpha_1, \ldots, \alpha_n$, there exists a unique $u \in U$ satisfying $u(x_i) = \alpha_i$, $i = 1, \ldots, n$.

Haar systems are so named because Haar [1918] proved that a finite-dimensional subspace U of $C(B)$ is a unicity space for $C(B)$ in the uniform norm if and only if U is a Haar system.

The term '*T*-system' has been generally reserved for Haar systems defined on an interval of $\mathbb{R}$. Different authors use these terms somewhat differently, but this is exactly the sense in which we will use it.

Definition 2. Let M be an interval in $\mathbb{R}$. The n-dimensional subspace U is a *T-system* on M if $U \subset C(M)$ and no $u \in U$, $u \neq 0$, has more than $n - 1$ distinct zeros in M.

As a simple consequence of the definition we have:

Proposition 1. *Let U be an n-dimensional subspace of $C(M)$, and $u_1, \ldots, u_n$ a basis for U. Then U is a T-system in $C(M)$ if and only if there exists an $\varepsilon \in \{-1, 1\}$, ε fixed, for which*

$$(A.1) \qquad \varepsilon \det(u_i(x_j))_{i,j=1}^n > 0$$

for all $x_1 < \cdots < x_n$ in M.

Remark. This is not true if U is a Haar system on B and B is not an interval in $\mathbb{R}$.

Proof. ($\Leftarrow$). Assume U is not a *T*-system. There then exists a $u \in U$, $u \neq 0$, and points $x_1, \ldots, x_n$ in M for which $u(x_i) = 0$, $i = 1, \ldots, n$. But then $\det(u_i(x_j))_{i,j=1}^n = 0$, contradicting (A.1).

($\Rightarrow$). Assume (A.1) does not hold. If $\det(u_i(x_j))_{i,j=1}^n = 0$ for some n distinct points $\{x_i\}_{i=1}^n$, there then exists a $u \in U$, $u \neq 0$, for which $u(x_i) = 0$, $i = 1, \ldots, n$, and U is not a *T*-system. We may therefore assume that $\det(u_i(x_j))_{i,j=1}^n \neq 0$ for any n distinct points $\{x_i\}_{i=1}^n$. Let $x_1 < \cdots < x_n$ and $y_1 < \cdots < y_n$ be all in M. Assume $\det(u_i(x_j))_{i,j=1}^n \cdot \det(u_i(y_j))_{i,j=1}^n < 0$. For each $\lambda \in [0, 1]$, let $z_j(\lambda) = \lambda x_j + (1 - \lambda) y_j$, $j = 1, \ldots, n$. Then $z_1(\lambda) < \cdots < z_n(\lambda)$ and the $z_j(\lambda)$ are all in M. Thus $\det(u_i(z_j(\lambda)))_{i,j=1}^n \neq 0$ for all $\lambda \in [0, 1]$. From continuity considerations, a contradiction immediately ensues. □

This equivalent definition based on the determinantal criterion (A.1) is very useful. In what follows, we will for ease of exposition set $\overline{M} = [a, b]$.

Proposition 2. *Let U be an n-dimensional T-system in $C(M)$. Given any $x_1 < \cdots < x_{n-1}$ in M, there exists a function $u \in U$ satisfying*

$$(A.2) \qquad u(x)(-1)^{i+1} > 0, \quad x \in (x_{i-1}, x_i),$$

$i = 1, \ldots, n$, where we set $x_0 = a$, $x_n = b$. Furthermore, if $v \in U$ satisfies $v(x_i) = 0$, $i = 1, \ldots, n - 1$, then $v = \alpha u$ for some $\alpha \in \mathbb{R}$.

It will be convenient to use the following notation. For any points $\{y_j\}_{j=1}^n$ (not necessarily ordered) in M, we set

$$U \begin{pmatrix} 1, & \cdots, & n \\ y_1, & \ldots, & y_n \end{pmatrix} = \det(u_i(y_j))_{i,j=1}^n.$$

Proof. Let $u_1, \ldots, u_n$ be any basis for U. Set

$$u(x) = \varepsilon U \begin{pmatrix} 1, & 2, & \cdots & , n \\ x, & x_1, & \cdots & , x_n \end{pmatrix}$$

where ε is as in (A.1). As a consequence of Proposition 1, u satisfies (A.2). Assume $v \in U$, $v \neq 0$, satisfies $v(x_i) = 0$, $i = 1, \ldots, n-1$. Choose $x_n \in M$ where $x_n \notin \{x_1, \ldots, x_{n-1}\}$. Then $u(x_n) \cdot v(x_n) \neq 0$. There exists an $\alpha \neq 0$ such that $(v - \alpha u)(x_n) = 0$. Thus $v - \alpha u$ vanishes at the n points $x_1, \ldots, x_n$. Since U is a T-system of dimension n, $v - \alpha u = 0$, i.e., $v = \alpha u$. This proves the claim of uniqueness.   □

Before continuing, let us introduce some additional notation. The reader will recall that for $f \in C(M)$, the zero set of f is denoted by $Z(f)$. In this appendix we abuse our earlier notation and let $|Z(f)|$ denote the *number* of distinct zeros of f in M. Thus U is an n-dimensional T-system in $C(M)$ if and only if $|Z(u)| \leq n - 1$ for every $u \in U$, $u \neq 0$. This result may be further strengthened if we distinguish between two types of zeros, namely zeros with sign changes and zeros without sign changes.

Definition 3. Let $f \in C(M)$. A zero of f at x_0 in the interior of M is said to be a *nonnodal zero* of f provided that f does not change sign at x_0, i.e., $f(x_0 - \delta)f(x_0 + \varepsilon) \geq 0$ for all ε and δ sufficiently small. All other zeros of f, including zeros at the endpoints of M (if such points exist) are called *nodal zeros*.

Definition 4. Let $f \in C(M)$. Then $|\widetilde{Z}(f)|$ counts the number of zeros of f, where nodal zeros are counted once and nonnodal zeros twice.

With these definitions we can now state.

Proposition 3. *Let U be an n-dimensional subspace of $C(M)$. Then U is a T-system in $C(M)$ if and only if $|\widetilde{Z}(u)| \leq n - 1$ for every $u \in U$, $u \neq 0$.*

Proof. ($\Leftarrow$). Since $|Z(f)| \leq |\widetilde{Z}(f)|$ for every $f \in C(M)$, we have $|Z(u)| \leq n-1$ for every $u \in U$, $u \neq 0$, and U is therefore a T-system.

($\Rightarrow$). Assume U is a T-system and $|\widetilde{Z}(u)| \geq n$ for some $u \in U$, $u \neq 0$. Let $Z(u) = \{x_1, \ldots, x_k\}$. Since U is a T-system, $k \leq n-1$. Since $|\widetilde{Z}(u)| \geq n$, at least one of these zeros is nonnodal. We add to the set $\{x_1, \ldots, x_k\}$ the points $x_j + \sigma$, $\sigma > 0$, σ small, for each nonnodal zero x_j, and also the point $x_i - \sigma$ for the first nonnodal zero x_i. This new set contains at least $n + 1$ points. We choose $n + 1$ points $y_1 < \cdots < y_{n+1}$ from this set with the property that $u(y_i)(-1)^i \delta \geq 0$, $i = 1, \ldots, n+1$, for some choice of $\delta \in \{-1, 1\}$. Such a choice is possible. Since $|Z(u)| \leq n-1$, strict inequality must hold for at least two of these points. Let $u_1, \ldots, u_n$ be any basis for U and consider the determinant

$$\begin{vmatrix} u(y_1) & \cdots & u(y_{n+1}) \\ u_1(y_1) & \cdots & u_1(y_{n+1}) \\ \vdots & & \vdots \\ u_n(y_1) & \cdots & u_n(y_{n+1}) \end{vmatrix}.$$

The determinant is zero since the first row is a linear combination of the other rows. However, expanding by the first row we obtain $0 = \sum_{i=1}^{n+1} a_i u(y_i)$, where $a_i(-1)^{i+1}\varepsilon > 0$, $i = 1, \ldots, n+1$, and ε is as in (A.1). Since $u(y_i)(-1)^i \delta \geq 0$, $i = 1, \ldots, n+1$, with at least two strict inequalities, a contradiction ensues. □

What about a converse result to Proposition 3? That is, given $s_1, \ldots, s_k$, $t_1, \ldots, t_l$ in M, all distinct, with $t_i \in$ int M, all i, and $k + 2l \leq n - 1$, does there exist a $u \in U$, $u \neq 0$, which has nodal zeros at the s_i, nonnodal zeros at the t_i, and no other zeros in M? For a general T-system, the answer is not necessarily yes. There may be additional zeros at the endpoints. However we have no need, in this work, for such a general result. The following theorem suffices (see Proposition 5.23).

Theorem 4. *Let U be an n-dimensional T-system in $C[a, b]$. Given $a \leq t_1 < \cdots < t_k \leq b$, let $\omega(t_i) = 2$ if $t_i \in (a, b)$ and $\omega(t_i) = 1$ if $t_i \in \{a, b\}$. Assume $\sum_{i=1}^n \omega(t_i) \leq n - 1$. There then exists a non-trivial non-negative $u \in U$ such that $u(t_i) = 0$, $i = 1, \ldots, k$.*

Proof. The idea of the proof is a simple one. However because of certain technical details, we divide the proof into a series of cases.

We first assume that $a < t_1 < \cdots < t_k < b$, and $n = 2m + 1$. (Thus $k \leq m$.) To the set $\{t_i\}_{i=1}^k$ we add the points $\{t_i + \delta\}_{i=1}^k$ for $\delta > 0$, δ small, and also the points $\{t_i'\}_{i=k+1}^m$ and $\{t_i' + \delta\}_{i=k+1}^m$ where, for convenience,

$$a < t_1 < t_1 + \delta < t_2 < \cdots < t_k + \delta < t_{k+1}' < t_{k+1}' + \delta < \cdots < t_m' < t_m' + \delta < b.$$

Let $\{s_i(\delta)\}_{i=1}^{2m}$ denote these $2m$ ordered points, and $u_1, \ldots, u_{2m+1}$ be any basis for U. Set

$$u^\delta(x) = \varepsilon U \begin{pmatrix} 1, & \cdots & , 2m & , 2m+1 \\ s_1(\delta), & \ldots, & s_{2m}(\delta) & , x \end{pmatrix}$$

where $\varepsilon \in \{-1, 1\}$ is as in (A.1).

From Proposition 2,

$$u^\delta(x) > 0, \quad x \in (s_{2i}(\delta), s_{2i+1}(\delta)),$$

$i = 0, 1, \ldots, m$, where $s_0(\delta) = a$, $s_{2m+1}(\delta) = b$ (and in fact $u^\delta(x) > 0$ for $x \in \{a, b\}$). Now

$$u^\delta(x) = \sum_{i=1}^{2m+1} a_i(\delta) u_i(x).$$

Normalize u^δ (multiply by a positive constant) so that $\sum_{i=1}^{2m+1} \left(a_i(\delta)\right)^2 = 1$. Letting $\delta \downarrow 0$ along an appropriate subsequence, we obtain a limit function

$$u(x) = \sum_{i=1}^{2m+1} a_i u_i(x)$$

for which $\sum_{i=1}^{2m+1} a_i^2 = 1$, i.e., u is non-trivial. Since $s_{2i}(\delta) - s_{2i-1}(\delta) \to 0$, $i = 1, \ldots, m$, u is non-negative and $u(t_i) = 0$, $i = 1, \ldots, k$. Thus the theorem is proved.

In all other cases the proof is essentially the same. Our only problem is in the choice of the set $\{s_i(\delta)\}_{i=1}^{n-1}$. We always start with the $\{t_i\}_{i=1}^{k}$, add the points $t_i + \delta$ for each $t_i \in (a,b)$, and also appropriate pairs of points t' and $t' + \delta$, $t' \in (a,b)$. Problems only arise because of the endpoints and the parity of n.

If $n = 2m+1$, and $t_1 = a$, $t_k = b$, then we use the set described above. If only one endpoint is in the set $\{t_i\}_{i=1}^{k}$, we can either add the other endpoint to the set, or add $a + \delta$ if a is in the set, or add $b - \delta$ if b is in the set.

If $n = 2m$, and $a < t_1$, $t_k < b$, then we add one of the endpoints to the set. If $t_1 = a$ and $t_k = b$, then we add either $a + \delta$ or $b - \delta$ to the set. If only one endpoint is among the $\{t_i\}_{i=1}^{k}$, then we use the set as described above. In all these cases the result follows.   □

Remark. From the above, it may be easily shown that we can always construct u as above, i.e., non-trivial, non-negative and vanishing at the $\{t_i\}_{i=1}^{k}$, with the additional property that u does not vanish at any other point in (a,b). (Construct u as above using the t_i', and $\tilde{u}$ using t_i'' different from the t_i', and take their sum.) In fact we can also construct u as above which vanishes on $[a,b]$ only at the $\{t_i\}_{i=1}^{k}$ except in the one particular case where $n = 2m+1$, and one of the endpoints is in the set $\{t_i\}_{i=1}^{k}$. In this case it may be that u necessarily vanishes at the other endpoint, cf. Zielke [1979, Chapter 10].

Part II. WT-Systems

We start with the definition of a weak Chebyshev (WT-) system. The definition to be presented depends upon a sign change count. As such, for each $f \in C[a,b]$ set

$$S(f) = \sup\{k : a \le x_1 < \cdots < x_{k+1} \le b, \ f(x_j)f(x_{j+1}) < 0, \ j = 1, \ldots, k\}.$$

Of course, if f is either non-negative or non-positive on $[a,b]$, then we set $S(f) = 0$. $S(f)$ counts the number of *sign changes* of f on $[a,b]$.

Definition 5. Let U be an n-dimensional subspace of $C[a,b]$. U is said to be a *weak Chebyshev (WT)-system* in $C[a,b]$ if $S(u) \le n - 1$ for every $u \in U$.

We could also define WT-systems on open or half-open intervals in an analogous fashion. However, since we will only consider functions continuous on $[a, b]$, these definitions are all equivalent. Note that the WT-system property is weaker than the T-system property. If $U \subset C[a, b]$ is a T-system on (a, b), then it is most certainly a WT-system on $[a, b]$.

Different equivalent definitions of WT-systems are sometimes given. Two additional equivalent definitions are the content of this next theorem. The first full statement of this theorem appeared in Jones, Karlovitz [1970]. However the proof therein was somewhat lacking. The proof given here is also partially based on Zielke [1979, Lemma 4.1].

Theorem 5. *Let U be an n-dimensional subspace of $C[a, b]$. Then the following are equivalent:*

1) $S(u) \le n - 1$ *for every $u \in U$.*
2) *Let $u_1, \ldots, u_n$ be any basis for U. There exists a $\sigma \in \{-1, 1\}$, σ fixed, such that*

$$\sigma \det(u_i(x_j))_{i,j=1}^n \ge 0$$

for every $a \le x_1 < \cdots < x_n \le b$.
3) *Given any $a = y_0 < y_1 < \cdots < y_{n-1} < y_n = b$, there exists a $u \in U$, $u \ne 0$, satisfying*

$$(-1)^j u(x) \ge 0, \ x \in (y_{j-1}, y_j), \ j = 1, \ldots, n.$$

In our proof of Theorem 5, we will use a 'smoothing' procedure. This is given by the following result.

Proposition 6. *Let $u_1, \ldots, u_n$ be linearly independent functions in $C[a, b]$. Assume that, for every $a \le x_1 < \cdots < x_n \le b$, we have $\det(u_i(x_j))_{i,j=1}^n \ge 0$. Then for each $\varepsilon > 0$ there exist functions $u_i^\varepsilon \in C[a, b]$, $i = 1, \ldots, n$, such that $\{u_i^\varepsilon\}_{i=1}^n$ is a T-system in $C[a, b]$, and each u_i^ε converges uniformly to u_i on $[a, b]$ as $\varepsilon \downarrow 0$, $i = 1, \ldots, n$.*

Proof. Let g be any fixed non-negative function in $C(\mathbb{R})$ with compact support and such that $g(x) = 1$ on $[a, b]$. Set

$$v_i(x) = \begin{cases} g(x)u_i(a), & x \le a \\ g(x)u_i(x), & a \le x \le b \\ g(x)u_i(b), & b \le x, \end{cases}$$

$i = 1, \ldots, n$.

The $\{v_i\}_{i=1}^n$ are linearly independent functions in $C(\mathbb{R})$, $v_i(x) = u_i(x)$ for all $x \in [a, b]$ and $i = 1, \ldots, n$, and as is easily checked

$$\det(v_i(x_j))_{i,j=1}^n \ge 0$$

for every choice of $x_1 < \cdots < x_n$.

For each $\varepsilon > 0$ and $i = 1, \ldots, n$, we define

$$u_i^\varepsilon(x) = \frac{1}{\varepsilon\sqrt{2\pi}} \int_{-\infty}^\infty e^{-\frac{(x-y)^2}{2\varepsilon^2}} v_i(y) \, dy.$$

These integrals are well defined since the v_i have compact support. We will show that the $\{u_i^\varepsilon\}_{i=1}^n$ satisfy the claims of the proposition.

We first prove that the $\{u_i^\varepsilon\}_{i=1}^n$ form a T-system on $[a,b]$. For given $a \leq x_1 < \cdots < x_n \leq b$, it follows from the basic composition formula (cf. Karlin [1968, p.17]) that

$$\det(u_i^\varepsilon(x_j))_{i,j=1}^n = \frac{1}{\varepsilon^n (2\pi)^{n/2}} \int \cdots \int_{y_1 < \cdots < y_n} \det(K_\varepsilon(x_j, y_k)) \cdot \det(v_i(y_k)) dy_1 \cdots dy_n,$$

where $K_\varepsilon(x,y) = \exp(-(x-y)^2/2\varepsilon^2)$. It is well known (see Karlin, Studden [1966, p.11]), that

$$\det(K_\varepsilon(x_j, y_k))_{j,k=1}^n > 0$$

for all $x_1 < \cdots < x_n$ and $y_1 < \cdots < y_n$. Furthermore $\det(v_i(y_k))_{i,k=1}^n \geq 0$ and is not identically zero on the domain of integration since the $\{v_i\}_{i=1}^n$ are linearly independent thereon. Thus

$$\det(u_i^\varepsilon(x_j))_{i,j=1}^n > 0,$$

i.e., the $\{u_i^\varepsilon\}_{i=1}^n$ form a T-system in $C[a,b]$.

It is a standard result in analysis that $u_i^\varepsilon \to v_i$ as $\varepsilon \downarrow 0$ uniformly on compact subsets of $\mathbb{R}$. Thus $u_i^\varepsilon \to u_i$ as $\varepsilon \downarrow 0$ uniformly on $[a,b]$ for each $i = 1, \ldots, n$. $\square$

Remark. The above is but one of several known 'smoothing' procedures which prove the proposition. It is the most commonly used. The $\{u_i^\varepsilon\}_{i=1}^n$ obtained actually form an extended Chebyshev (ET)-system on $[a,b]$. We have, however, no need for this stronger result.

We now prove Theorem 5.

Proof of Theorem 5. $(2) \Rightarrow (1)$. Assume that $\det(u_i(x_j))_{i,j=1}^n \geq 0$ for all $a \leq x_1 < \cdots < x_n \leq b$. We must prove that $S(u) \leq n-1$ for all $u \in U$, $u \neq 0$. Assume $u = \sum_{i=1}^n a_i u_i$, $u \neq 0$, and $S(u) \geq n$. For each $\varepsilon > 0$, let $\{u_i^\varepsilon\}_{i=1}^n$ be as given in Proposition 6. Set $u^\varepsilon(x) = \sum_{i=1}^n a_i u_i^\varepsilon$. Since $S(u) \geq n$, we have $S(u^\varepsilon) \geq n$ for all ε sufficiently small. But $\{u_i^\varepsilon\}_{i=1}^n$ is a T-system in $C[a,b]$ and $S(u^\varepsilon) \leq Z(u^\varepsilon) \leq n-1$. This is a contradiction.

$(2) \Rightarrow (3)$. We are given $a = y_0 < y_1 < \cdots < y_{n-1} < y_n = b$. For $\varepsilon > 0$, let $\{u_i^\varepsilon\}_{i=1}^n$ be as given in Proposition 6. Since $\{u_i^\varepsilon\}_{i=1}^n$ is a T-system in $C[a,b]$, there exists a unique

$$u^\varepsilon = \sum_{i=1}^n a_i(\varepsilon) u_i^\varepsilon$$

satisfying $\sum_{i=1}^n (a_i(\varepsilon))^2 = 1$, and

$$(-1)^j u^\varepsilon(x) > 0, \quad x \in (y_{j-1}, y_j),$$

$j = 1, \ldots, n$ (see Proposition 2). On some subsequence $\{\varepsilon_m\}_1^\infty$ tending to zero, $a_i(\varepsilon_m) \to b_i$, $i = 1, \ldots, n$. Set $u = \sum_{i=1}^n b_i u_i$. u is non-trivial since $\sum_{i=1}^n b_i^2 = 1$. Furthermore u^{ε_m} converges uniformly to u on $[a, b]$ and therefore

$$(-1)^j u(x) \geq 0, \quad x \in (y_{j-1}, y_j), \quad j = 1, \ldots, n.$$

We will prove that each of (1) and (3) imply (2). To this end, let us assume that (2) does not hold. As such there exist $a \leq z_1 < \cdots < z_n \leq b$ and $a \leq y_1 < \cdots < y_n \leq b$ satisfying

$(A.3)$ $\qquad \det(u_i(z_j))_{i,j=1}^n \cdot \det(u_i(y_j))_{i,j=1}^n < 0.$

We first claim that we may choose the $\{z_j\}_{j=1}^n$ and $\{y_j\}_{j=1}^n$ satisfying (A.3), and such that the two sequences differ in only one element. That is, there exist $\{w_j\}_{j=1}^{n-1}$, z and y (all distinct) for which

$(A.4)$ $\qquad \{z_j\}_{j=1}^n = \{w_j\}_{j=1}^{n-1} \cup \{z\}$ and $\{y_j\}_{j=1}^n = \{w_j\}_{j=1}^{n-1} \cup \{y\}.$

We prove this claim as follows. Assume $\det(u_i(z_j))_{i,j=1}^n > 0$. Replace z_1 by any y_{i_1} so that the resulting determinant is non-zero. This is certainly possible. Reorder the columns of this matrix so that the points are once again in increasing order of magnitude. If the determinant of this 'new' matrix is negative, we are finished. We take $\{w_j\}_{j=1}^{n-1} = \{z_j\}_{j=2}^n$, $z = z_1$, and $y = y_{i_1}$. If the determinant is positive we continue this process. That is, at each step k we replace z_k by some y_{i_k} from the remaining $\{y_j\}_{j=1}^n$. If at some stage the associated determinants change sign from positive to negative, then we choose the $\{w_j\}_{j=1}^{n-1}$, z and y, analogously to the above. That this must happen at some stage follows from the fact that after n steps we have eliminated all the $\{z_j\}_{j=1}^n$, and $\det(u_i(y_j))_{i,j=1}^n < 0$.

We therefore assume that (A.3) and (A.4) hold, and

$$w_1 < \cdots < w_{p-1} < z < w_p < \cdots < w_{n-1}$$

$$w_1 < \cdots < w_{q-1} < y < w_q < \cdots < w_{n-1}$$

for some $p, q \in \{1, \ldots, n\}$, and $y < z$ ($q \leq p$). Since $\det(u_i(z_j))_{i,j=1}^n \neq 0$, there exists a unique $u^* \in U$ satisfying $u^*(w_j) = 0$, $j = 1, \ldots, n-1$, and $u^*(z) = 1$. From (A.3) and simple linear algebra it follows that $(-1)^{p+q+1} u^*(y) > 0$.

$(3) \Rightarrow (2)$. Assume that (3) holds, (2) does not hold, and the $\{w_j\}_{j=1}^{n-1}$, z, y and u^* are as above. From continuity considerations, we may assume that $a < w_1$, and $w_{n-1} < b$. From (3), there exists a $\tilde{u} \in U$, $\tilde{u} \neq 0$, satisfying $\tilde{u}(x)(-1)^{i+p} \geq 0$ for $x \in (w_{i-1}, w_i)$, $i = 1, \ldots, n$, where $w_0 = a$ and $w_n = b$. Since U restricted to the points $\{w_j\}_{j=1}^{n-1}$ is of dimension $n-1$, we have that if $u \in U$ satisfies $u(w_j) = 0$, $j = 1, \ldots, n-1$, then $u = \alpha \tilde{u}$ for some $\alpha \in \mathbb{R}$. Since $u^*(w_j) = 0$, $j = 1, \ldots, n-1$, and $u^*(z) = 1$, $z \in (w_{p-1}, w_p)$, it follows that $u^* = \alpha \widetilde{u}$ for some $\alpha > 0$. Because $y \in (w_{q-1}, w_q)$ we therefore have $u^*(y)(-1)^{p+q} \geq 0$. But $u^*(y)(-1)^{p+q+1} > 0$. This is a contradiction.

(1)$\Rightarrow$(2). We assume that (2) does not hold, and the $\{w_j\}_{j=1}^{n-1}$, z, y and u^* are as above. Since $\det(u_i(z_j))_{i,j=1}^n \neq 0$, there exists a unique $v \in U$ satisfying

$$v(z_j) = \begin{cases} (-1)^{j+p+1}, & j = 1, \ldots, q-1 \\ (-1)^{j+p}, & j = q, \ldots, n. \end{cases}$$

Note that v has at least $n - 2$ sign changes, one in each of (z_{j-1}, z_j) for $j \in \{2, \ldots, n\} \setminus \{q\}$. Let $\widetilde{u} = v + \alpha u^*$ where $\alpha > 0$.

Now $\widetilde{u}(w_j) = v(w_j) + \alpha u^*(w_j) = v(w_j)$ for each $j = 1, \ldots, n-1$ and $\widetilde{u}(z) = v(z) + \alpha u^*(z) = v(z) + \alpha$. Since $z = z_p$, we have $v(z) = 1$ and therefore $\operatorname{sgn} \widetilde{u}(z_j) = \operatorname{sgn} v(z_j)$ for all $j = 1, \ldots, n$ and $\alpha > 0$. Since $u^*(y)(-1)^{p+q+1} > 0$, we can choose $\alpha > 0$, sufficiently large, so that $\operatorname{sgn} \widetilde{u}(y) = (-1)^{p+q+1}$. Now $y \in (z_{q-1}, z_q)$, by construction. It therefore follows that $\widetilde{u}$ has at least one sign change in each (z_{j-1}, z_j), $j \in \{2, \ldots, n\} \setminus \{q\}$ and two sign changes in (z_{q-1}, z_q). Thus $S(\widetilde{u}) \geq n$. This contradicts (1).   □

We now consider various additional properties enjoyed by WT-systems. This next result is due to Sommer [1983a, Theorem 1.4].

Proposition 7. *Let U be a WT-system in $C[a, b]$. Then, for every $a \leq c < d \leq b$, U is a WT-system in $C[c, d]$.*

Proof. Assume $\dim U = n$. If $\dim U|_{[c,d]} = n$ then this result is an immediate consequence of any of the equivalent definitions of a WT-system. We therefore assume that $\dim U|_{[c,d]} = m < n$ (and $m \geq 1$). Assume U is not a WT-system on $[c, d]$. There therefore exist $c \leq x_1 < \cdots < x_{m+1} \leq d$ and a $u^* \in U$ such that $u^*(x_i)(-1)^i > 0$, $i = 1, \ldots, m+1$. (This follows from the definition of a WT-system.) Set

$$V = \{v : v \in U, v(x) = 0, \text{ all } x \in [c, d]\}.$$

Since $\dim U|_{[c,d]} = m$, we have $\dim V = n - m$. Let $\{y_j\}_{j=1}^{n-m}$ be points in $[a, b]$ over which V is linearly independent. Thus $y_j \notin [c, d]$, $j = 1, \ldots, n-m$. Assume

$$y_1 < \cdots < y_k < c < d < y_{k+1} < \cdots < y_{n-m}.$$

Let $v \in V$ satisfy $v(y_j) = (-1)^{j+k}$, $j = 1, \ldots, k$, and $v(y_j) = (-1)^{j+m+k+1}$, $j = k+1, \ldots, n-m$. Set $u = u^* + \alpha v$, where $\alpha > 0$ is chosen sufficiently large so that $\operatorname{sgn} u(y_j) = \operatorname{sgn} v(y_j)$ for all j. Since $v(x_i) = 0$, $i = 1, \ldots, m+1$, it follows that $S(u) \geq n$. This contradicts our hypothesis that U is a WT-system on $[a, b]$.   □

WT-systems are characterized in terms of sign changes while T-systems are characterized in terms of zeros. We want to bridge the gap somewhat between WT- and T-systems.

Definition 6. Let U be a finite-dimensional subspace of $C[a,b]$. We will say that $x \in [a,b]$ is *essential* if there exists a $u \in U$ for which $u(x) \neq 0$.

In other words, x is essential if $x \in \operatorname{supp} U$.

Definition 7. Let U be a finite-dimensional subspace of $C[a,b]$. Assume $u^* \in U$ and $u^*(x_i) = 0$, $i = 1, \ldots, k$ for some $a \leq x_1 < \cdots < x_k \leq b$. The $\{x_i\}_{i=1}^k$ will be called *separated zeros* of u^* if there exist $\{y_i\}_{i=1}^{k-1}$ satisfying $y_i \in (x_i, x_{i+1})$, $i = 1, \ldots, k-1$, for which $u^*(y_i) \neq 0$.

With these definitions we can now prove this next result due to Stockenberg [1977a].

Theorem 8. *Let U be an n-dimensional WT-system in $C[a,b]$. Assume $u^* \in U$ has n separated essential zeros $\{x_i\}_{i=1}^n$ where $a \leq x_1 < \cdots < x_n \leq b$. Then $u^*(x) = 0$ for all $x \in [a, x_1] \cup [x_n, b]$.*

Proof. We will prove that $u^*(x) = 0$ for all $x \in [x_n, b]$. The proof of the fact that u^* vanishes identically on $[a, x_1]$ is totally analogous. The n zeros $\{x_i\}_{i=1}^n$ of u^* are separated. Let $\{y_i\}_{i=1}^{n-1}$ satisfy $y_i \in (x_i, x_{i+1})$, and $u^*(y_i) \neq 0$, $i = 1, \ldots, n-1$. Assume contrary to our claim that there exists a $y_n \in (x_n, b]$ with $u^*(y_n) \neq 0$.

We first claim the existence of a $\tilde{u} \in U$ for which $\tilde{u}(x_i) \neq 0$, $i = 1, \ldots, n$. This follows from the fact that the $\{x_i\}_{i=1}^n$ are essential, and an induction argument. Since x_j is essential there exists a $u_j \in U$ satisfying $u_j(x_j) \neq 0$. Assume there exists a $v \in U$ for which $v(x_i) \neq 0$, $i = 1, \ldots, k-1$ ($k \geq 2$). If $v(x_k) \neq 0$, then we immediately advance our induction. If $v(x_k) = 0$, let $w = v + \varepsilon u_k$. For all but a finite number of $\varepsilon \neq 0$, we have $w(x_i) \neq 0$, $i = 1, \ldots, k$. This proves the induction step. Thus $\tilde{u}$ exists as claimed.

Now $|u^*(y_i)| > 0$, $i = 1, \ldots, n$. We can therefore choose $\tilde{u}$ as above satisfying

$$\max_{i=1,\ldots,n} |\tilde{u}(y_i)| < \min_{i=1,\ldots,n} |u^*(y_i)|.$$

We consider both $\tilde{u} - u^*$ and $\tilde{u} + u^*$. By construction $(\tilde{u} \pm u^*)(x_i) = \tilde{u}(x_i) \neq 0$ for each $i = 1, \ldots, n$, while $\operatorname{sgn}((\tilde{u} - u^*)(y_i)) = -\operatorname{sgn} u^*(y_i) \neq 0$, and $\operatorname{sgn}((\tilde{u} + u^*)(y_i)) = \operatorname{sgn} u^*(y_i) \neq 0$, for each $i = 1, \ldots, n$. Thus on (x_i, x_{i+1}) the number of sign changes of $\tilde{u} - u^*$ plus the number of sign changes of $\tilde{u} + u^*$ is at least two. It therefore follows that on (x_1, x_n), the number of sign changes of $\tilde{u} - u^*$ plus the number of sign changes of $\tilde{u} + u^*$ is at least $2(n-1)$.

Since U is a WT-system of dimension n, we see that each of $\tilde{u} \pm u^*$ has exactly $n-1$ sign changes on (x_1, x_n). But $(\tilde{u} \pm u^*)(x_n) = \tilde{u}(x_n) \neq 0$, while $\operatorname{sgn}((\tilde{u} - u^*)(y_n)) = -\operatorname{sgn} u^*(y_n) \neq 0$ and $\operatorname{sgn}((\tilde{u} + u^*)(y_n)) = \operatorname{sgn} u^*(y_n) \neq 0$. Thus either $\tilde{u} - u^*$ or $\tilde{u} + u^*$ has an additional sign change in (x_n, y_n). This is a contradiction. □

From Theorem 8 we obtain various consequences used in Chapter 4.

Corollary 9. *Assume U is an n-dimensional WT-system in $C[a,b]$, and every point of (a,b) is essential. Let $u^* \in U$, $u^* \neq 0$. Then*

$$Z(u^*) = \bigcup_{i=1}^{m} [c_i, d_i],$$

where $a \leq c_1 \leq d_1 < c_2 \leq d_2 < \cdots < c_m \leq d_m \leq b$, and $m \leq n+1$. Furthermore, if $a < d_1$ or $c_m < b$, then $m \leq n$. Also, if u^* has no zero interval in $[a,b]$, then u^* has at most $n-1$ zeros in (a,b).

Corollary 10. *Assume U is an n-dimensional WT-system in $C[a,b]$, every point of (a,b) is essential, and no $u \in U$, $u \neq 0$, has a zero interval in $[a,b]$. Then U is a T-system on (a,b).*

We end this appendix with the following generic property of WT-systems.

Theorem 11. *Let U be an n-dimensional WT-system in $C[a,b]$, $n \geq 2$. Then U contains an $(n-1)$-dimensional WT-system in $C[a,b]$.*

In other words, given U as above, there exist $U_1 \subset \cdots \subset U_{n-1} \subset U$ such that U_k is a k-dimensional WT-system in $C[a,b]$, $k = 1, \ldots, n-1$. Theorem 11 was independently and simultaneously proved by Sommer, Strauss [1977] and Stockenberg [1977b]. Our proof follows that found in Zielke [1979, Theorem 7.6]. Before proving Theorem 11, let us note an immediate consequence of this result and (3) of Theorem 5.

Corollary 12. *Let U be an n-dimensional WT-system in $C[a,b]$. Given any $y_0 = a < y_1 < \cdots < y_k < b = y_{k+1}$, $k \leq n-1$, there exists a $u^* \in U$, $u^* \neq 0$, satisfying*

$$(-1)^i u^*(x) \geq 0, \quad x \in (y_{i-1}, y_i),$$

$i = 1, \ldots, k+1$.

Proof of Theorem 11. Let $P = \operatorname{supp} U$, i.e., P is the set of all essential points in $[a,b]$. For each $x \in [a,b]$, set

$$U_x = \{u : u \in U, u(x) = 0\}.$$

U_x is a subspace of U. If $x \in P$, then $\dim U_x = n-1$. If $x \notin P$, then $U_x = U$. For each $x \in [a,b)$, let

$$A_x = \{u : u \in U, u \text{ has } n-1 \text{ sign changes in } [x,b]\}.$$

That is, $u \in A_x$ provided that there exist points $x \leq y_1 < \cdots < y_n \leq b$ for which $u(y_i)u(y_{i+1}) < 0$, $i = 1, \ldots, n-1$. (Also set $A_b = \emptyset$.) Let $B_x = U \backslash A_x$. We separate out certain facts in this next lemma.

Lemma 13. *B_x is closed and, for each $x \in P$, $U_x \subseteq B_x$.*

Proof. We show that B_x is closed by proving that A_x is open. Let $u \in A_x$, and $x \leq y_1 < \cdots < y_n \leq b$ where $u(y_i)u(y_{i+1}) < 0$, $i = 1, \ldots, n-1$. Set $\alpha = \min\{|u(y_i)| : i = 1, \ldots, n\}$, and

$$C = \{w : w \in U, \|w\|_\infty < \alpha\}.$$

For each $w \in C$, $u + w \in A_x$ since $u(y_i)(u + w)(y_i) > 0$, $i = 1, \ldots, n$. Thus A_x is open and B_x is closed.

Assume $x \in P$ and $U_x \not\subseteq B_x$. Then $U_x \cap A_x \neq \emptyset$. There exists a $u \in U$ and points $x < y_1 < \cdots < y_n \leq b$ such that $u(y_i)(-1)^i > 0$, $i = 1, \ldots, n$, and $u(x) = 0$. Since $x \in P$, i.e., x is essential, there exists a $v \in U$ for which $v(x) > 0$. Set $w = u + \varepsilon v$. For $\varepsilon > 0$, ε sufficiently small, $w(y_i)(-1)^i > 0$, $i = 1, \ldots, n$, and $w(x) > 0$. Thus $S(w) \geq n$. This contradicts the fact that U is a WT-system. □

Proof of Theorem 11 (continued). If $a \in P$, then set $V = U_a$. From Lemma 13, it follows that, for all $v \in V$, $S(v) \leq n-2$, i.e., V is an $(n-1)$-dimensional WT-system in $C[a, b]$. (Similarly, if $b \in P$ we can easily construct the desired V.) We therefore assume that $a \notin P$, i.e., $u(a) = 0$ for all $u \in U$. We will, without loss of generality, assume that $a = \inf P$. (Otherwise replace a by $\inf P$ since all $u \in U$ vanish identically on $[a, \inf P]$.)

Let $\{x_k\}_{i=1}^\infty$ be a sequence of decreasing points in P satisfying $\lim_{k\to\infty} x_k = a$. For ease of exposition, set $B_k = B_{x_k}$, $U_k = U_{x_k}$, and $A_k = A_{x_k}$. From our previous results and the definition of B_k, we have that B_k is closed; $B_{k+1} \subseteq B_k$; $U_k \subseteq B_k$; and $\dim U_k = n - 1$, all k.

Since U is finite-dimensional, all norms on U are equivalent. For convenience, let $\| \cdot \|$ be any norm on U obtained from an inner product. Let $u_1^k, \ldots, u_{n-1}^k$ be any orthonormal basis for U_k and set

$$S = \{u : u \in U, \|u\| = 1\}.$$

Thus $u_i^k \in S$ for all i and k. Because S is compact there exists a subsequence of $\{k\}$, again denoted by $\{k\}$, such that $\lim_{k\to\infty} u_i^k = u_i$, $i = 1, \ldots, n-1$. As is easily checked, the $\{u_i\}_{i=1}^{n-1}$ are linearly independent elements of U. Set $V = \text{span}\{u_1, \ldots, u_{n-1}\}$. We will prove that V is a WT-system in $C[a, b]$.

Set $B = \bigcap_{j=1}^\infty B_j$. Since B_j is closed, so is B. Furthermore $u_i^k \in U_k \subseteq B_k$, and $B_k \subseteq B_j$ for all $j \leq k$. Thus $u_i^k \in \bigcap_{j=1}^k B_j$, implying that $u_i \in B$, $i = 1, \ldots, n-1$. Because $U_k \subseteq B_k$ for all k, it follows that $V \subseteq B$, i.e., linear combinations of the u_i are in B. Let $v \in V$. Since $V \subseteq U$, $S(v) \leq n-1$. Assume $S(v) = n-1$. Thus there exist $a < y_1 < \cdots < y_n \leq b$ such that $v(y_i)v(y_{i+1}) < 0$, $i = 1, \ldots, n-1$. (We have $a < y_1$ since $a \in P$.) Because $\lim_{k\to\infty} x_k = a$, it follows that $v \in A_k$ for all k sufficiently large, i.e., $x_k < y_1$. Thus $v \notin B_k$ and $v \notin B$. But $V \subseteq B$ and therefore $v \notin V$. This is a contradiction. Thus $S(v) \leq n-2$ for all $v \in V$. □

This last result is included because it is such an elegant consequence of Theorem 11. It is generally proven by different methods.

Proposition 13. *Let U be an n-dimensional subspace of $C[a, b]$, $n \geq 2$. Assume that U is a T-system on (a, b). Then U contains an $(n-1)$-dimensional subspace which is also a T-system on (a, b).*

Proof. U is a WT-system in $C[a, b]$. From Theorem 11 there exists an $(n-1)$-dimensional subspace V which is a WT-system in $C[a, b]$. Assume some $v \in V$, $v \neq 0$, has $n-1$ zeros in (a, b). (It can have no more since $v \in U$.) Because U is a T-system of dimension n, we have from Proposition 2 that v has $n-1$ sign changes in (a, b). This is a contradiction. Thus V is a T-system on (a, b).   □

Remark. Proposition 13 is not valid if we replace the open interval by the closed or half-open interval, see Zielke [1979, p.43–45].

Appendix B
Convexity Cones and L^1-Approximation

If U is a WT-system on $[a, b]$ and f is in the convexity cone generated by U, i.e., $V = \text{span}\{U, f\}$ is a WT-system, then the problem of one- and two-sided $L^1(\mu)$-approximation to f from U takes on a special character. In the problem of two-sided $L^1(\mu)$-approximation, a best approximant is obtained by interpolation at certain fixed points dependent only on U and μ, and independent of the specific f in the convexity cone. It is this theory which we develop in Part I of this appendix.

In Part II we present the somewhat analogous theory for the problem of one-sided $L^1(\mu)$-approximation. There we highlight the connection between our problem and a moment problem. This leads us to a consideration of 'principal representations'. We shall prove that a best one-sided $L^1(\mu)$ approximant to each f in the convexity cone generated by U is also obtained by interpolation at certain fixed points.

Part I. Two-Sided L^1-Approximation

Let $K \subset I\!\!R^d$ be compact, satisfying $K = \overline{\text{int } K}$. Let U be a finite-dimensional subspace of $C(K)$ and μ an 'admissible' measure on K. Assume that the function $h \in L^\infty(K, \mu)$ satisfies

$$1) \ |h| = 1 \ \text{on} \ K$$

$$2) \int_K hu \, d\mu = 0, \ \text{all} \ u \in U.$$

If $f \in C(K)$ and there exists a $u^* \in U$ satisfying

$$3) \ h(f - u^*) \geq 0 \ \text{on} \ K,$$

then $u^* \in P_U(f)$. This follows from Theorem 2.3 since (1) and (3) imply

$$\int_K h(f - u^*) d\mu = \|f - u^*\|_1.$$

Our problem is, of course, in finding the appropriate h and u^* for a given f. In certain special cases this problem has a simple elegant solution. This occurs when $K = [a, b] \subset I\!\!R$, U is a WT-system, and f is in the convexity cone generated by U. In this case a best approximant is obtained by interpolation at certain fixed points, dependent only on U and μ.

In developing this theory, we start with a general result due to Hobby, Rice [1965].

Theorem 1. *Let ν be a finite non-atomic positive measure on $[a, b]$, and U an n-dimensional subspace of $L^1([a, b], \nu)$. There exist points $a = x_0 < x_1 < \cdots < x_m < x_{m+1} = b$ such that $m \leq n$, and*

$$\sum_{j=1}^{m+1} (-1)^j \int_{x_{j-1}}^{x_j} u \, d\nu = 0,$$

for all $u \in U$.

Our proof of Theorem 1 (Pinkus [1976b]) is based on the Borsuk Antipodality Theorem (Borsuk [1933]).

Theorem 2 (Borsuk). *Let Ω be a bounded open symmetric neighborhood of 0 in $\mathbb{R}^{n+1}$, and T a continuous map of $\partial\Omega$ (the boundary of Ω) into $\mathbb{R}^n$, with T odd on $\partial\Omega$, i.e., $T(-\mathbf{t}) = -T(\mathbf{t})$ for all $\mathbf{t} \in \partial\Omega$. There then exists a $\mathbf{t}^* \in \partial\Omega$ for which $T(\mathbf{t}^*) = 0$.*

Proof of Theorem 1. Let

$$\Xi_{n+1} = \Big\{ \mathbf{t} : \mathbf{t} = (t_1, \ldots, t_{n+1}), \ \sum_{i=1}^{n+1} |t_i| = b - a \Big\}.$$

For each $\mathbf{t} \in \Xi_{n+1}$, set $x_0(\mathbf{t}) = a$, and $x_j(\mathbf{t}) = a + \sum_{i=1}^{j} |t_i|$, $j = 1, \ldots, n+1$. Thus $a = x_0(\mathbf{t}) \leq x_1(\mathbf{t}) \leq \cdots \leq x_{n+1}(\mathbf{t}) = b$. Let $u_1, \ldots, u_n$ be any basis for U. To each $\mathbf{t} \in \Xi_{n+1}$, define

$$c_i(\mathbf{t}) = \sum_{j=1}^{n+1} (\operatorname{sgn} t_j) \int_{x_{j-1}(\mathbf{t})}^{x_j(\mathbf{t})} u_i d\nu,$$

and $C(\mathbf{t}) = (c_1(\mathbf{t}), \ldots, c_{n+1}(\mathbf{t}))$. Since $x_j(\mathbf{t}) = x_{j-1}(\mathbf{t})$ if and only if $t_j = 0$, it follows that $C(\mathbf{t})$ is a continuous map of Ξ_{n+1} into $\mathbb{R}^n$. Furthermore $C(\mathbf{t})$ is odd. Thus, from Theorem 2, there exists a $\mathbf{t}^* \in \Xi_{n+1}$ for which $c_i(\mathbf{t}^*) = 0$, $i = 1, \ldots, n$. From the $\{x_j(\mathbf{t}^*)\}_{j=0}^{n+1}$, we easily choose the $\{x_j\}_{j=0}^{m+1}$ so that Theorem 1 holds.   □

We shall use Theorem 1 where U is a T- or WT-system. Recall that, in Appendix A we defined $S(f)$ for every $f \in C[a, b]$. Let us generalize this notion to $f \in L^1([a, b], \mu)$.

Definition 1. *Let $f \in L^1([a, b], \mu)$. We say that f has k sign changes on $[a, b]$, denoted $S(f) = k$, if there exist $k + 1$ disjoint ordered intervals $I_1, \ldots, I_{k+1}$ (by ordered we mean that $x < y$ for all $x \in I_i$, $y \in I_{i+1}$) whose union is $[a, b]$, and where $\varepsilon(-1)^i f \geq 0$ μ a.e. on I_i, $i = 1, \ldots, k+1$, with $\varepsilon \in \{-1, 1\}$, ε fixed, and $\mu\{x : x \in I_i, f(x) \neq 0\} > 0$, all i. If no such k exists, we set $S(f) = \infty$.*

Proposition 3. *Let U be an n-dimensional subspace of $C[a, b]$, and μ an 'admissible' measure. Assume $g \in L^1([a, b], \mu)$, $g \neq 0$, satisfies*

$$\int_a^b gu \, d\mu = 0,$$

for all $u \in U$. If U is a T-system, then $S(g) \geq n$. If U is a WT-system, and $\mu(Z(g)) = 0$, then $S(g) \geq n$.

Proof. Assume $S(g) = m < n$, and let $I_1, \ldots, I_{m+1}$ be as in Definition 1. Set $\bar{I}_j = [c_j, d_j]$, $j = 1, \ldots, m+1$. Thus $d_j = c_{j+1}$, $j = 1, \ldots, m+1$, and $a = c_1 < c_2 < \cdots < c_{m+2} = b$. Since $m < n$, and U is in either case a WT-system on $[a, b]$, there exists from Corollary 12 of Appendix A a $u^* \in U$, $u^* \neq 0$, satisfying $\varepsilon(-1)^j u^*(x) \geq 0$ for all $x \in (c_j, c_{j+1})$, $j = 1, \ldots, m+1$, where $\varepsilon \in \{-1, 1\}$ is chosen so that $gu^* \geq 0$ μ a.e.. Thus

$$0 = \int_a^b gu^* d\mu = \int_a^b |gu^*| d\mu.$$

Now, if U is a T-system, then $\mu(Z(u^*)) = 0$, and a contradiction ensues. If U is a WT-system, then we may have $\mu(Z(u^*)) > 0$. But by assumption $\mu(Z(g)) = 0$, and a contradiction again ensues. □

As a consequence of Proposition 3, we have:

Corollary 4. *Let U be an n-dimensional WT-system on $[a, b]$, and μ an 'admissible' measure. Assume there exist $a = x_0 < x_1 < \cdots < x_{m+1} = b$ for which*

$$\sum_{j=1}^{m+1} (-1)^j \int_{x_{j-1}}^{x_j} u \, d\mu = 0$$

for all $u \in U$. Then $m \geq n$.

The above result, together with Theorem 1, implies the existence of $a = x_0 < x_1 < \cdots < x_{n+1} = b$ satisfying

$$(B.1) \qquad \sum_{j=1}^{n+1} (-1)^j \int_{x_{j-1}}^{x_j} u \, d\mu = 0$$

for all $u \in U$. The question of the uniqueness of these points we defer until after a discussion of their use.

Definition 2. *Let U be an n-dimensional subspace of $C[a, b]$, and $u_1, \ldots, u_n$ a basis for U. Set*

$$\mathcal{C}^+(U) = \{u_{n+1} : u_{n+1} \in C[a, b], \det(u_i(x_j))_{i,j=1}^{n+1} \geq 0 \text{ for all choices}$$
$$\text{of } a \leq x_1 < \cdots < x_{n+1} \leq b\}.$$

and

$$\mathcal{C}^-(U) = \{u_{n+1} : -u_{n+1} \in \mathcal{C}^+(U)\}.$$

Each $\mathcal{C}^{\pm}(U)$ is a convex cone. We say that $\mathcal{C}(U) = \mathcal{C}^+(U) \cup \mathcal{C}^-(U)$ is the *convexity cone* generated by U. Alternatively, $\mathcal{C}(U)$ is the set of all $f \in C[a, b]$ for which either $f \in U$, or $V = U \cup \{f\}$ is a WT-system of dimension $n + 1$.

Note that $\mathcal{C}(U)$, unlike $\mathcal{C}^{\pm}(U)$, is independent of the choice of basis for U, and is not in general convex.

Theorem 5. *Let U be an n-dimensional WT-system on $[a,b]$, and μ an 'admissible' measure. Let $a = x_0 < x_1 < \cdots < x_{n+1} = b$ satisfy (B.1). To each $f \in \mathcal{C}(U)$ there exists a $u^* \in U$ for which*

$$(B.2) \qquad \varepsilon(-1)^j (f - u^*)(x) \geq 0, \quad x \in (x_{j-1}, x_j),$$

$j = 1, \ldots, n+1$, for some $\varepsilon \in \{-1, 1\}$, ε fixed. Furthermore $u^* \in P_U(f)$.

Proof. Assume $u^* \in U$ exists satisfying (B.2). Set $h(x) = (-1)^j \varepsilon$, for $x \in (x_{j-1}, x_j]$, $j = 1, \ldots, n+1$ $(h(a) = \varepsilon)$. Then

$$1)\ |h| = 1 \text{ on } [a,b]$$

$$2)\ \int_a^b h u \, d\mu = 0, \text{ all } u \in U$$

$$3)\ \int_a^b h(f - u^*) d\mu = \|f - u^*\|_1.$$

From Theorem 2.3, $u^* \in P_U(f)$. It remains to prove the existence of $u^* \in U$ satisfying (B.2).

Assume $f \in \mathcal{C}(U)$. If $f \in U$ we take, for fairly obvious reasons, $u^* = f$. We therefore assume that $f \notin U$. Set $V = U \cup \{f\}$. V is a WT-system of dimension $n + 1$. From Theorem 5 (or Corollary 12) of Appendix A, there exists a $v \in V$, $v \neq 0$, such that

$$(-1)^j v(x) \geq 0, \quad x \in (x_{j-1}, x_j),$$

$j = 1, \ldots, n+1$. Now $v = \alpha f - u$ for some $u \in U$ and $\alpha \in \mathbb{R}$. If $\alpha \neq 0$ we are finished, i.e., $u^* = u/\alpha$. We have $\alpha \neq 0$ since, from (B.1),

$$0 < \|v\|_1 = \sum_{j=1}^{n+1} (-1)^j \int_{x_{j-1}}^{x_j} v \, d\mu = \alpha \Big(\sum_{j=1}^{n+1} (-1)^j \int_{x_{j-1}}^{x_j} f \, d\mu \Big). \qquad \square$$

Remark. If $u_1, \ldots, u_n$ is any basis for U, and $\det(u_i(x_j))_{i,j=1}^n \neq 0$, then the u^* of Theorem 5 is uniquely determined by $u^*(x_j) = f(x_j)$, $j = 1, \ldots, n$. If $\det(u_i(x_j))_{i,j=1}^n = 0$, then to each $f \in \mathcal{C}(U)$ there always exist $u \in U$ satisfying $u(x_j) = f(x_j)$, $j = 1, \ldots, n$. There is, moreover, at least one for which (B.2) holds. In this latter case the u^* need not be uniquely determined. As an example, set $U = \text{span}\{|x|\}$ on $[-1,1]$, μ Lebesgue measure, and $f = x_+$. Then $f \in \mathcal{C}(U)$, $x_1 = 0$, and we can take $u^*(x) = \alpha|x|$ for any $0 \leq \alpha \leq 1$.

Proposition 6. *Let U be an n-dimensional WT-system on $[a,b]$, and μ an 'admissible' measure. Let $a = x_0 < x_1 < \cdots < x_{n+1} = b$ satisfy (B.1). Assume $u_1, \ldots, u_n$ is any basis for U, and $\det(u_i(x_j))_{i,j=1}^n \neq 0$. Given $f \in \mathcal{C}(U)$, the u^* of (B.2) is the unique best approximant to f from U.*

Proof. Let $u^* \in U$ satisfy (B.2). Assume $u \in U$. Then

$$\|f - u^*\|_1 = \varepsilon \sum_{j=1}^{n+1} (-1)^j \int_{x_{j-1}}^{x_j} (f - u^*) d\mu$$

$$= \varepsilon \sum_{j=1}^{n+1} (-1)^j \int_{x_{j-1}}^{x_j} (f - u) d\mu \leq \|f - u\|_1.$$

For equality to hold, we must have

$$\varepsilon(-1)^j (f - u)(x) \geq 0$$

for $x \in (x_{j-1}, x_j)$, $j = 1, \ldots, n+1$. It is therefore necessary that $u(x_j) = f(x_j)$, $j = 1, \ldots, n$, which implies that $u(x_j) = u^*(x_j)$, $j = 1, \ldots, n$. Since $\det(u_i(x_j))_{i,j=1}^n \neq 0$ we have $u = u^*$. □

If U is a T-system, then the conditions of Proposition 6 are certainly satisfied. In this case we also have the uniqueness of the points $\{x_j\}_{j=1}^n$.

Proposition 7. *Let U be an n-dimensional T-system on $[a,b]$, and μ an 'admissible' measure. There exist a unique set of points $a = x_0 < x_1 < \cdots < x_{n+1} = b$ satisfying (B.1).*

Proof. Assume $a = x_0 < x_1 < \cdots < x_{n+1} = b$ and $a = y_0 < y_1 < \cdots < y_{n+1} = b$ satisfy (B.1), with $(x_1, \ldots, x_n) \neq (y_1, \ldots, y_n)$. Set $h_{\mathbf{x}}(x) = (-1)^j$, $x \in (x_{j-1}, x_j]$, $j = 1, \ldots, n+1$, and $h_{\mathbf{y}}(x) = (-1)^j$, $x \in (y_{j-1}, y_j]$, $j = 1, \ldots, n+1$. Then

$$\int_a^b (h_{\mathbf{x}} - h_{\mathbf{y}}) u \, d\mu = 0$$

for all $u \in U$. From Proposition 3, $S(h_{\mathbf{x}} - h_{\mathbf{y}}) \geq n$.

Now assume without loss of generality that $x_1 \leq y_1$. Then $(h_{\mathbf{x}} - h_{\mathbf{y}})(x) = 0$, $x \in (a, x_1]$, and $(-1)^j(h_{\mathbf{x}} - h_{\mathbf{y}})(x) \geq 0$, $x \in (x_{j-1}, x_j]$, $j = 2, \ldots, n+1$. Thus $S(h_{\mathbf{x}} - h_{\mathbf{y}}) \leq n - 1$. This is a contradiction. □

For a general WT-system, it is not necessary that the $\{x_j\}_{j=1}^n$ satisfying (B.1) are uniquely defined. Let $U = \text{span}\{(|x| - 1)_+\}$ on $[-2, 2]$, and μ be Lebesgue measure. U is a WT-system. We are free to choose x_1 to be any point in $[-1, 1]$. Note that this implies (from Theorem 5) that every $f \in C(U)$ vanishes identically on $[-1, 1]$.

For a WT-system we ask two questions. Firstly, when is there a unique set of $\{x_j\}_{j=1}^n$ satisfying (B.1)? Secondly, when do we have $\det(u_i(x_j))_{i,j=1}^n \neq 0$? These questions are not totally unrelated. We have from Sommer [1979, Theorem 2.2(ii)],

Proposition 8. *Let U be an n-dimensional WT-system on $[a,b]$, and μ an 'admissible' measure. There do not exist two sets of points $a = x_0 < x_1 <$*

$\cdots < x_{n+1} = b$ and $a = y_0 < y_1 < \cdots < y_{n+1} = b$ which both satisfy (B.1) and such that $\det(u_i(x_j))_{i,j=1}^n \cdot \det(u_i(y_j))_{i,j=1}^n \neq 0$.

Proof. Assume to the contrary the existence of distinct $\{x_j\}_{j=1}^n$ and $\{y_j\}_{j=1}^n$ as above, satisfying (B.1), and such that $\det(u_i(x_j))_{i,j=1}^n \cdot \det(u_i(y_j))_{i,j=1}^n \neq 0$. With no loss of generality, we assume that $x_i = y_i$, $i = 1, \ldots, k-1$, and $x_k < y_k$. Set $h_x(x) = (-1)^j$, $x \in (x_{j-1}, x_j]$, $j = 1, \ldots, n+1$, and $h_y(x) = (-1)^j$, $x \in (y_{j-1}, y_j]$, $j = 1, \ldots, n+1$. There then exists an $\varepsilon > 0$ such that

$$(h_x - h_y)(x) = \begin{cases} 0, & a < x \leq x_k \\ 2(-1)^{k+1}, & x_k < x < x_k + \varepsilon, \end{cases}$$

and $(-1)^j(h_x - h_y)(x) \geq 0$, $x \in (x_{j-1}, x_j]$, $j = k+1, \ldots, n+1$. Let $\tilde{u} \in U$ satisfy $\tilde{u}(x_j) = (-1)^{k+1}\delta_{kj}$, $j = 1, \ldots, n$. Such a $\tilde{u}$ is uniquely defined since $\det(u_i(x_j))_{i,j=1}^n \neq 0$. Furthermore, it is easily seen that

$$(-1)^j\tilde{u}(x) \geq 0, \quad x \in (x_{j-1}, x_j],$$

$j = k+1, \ldots, n+1$. Now $\int_a^b (h_x - h_y)u\, d\mu = 0$ for all $u \in U$, but by construction $\int_a^b (h_x - h_y)\tilde{u}\, d\mu > 0$. This contradiction proves the proposition.   □

There are classes of WT-systems for which the $\{x_j\}_{j=1}^n$ satisfying (B.1) are unique, and $\det(u_i(x_j))_{i,j=1}^n \neq 0$. From Sommer [1979, Theorem 2.2],

Proposition 9. *Let U be an n-dimensional WT-system on $[a,b]$, and μ an 'admissible' measure. Assume that U is a unicity space for $C_1([a,b], \mu)$. Then the set of points $\{x_j\}_{j=1}^n$ satisfying (B.1) is unique, and $\det(u_i(x_j))_{i,j=1}^n \neq 0$.*

Proof. Let $\{x_j\}_{j=1}^n$ satisfy (B.1). It suffices, on the basis of Proposition 8, to prove that $\det(u_i(x_j))_{i,j=1}^n \neq 0$. Set $h_x(x) = (-1)^j$, $x \in (x_{j-1}, x_j]$, $j = 1, \ldots, n+1$ ($h(a) = -1$). Assume that $\det(u_i(x_j))_{i,j=1}^n = 0$. There then exists a $u^* \in U$, $u^* \neq 0$, satisfying $u^*(x_j) = 0$, $j = 1, \ldots, n$. Thus

1) $|h_x| = 1$, on all of $[a,b]$

2) $\int_a^b h_x u\, d\mu = 0$, all $u \in U$

3) $h_x|u^*|$ is continuous.

From Theorem 3.1, U is not a unicity space for $C_1([a,b], \mu)$.   □

The condition of Proposition 9 is, of course, measure dependent and therefore difficult to verify. If U satisfies Property A then for every 'admissible' measure, the set of points $\{x_j\}_{j=1}^n$ satisfying (B.1) is unique and $\det(u_i(x_j))_{i,j=1}^n \neq 0$. It is not clear whether Property A is a necessary condition for this conclusion to hold. From Kroó [1988],

Theorem 10. *Let U be an n-dimensional WT-system on $[a,b]$. Then the following are equivalent.*

1) For every 'admissible' measure μ, the set of points $\{x_j\}_{j=1}^n$ satisfying (B.1) is unique and $\det(u_i(x_j))_{i,j=1}^n \neq 0$.

2) For every $a = x_0 < x_1 < \cdots < x_{n+1} = b$ such that there exists a $u^* \in U$, $u^* \neq 0$, satisfying $u^*(x_j) = 0$, $j = 1, \ldots, n$, there also exists a $\tilde{u} \in U$, $\tilde{u} \neq 0$, for which $(-1)^j \tilde{u}(x) \geq 0$, $x \in (x_{j-1}, x_j)$, $j = 1, \ldots, n+1$.

Proof. $(1) \Rightarrow (2)$. Assume (2) does not hold. There then exist $a = x_0 < x_1 < \cdots < x_{n+1} = b$ and a $u^* \in U$, $u^* \neq 0$, satisfying $u^*(x_j) = 0$, $j = 1, \ldots, n$, while there does not exist a $\tilde{u} \in U$, $\tilde{u} \neq 0$, for which $(-1)^j \tilde{u}(x) \geq 0$, $x \in (x_{j-1}, x_j)$, $j = 1, \ldots, n+1$. Let $h(x) = (-1)^j$ for $x \in (x_{j-1}, x_j]$, $j = 1, \ldots, n+1$. Then

$$W = \{hu : u \in U\}$$

is an n-dimensional subspace of $L^\infty[a, b]$ which does not contain a non-negative non-trivial function. From Corollary 4.4 there exists an 'admissible' measure μ satisfying

$$\int_a^b hu \, d\mu = 0$$

for all $u \in U$. In other words, the above $\{x_j\}_{j=1}^n$ satisfy (B.1) for the measure μ. Since $u^*(x_j) = 0$, $j = 1, \ldots, n$, we have that $\det(u_i(x_j))_{i,j=1}^n = 0$, contradicting (1).

$(2) \Rightarrow (1)$. Let $a = x_0 < x_1 < \cdots < x_{n+1} = b$ satisfy (B.1) for some 'admissible' measure μ. It suffices, on the basis of Proposition 8, to prove that $\det(u_i(x_j))_{i,j=1}^n \neq 0$. If $\det(u_i(x_j))_{i,j=1}^n = 0$, then there exists from (2) a $\tilde{u} \in U$, $\tilde{u} \neq 0$, satisfying $(-1)^j \tilde{u}(x) \geq 0$, $x \in (x_{j-1}, x_j)$, $j = 1, \ldots, n+1$. But then

$$\sum_{j=1}^{n+1} (-1)^j \int_{x_{j-1}}^{x_j} \tilde{u} \, d\mu = \|\tilde{u}\|_1 > 0,$$

contradicting (B.1). Thus $\det(u_i(x_j))_{i,j=1}^n \neq 0$. □

Another class of WT-systems for which the conclusions of Proposition 9 are valid depend upon $\mathcal{C}(U)$. For each $a \leq t_1 < \cdots < t_n \leq b$, set

$$U[t_1, \ldots, t_n] = \{(f(t_1), \ldots, f(t_n)) : f \in \mathcal{C}(U)\}.$$

We define the *dimension* of $U[t_1, \ldots, t_n]$, denoted $\dim U[t_1, \ldots, t_n]$, as the dimension of the smallest linear subspace in $\mathbb{R}^n$ containing $U[t_1, \ldots, t_n]$. From Micchelli [1977, Lemma 3],

Proposition 11. Let U be an n-dimensional WT-system on $[a, b]$. Assume that for every choice of $a < t_1 < \cdots < t_n < b$ we have $\dim U[t_1, \ldots, t_n] = n$. Then for any 'admissible' measure μ, the set of points $\{x_j\}_{j=1}^n$ satisfying (B.1) is unique, and $\det(u_i(x_j))_{i,j=1}^n \neq 0$.

Proof. While there is a simple direct proof to this proposition, it also follows as a consequence of (2) of Theorem 10. For assume we are given $a = x_0 <$

$x_1 < \cdots < x_{n+1} = b$, and there exists a $u^* \in U$, $u^* \neq 0$, satisfying $u^*(x_j) = 0$, $j = 1, \ldots, n$. There then exists a vector $c = (c_1, \ldots, c_n) \neq 0$ such that $\sum_{j=1}^n c_j u(x_j) = 0$ for all $u \in U$. Since dim $U[x_1, \ldots, x_n] = n$, there exists an $f \in C(U)$ for which $\sum_{j=1}^n c_j f(x_j) \neq 0$. Because span$\{f, U\}$ is a WT-system of dimension $n + 1$, there exists a $\tilde{v} = \alpha f + \tilde{u}$, $\alpha \in \mathbb{R}$, $\tilde{u} \in U$, satisfying $(-1)^j \tilde{v}(x) \geq 0$, $x \in (x_{j-1}, x_j)$, $j = 1, \ldots, n+1$. Thus $\tilde{v}(x_j) = 0$, $j = 1, \ldots, n$. Since $0 = \sum_{j=1}^n c_j \tilde{v}(x_j) = \alpha \sum_{j=1}^n c_j f(x_j)$, this implies that $\alpha = 0$. Therefore $\tilde{v} = \tilde{u} \in U$ satisfies (2) of Theorem 10.   □

Of course, Proposition 11 begs the question of when dim $U[t_1, \ldots, t_n] = n$ for every $a < t_1 < \cdots < t_n < b$. We shall, however, not consider this complicated problem.

Let U be a WT-system on $[a, b]$. Assume there are unique $\{x_j\}_{j=1}^n$ satisfying (B.1), and $\det(u_i(x_j))_{i,j=1}^n \neq 0$. It may be that U is not a unicity space for $C_1([a, b], \mu)$, and dim $U[t_1, \ldots, t_n] < n$, for some $a < t_1 < \cdots < t_n < b$. As an example, set $U = \text{span}\{u\}$ on $[-2, 2]$, where

$$u(x) = \begin{cases} |x + 1|, & -2 \leq x \leq 0, \\ |x - 1|, & 0 \leq x \leq 2, \end{cases}$$

and μ is Lebesgue measure. Then necessarily $x_1 = 0$, and $u(x_1) = 1 \neq 0$. U is not a unicity space since

$$h(x) = \begin{cases} 1, & 1 \leq |x| \leq 2, \\ -1, & |x| < 1, \end{cases}$$

satisfies $\int_{-2}^2 hu\, d\mu = 0$, and $h|u| \in C[-2, 2]$. Furthermore, every $f \in C(U)$ must satisfy $f(1) = f(-1) = 0$, so that dim $U[\pm 1] = 0$.

It is an open question as to whether there may exist distinct $\{x_j\}_{j=1}^n$ and $\{y_j\}_{j=1}^n$ satisfying (B.1) with $\det(u_i(x_j))_{i,j=1}^n \neq 0$ and $\det(u_i(y_j))_{i,j=1}^n = 0$, where U is an n-dimensional WT-system.

Part II. One-Sided L^1-Approximation

The results of the previous section are somewhat related to certain problems in moment theory (the so-called $(-L, L)$-moment problem of Markov), but we did not use this connection. The results of this section are very much related to moment theory. We shall herein consider principal representations of moments, the Markov-Krein Inequality, and other associated results. An extensive treatment of these problems, as they pertain to T-systems (but not to WT-systems), can be found in the first few chapters of Karlin, Studden [1966], and Krein, Nudel'man [1977] (we shall follow more closely the former). Both works are heavily based on Krein [1951]. We shall, however, be very selective in what we present from this general theory since we have a particular goal in mind. Thus many of the results to be presented are special cases of a more general theory.

To see how the one-sided L^1-approximation problem is connected to a moment theory problem, let us recall Theorem 5.2. For ease of exposition, we shall always assume that $K = [a, b]$, μ is an 'admissible' measure, and U is an n-dimensional subspace of $C[a, b]$ which contains a strictly positive function. From Theorem 5.2, $u^* \in U$ is a best one-sided (from below) $L^1(\mu)$ approximant to $f \in C[a, b]$ if and only if $u^* \in \mathcal{U}(f)$, i.e., $u^* \le f$, and there exist distinct points $\{x_i\}_{i=1}^{k}$ in $[a, b]$, $1 \le k \le n$, and positive values $\{\lambda_i\}_{i=1}^{k}$ for which

$$a)\ (f - u^*)(x_i) = 0,\ i = 1, \ldots, k$$

$$b)\ \int_a^b u\, d\mu = \sum_{i=1}^{k} \lambda_i u(x_i),\ \text{all } u \in U.$$

We also recall (see (5.2)) that u^* is a best approximant to f from $\mathcal{U}(f)$ if and only if it satisfies

$$\int_a^b u^*\, d\mu = \max\left\{ \int_a^b u\, d\mu\ :\ u \in \mathcal{U}(f) \right\}.$$

Obviously, for any $u \in \mathcal{U}(f)$, we have

$$\int_a^b u\, d\mu \le \int_a^b f\, d\mu.$$

Furthermore, if $\tilde{\mu}$ is any non-negative measure for which $\int_a^b u\, d\mu = \int_a^b u\, d\tilde{\mu}$ for all $u \in U$, then

$$\int_a^b u\, d\mu \le \int_a^b f\, d\tilde{\mu}$$

for all $u \in \mathcal{U}(f)$. As such we obtain

$$\max\left\{ \int_a^b u\, d\mu : u \in \mathcal{U}(f) \right\}$$

$$\le \inf\left\{ \int_a^b f\, d\tilde{\mu} : \tilde{\mu} \ge 0 \text{ and } \int_a^b u\, d\mu = \int_a^b u\, d\tilde{\mu} \text{ for all } u \in U \right\}.$$

Consider the 'dual' problem given on the right of this inequality. We can reinterpret Theorem 5.2 as simply saying that the above infimum is attained and its exact value is $\int_a^b u^*\, d\mu$. It is attained for the measure $d\tilde{\mu} = \sum_{i=1}^{k} \lambda_i \delta_{x_i}$, where δ_{x_i} is the Dirac-Delta measure with mass one at x_i. (Note that equality in the above does not always hold without the assumption of the existence of a strictly positive function in U, see e.g. the example after Corollary 5.6.)

In other words, if $U = \text{span}\{u_1, \ldots, u_n\}$ and

$$\int_a^b u_j\, d\mu = c_j, \qquad j = 1, \ldots, n,$$

then the dual problem is of the form

$$(B.3) \qquad \min\left\{ \int_a^b f\, d\tilde{\mu} : \tilde{\mu} \ge 0,\ \int_a^b u_j\, d\tilde{\mu} = c_j,\ j = 1, \ldots, n \right\}.$$

We shall prove that if U is a WT-system, satisfying the previous assumptions, and if f is in the convexity cone of U, then this problem has a solution $\tilde{\mu}$ independent of f. (To be precise, it does depend upon whether $f \in C^+(U)$ or $f \in C^-(U)$.)

We initiate our study of this related moment problem under the more restrictive assumption that U is a T-system of dimension n on [a,b].

Let $u_1, \ldots, u_n$ be a basis for the T-system U on [a,b]. The *moment space* $\mathcal{M}_n$ is defined as follows:

$$\mathcal{M}_n = \{ \mathbf{c} = (c_1, \ldots, c_n) : c_j = \int_a^b u_j d\sigma, \ j = 1, \ldots, n \},$$

where σ varies over all non-decreasing right continuous functions of bounded variation on [a,b]. The vector $\mathbf{c} \in \mathcal{M}_n$ is said to be a *moment* of U with associated measure σ. Obviously $\mathcal{M}_n$ is a convex cone. Since U contains a strictly positive function (U is a T-system) it also follows that $\mathcal{M}_n$ is closed.

An equivalent formulation of $\mathcal{M}_n$ is given by the following. The vectors

$$D_n = \{ (u_1(x), \ldots, u_n(x)) : a \leq x \leq b \}$$

form a curve in $\mathbb{R}^n$. Let $\mathcal{D}(D_n)$ denote the smallest convex cone containing D_n. Then, as is readily shown, $\mathcal{M}_n = \mathcal{D}(D_n)$.

To obtain 'representations' for the moments of U, we first characterize the boundary of $\mathcal{M}_n = \mathcal{D}(D_n)$. Before doing so, let us recall (see Section 6 of Chapter 5) the definition of the *index* of a set of distinct points $\{x_1, \ldots, x_r\}$ in [a,b]. It shall be denoted $I(x_1, \ldots, x_r)$ and is equal to $\sum_{i=1}^r w(x_i)$, where $w(x) = 2$ if $x \in (a, b)$, while $w(a) = w(b) = 1$.

Theorem 12. $\mathbf{c}^* \in \mathcal{M}_n$ *is a boundary point of* $\mathcal{M}_n$ *if and only if* $\mathbf{c}^*$ *admits a representation of the form*

$$(B.4) \qquad c_j^* = \sum_{i=1}^r \lambda_i u_j(x_i), \qquad j = 1, \ldots, n,$$

where $\lambda_i > 0$, $i = 1, \ldots, r$, *and* $I(x_1, \ldots, x_r) \leq n - 1$.

Proof. ($\Rightarrow$). Assume $\mathbf{c}^*$ is a boundary point of $\mathcal{M}_n$, $\mathbf{c}^* \neq \mathbf{0}$. Since $\mathcal{M}_n$ is a closed convex cone, there exists a supporting hyperplane to $\mathcal{M}_n$ at $\mathbf{c}^*$ which contains the origin. Thus there exists an $\mathbf{a} = (a_1, \ldots, a_n) \neq \mathbf{0}$ for which

$$a) \sum_{j=1}^n a_j c_j \geq 0, \text{ all } \mathbf{c} \in \mathcal{M}_n$$

$$b) \sum_{j=1}^n a_j c_j^* = 0.$$

Set $u^* = \sum_{j=1}^n a_j u_j$. Thus $u^* \neq 0$. Because $(u_1(x), \ldots, u_n(x)) \in \mathcal{M}_n$ for each $x \in [a, b]$, it follows from (a) that $u^* \geq 0$. Since $\mathbf{c}^* \in \mathcal{M}_n$, there exists a non-decreasing right continuous function σ^* of bounded variation such that

$$c_j^* = \int_a^b u_j d\sigma^*, \qquad j = 1, \ldots, n.$$

From (b), we have

$$\int_a^b u^* d\sigma^* = 0.$$

For equality to hold ($u^* \geq 0$, $d\sigma^* \geq 0$) it is necessary that u^* vanish at every point of increase of σ^*. Since U is a T-system, u^* has at most a finite number of zeros. Furthermore, since $u^* \geq 0$, we have from Proposition 3 of Appendix A that $|\widetilde{Z}(u^*)| \leq n-1$. Thus (B.4) holds with $I(x_1, \ldots, x_r) \leq |\widetilde{Z}(u^*)| \leq n-1$.

($\Leftarrow$). We now assume that $\mathbf{c}^*$ admits a representation of the form (B.4). Since $I(x_1, \ldots, x_r) \leq n - 1$, there exists, from Theorem 4 of Appendix A, a $u^* = \sum_{j=1}^n a_j u_j$ which is non-negative, non-trivial, and vanishes at the x_i, $i = 1, \ldots, r$. Thus

$$\sum_{j=1}^n a_j c_j^* = \sum_{i=1}^r \lambda_i u^*(x_i) = 0.$$

Furthermore, since $u^* \geq 0$,

$$\sum_{j=1}^n a_j c_j \geq 0$$

for all $\mathbf{c} \in D_n$ and thus for all $\mathbf{c} \in \mathcal{D}(D_n) = \mathcal{M}_n$. We have constructed a supporting hyperplane to $\mathcal{M}_n$ which touches at $\mathbf{c}^*$. Thus $\mathbf{c}^*$ is a boundary point of $\mathcal{M}_n$. □

Interior points of $\mathcal{M}_n$ may have many different representations. Prior to obtaining our main theorem on 'principal' representations, we present the following ancillary result.

Proposition 13. *Let $\mathbf{c}^* \in \operatorname{int} \mathcal{M}_n$. For each $x^* \in \{a, b\}$ there exists a representation of the form*

$$(B.5) \qquad c_j^* = \sum_{i=1}^r \lambda_i u_j(x_i), \qquad j = 1, \ldots, n,$$

where $\lambda_i > 0$, $i = 1, \ldots, r$, $x^* \in \{x_1, \ldots, x_r\}$, and $I(x_1, \ldots, x_r) = n$.

Proof. Assume $n \geq 2$. Set $\mathbf{c}^o = (\lambda u_1(x^*), \ldots, \lambda u_n(x^*))$, $\lambda > 0$. From Theorem 12, $\mathbf{c}^o \in \partial \mathcal{M}_n$ and thus $\mathbf{c}^o \neq \mathbf{c}^*$. Let L denote the straight line in $\mathbb{R}^n$ containing $\mathbf{c}^o$ and $\mathbf{c}^*$. As we traverse L from $\mathbf{c}^o$ through $\mathbf{c}^*$ we must eventually hit the boundary of $\mathcal{M}_n$ again for some $\lambda > 0$ sufficiently large (choose any such λ) at some point $\widetilde{\mathbf{c}}$. Thus

$$\mathbf{c}^* = \alpha \mathbf{c}^o + (1 - \alpha)\widetilde{\mathbf{c}}$$

for some $\alpha \in (0,1)$. Since $\tilde{c}$ is on the boundary of $\mathcal{M}_n$, it has a representation of the form (B.4) of index at most $n-1$. An additional point of index one comes from c°. Thus c^* has a representation of the form (B.5) with index $I(x_1, \ldots, x_r) \leq n$. If $I(x_1, \ldots, x_r) \leq n-1$, then from Theorem 12 c^* is on the boundary of $\mathcal{M}_n$, a contradiction. Thus $I(x_1, \ldots, x_r) = n$. □

If $c^* \in \operatorname{int} \mathcal{M}_n$ has a representation

$$c_j^* = \sum_{i=1}^{r} \lambda_i u_j(x_i), \qquad j = 1, \ldots, n,$$

then, from Theorem 12, $I(x_1, \ldots, x_r) \geq n$. If $I(x_1, \ldots, x_r) = n$, then we say that the representation is *principal*. We divide principal representations into two sets, *lower* and *upper*. A principal representation, using the points $x_1, \ldots, x_r$, is said to be *upper* if $b \in \{x_1, \ldots, x_r\}$. Otherwise it is called *lower*. (We shall soon motivate the reason for this convention.) The explicit form of these principal representations depends upon the parity of n. Assume $c^* \in \operatorname{int} \mathcal{M}_n$ has a principal representation using the points (nodes) $x_1, \ldots, x_r$. We adopt the convention that $a \leq x_1 < \cdots < x_r \leq b$.

If $n = 2m$, then for a lower principal representation $m = r$, and

$$a < x_1 < \cdots < x_m < b.$$

For an upper principal representation $m + 1 = r$, and

$$a = x_1 < \cdots < x_{m+1} = b.$$

If $n = 2m + 1$, then for a lower principal representation $m + 1 = r$, and

$$a = x_1 < \cdots < x_{m+1} < b.$$

For an upper principal representation $m + 1 = r$, and

$$a < x_1 < \cdots < x_{m+1} = b.$$

We can now state:

Theorem 14. *To each $c^* \in \operatorname{int} \mathcal{M}_n$ there exist exactly two principal representations, one lower and one upper.*

Proof. There are two claims made in the statement of the theorem, namely existence and uniqueness. Proposition 13 provides us with a proof of the existence of an upper principal representation (take $x^* = b$). For $n = 2m+1$ we also obtain the existence of a lower principal representation (take $x^* = a$). We complete the existence portion of the proof by constructing a lower principal representation in the case $n = 2m$.

For each $d \in [a, b]$, replace the interval $[a, b]$ by $[d, b]$. Let $\mathcal{M}_n(d)$ denote the moment space thereon (using the functions $u_1, \ldots, u_n$ restricted to $[d, b]$).

Let $\mathbf{c}^* \in \text{int}\,\mathcal{M}_n = \text{int}\,\mathcal{M}_n(a)$. Now $\mathbf{c}^* \notin \mathcal{M}_n(b) = \{(\lambda u_1(b), \ldots, \lambda u_n(b)) : \lambda \geq 0\}$, since otherwise it is, by Theorem 12, on the boundary of $\mathcal{M}_n$. From continuity considerations there exists a $d^* \in (a, b)$ with the property that $\mathbf{c}^* \in \partial\mathcal{M}_n(d^*)$. Applying Theorem 12, we see that $\mathbf{c}^*$ has a representation of the form (B.4) with respect to the interval $[d^*, b]$. That is, of index at most $n - 1$ with respect to $[d^*, b]$, and thus of index at most n with respect to $[a, b]$. Since $\mathbf{c}^* \in \text{int}\,\mathcal{M}_n$, it must be of exact index n with respect to $[a, b]$, and exact index $n - 1$ with respect to $[d^*, b]$. Since $n = 2m$, this implies that neither a (by construction) nor b (by a parity argument) is a node. This representation is therefore a lower principal representation.

It remains to prove the uniqueness of each of the principal representations. The argument is essentially the same in all four possible cases (upper or lower, n even or odd). We shall therefore prove, for example, the uniqueness of the lower principal representation for $n = 2m + 1$.

Let $\mathbf{c}^* \in \text{int}\,\mathcal{M}_n$, $n = 2m + 1$, and assume

$$c_j^* = \sum_{i=1}^{m+1} \lambda_i u_j(x_i) = \sum_{i=1}^{m+1} \alpha_i u_j(y_i), \qquad j = 1, \ldots, 2m + 1,$$

where $\lambda_i, \alpha_i > 0$, $i = 1, \ldots, m + 1$, and $a = x_1 < \cdots < x_{m+1} < b$; $a = y_1 < \cdots < y_{m+1} < b$. Thus

$$\sum_{i=1}^{m+1} \lambda_i u(x_i) = \sum_{i=1}^{m+1} \alpha_i u(y_i)$$

for all $u \in U$. Assume $y_k \notin \{x_1, \ldots, x_{m+1}\}$. Let $u \in U$ satisfy $u(x_i) = 0$, $i = 1, \ldots, m + 1$, and $u(y_i) = \delta_{ik}$, $i = 1, \ldots, m + 1$. Such a u exists since U is a T-system of dimension $2m + 1$ and we have specified values at no more than $2m + 1$ points (recall that $x_1 = y_1$). But this implies that $\alpha_k = 0$, a contradiction. Thus $y_k \in \{x_1, \ldots, x_{m+1}\}$ for all $k = 1, \ldots, m + 1$, and therefore $x_i = y_i$, $i = 1, \ldots, m + 1$. If $\lambda_k \neq \alpha_k$ for some k, let $u \in U$ satisfy $u(x_i) = \delta_{ik}$, $i = 1, \ldots, m + 1$. A contradiction ensues as above. This proves the uniqueness. $\quad\square$

Quadrature formulae for U with positive coefficients are representations of finite index. For given $\mathbf{c}^* \in \text{int}\,\mathcal{M}_n$ with $n = 2m$, the lower principal representation is often referred to as a quadrature formula of *Gaussian type*. The upper principal representation is sometimes said to be a quadrature formula of *Lobatto type*. For $\mathbf{c}^* \in \text{int}\,\mathcal{M}_n$ with $n = 2m + 1$, both the lower and upper principal representations are referred to as quadrature formulae of *Radau type*.

The principal representations are solutions to various extremal problems. One of these problems is directly related to our study of one-sided L^1-approximation.

In the first part of this appendix, we defined $C^+(U)$ and $C^-(U)$. We recall that

$$C^+(U) = \{u_{n+1} : u_{n+1} \in C[a,b], \det(u_i(x_j))_{i,j=1}^{n+1} \geq 0 \text{ for all choices}$$
$$\text{of } a \leq x_1 < \cdots < x_{n+1} \leq b\}.$$

and

$$C^-(U) = \{u_{n+1} : -u_{n+1} \in C^+(U)\}.$$

If, in the definition of $C^{\pm}(U)$, we replace some u_i by $-u_i$, or interchange the order of u_i and u_{i+1}, then $C^+(U)$ and $C^-(U)$ are interchanged. There are always these two distinct sets. But which is denoted $C^+(U)$ and which is denoted $C^-(U)$ depends on our choice of basis for U. To fix, once and for all, these two convex cones, we introduce the following notation. For an n-dimensional T-system $U = \text{span}\{u_1,\ldots,u_n\}$, we say that $\{u_1,\ldots,u_n\}$ is a T^+-*system* if $\det(u_i(x_j))_{i,j=1}^n > 0$ for all choices of $a \leq x_1 < \cdots < x_n \leq b$. Similarly, for an n-dimensional WT-system $U = \text{span}\{u_1,\ldots,u_n\}$, we say that $\{u_1,\ldots,u_n\}$ is a WT^+-*system* if $\det(u_i(x_j))_{i,j=1}^n \geq 0$ for all choices of $a \leq x_1 < \cdots < x_n \leq b$. We now always define $C^{\pm}(U)$ as previously where $\{u_1,\ldots,u_n\}$ is either a T^+- or WT^+-system, as appropriate. The sets $C^+(U)$ and $C^-(U)$ are now dependent only on U and not on the choice of basis for U.

For a given 'admissible' measure μ, set

$$c_j = \int_a^b u_j d\mu, \qquad j = 1,\ldots,n.$$

The vector $\mathbf{c} = (c_1,\ldots,c_n)$ lies in the interior of $\mathcal{M}_n$ (Theorem 12). For such a measure μ, we let $\underline{\mu}$ and $\overline{\mu}$ denote the unique lower and upper principal representations, respectively. We then have:

Theorem 15. *Let U be an n-dimensional T-system on $[a,b]$, and μ an 'admissible' measure. For each $f \in C^+(U)$ there exists a $u^* \in U$ satisfying $u^* \leq f$ and*

$$\int_a^b (f - u^*) d\underline{\mu} = 0.$$

Furthermore, u^* is a best one-sided $L^1(\mu)$ approximant from U to f from below.

Analogously we also have:

Corollary 16. *Let U be an n-dimensional T-system on $[a,b]$, and μ an 'admissible' measure. For each $f \in C^-(U)$ there exists a $u^* \in U$ satisfying $u^* \leq f$ and*

$$\int_a^b (f - u^*) d\overline{\mu} = 0.$$

Furthermore, u^* is a best one-sided $L^1(\mu)$ approximant to f from below.

Let us also note one immediate consequence of Theorem 15 and Corollary 16, and (B.3).

Theorem 17 (Markov-Krein Inequality). *Let U be an n-dimensional T-system on $[a, b]$, μ an 'admissible' measure, and $f \in C^+(U)$. Then*

$$\int_a^b f \, d\underline{\mu} \le \int_a^b f \, d\mu \le \int_a^b f \, d\overline{\mu}.$$

Remark. In the above, it is not necessary that μ be an 'admissible' measure, but only that the associated moment be in the interior of $\mathcal{M}_n$.

Proof of Theorem 15. If $u^* \in U$ exists satisfying $u^* \le f$ with

$$\int_a^b (f - u^*) d\underline{\mu} = 0,$$

then from (B.3) or Theorem 5.2, we see that u^* is the desired best approximant. For completeness we simply note that, if $u \in U$ satisfies $u \le f$, then

$$\int_a^b u \, d\mu = \int_a^b u \, d\underline{\mu} \le \int_a^b f \, d\underline{\mu} = \int_a^b u^* d\underline{\mu} = \int_a^b u^* d\mu.$$

Thus $\int_a^b u \, d\mu \le \int_a^b u^* d\mu$ for all $u \in U$ satisfying $u \le f$.

It remains to prove the existence of the desired u^*. If $f \in U$, there is nothing to prove. We therefore assume that $\{u_1, \ldots, u_n\}$ is a T^+-system and $\{u_1, \ldots, u_n, f\}$ is a WT^+-system. We use an argument to be found in Theorem 4 of Appendix A.

For convenience set $f = u_{n+1}$ and assume that $n = 2m$, and

$$\int_a^b u \, d\underline{\mu} = \sum_{i=1}^m \lambda_i u(x_i)$$

for all $u \in U$, where $a < x_1 < \cdots < x_m < b$. We shall also assume that $\{u_1, \ldots, u_n, f\}$ is a T^+-system of $[a, b]$. If $\{u_1, \ldots, u_n, f\}$ is only a WT^+-system, we 'smooth' it (see Proposition 6 of Appendix A), obtain the desired result, and then limit back. This is easily done.

For $\varepsilon > 0$, ε small, set

$$v^\varepsilon(x) = cU \begin{pmatrix} 1, & & \cdot & \cdot & \cdot & , 2m+1 \\ x_1, & x_1 + \varepsilon, & \ldots & , x_m & , x_m + \varepsilon & , x \end{pmatrix}$$

where $c > 0$, to be chosen later. Since $\{u_1, \ldots, u_{2m+1}\}$ is a T^+-system, we have $v^\varepsilon(x_i) = v^\varepsilon(x_i + \varepsilon) = 0$, $i = 1, \ldots, m$, and $v^\varepsilon(x) > 0$ for $x \in (x_{i-1} + \varepsilon, x_i)$, $i = 1, \ldots, m+1$, where $x_0 + \varepsilon = a$ and $x_{m+1} = b$. Let $v^\varepsilon = \sum_{i=1}^{2m+1} c a_i^\varepsilon u_i$. Choose $c > 0$ so that $\sum_{i=1}^{2m+1} (c a_i^\varepsilon)^2 = 1$. Letting $\varepsilon \downarrow 0$ along a subsequence, we obtain a $v = \sum_{i=1}^{2m+1} a_i u_i$ which is non-negative, non-trivial, and vanishes at the x_i, $i = 1, \ldots, m$. Now $v = \alpha f - \tilde{u}$ for some $\alpha \in \mathbb{R}$ and $\tilde{u} \in U$, where $\alpha = a_{2m+1}$. By construction,

$$a_{2m+1}^\varepsilon = cU \begin{pmatrix} 1, & . & . & . & , 2m \\ x_1, & x_1 + \varepsilon, & \ldots, & x_m, & x_m + \varepsilon \end{pmatrix}.$$

Since $\{u_1, \ldots, u_{2m}\}$ is a T^+-system, we have $a_{2m+1}^\varepsilon > 0$. Thus $\alpha \geq 0$. We have therefore constructed an $\alpha \geq 0$ and $\tilde{u} \in U$ such that $\alpha f - \tilde{u} \geq 0$. Furthermore $\alpha f - \tilde{u} \neq 0$ and

$$\int_a^b (\alpha f - \tilde{u}) d\mu = 0.$$

If $\alpha = 0$, then $-\tilde{u} \geq 0$, $(-\tilde{u} \neq 0)$, and

$$0 < \int_a^b -\tilde{u} \, d\mu = \int_a^b -\tilde{u} \, d\mu = 0,$$

a contradiction. Thus $\alpha > 0$. Setting $u^* = \tilde{u}/\alpha$ proves our claim. □

Remark. We make no claim as to the uniqueness of u^*. For example, let μ be Lebesgue measure on $[-1, 1]$, $U = \text{span}\{1, x\}$, and $f(x) = x_+$. Then $f \in \mathcal{C}^+(U)$ and

$$\int_{-1}^1 u \, d\mu = 2u(0).$$

The functions $u^*(x) = \alpha x$, $0 \leq \alpha \leq 1$, all satisfy the conditions of Theorem 15.

Before turning to a discussion of the uniqueness problem, let us extend the above results to WT-systems. The above theory depended on the existence of principal representations. It is *not* true that every WT-system has principal representations with respect to every interior point of the moment space. For example, if U is an n-dimensional WT-system with a basis of non-negative functions with disjoint support, then every representation of the moments generated by any 'admissible' measure must contain at least n distinct points.

To obtain principal representations for a WT-system U, it is sufficient (but certainly not necessary) that U contain a strictly positive function. From Micchelli, Pinkus [1977, Theorem 2.2], we have:

Proposition 18. *Let U be an n-dimensional WT-system on $[a, b]$, containing a strictly positive function. Assume μ is an 'admissible' measure and $\mathbf{c} = (c_1, \ldots, c_n)$, where*

$$c_j = \int_a^b u_j \, d\mu, \qquad j = 1, \ldots, n.$$

There exist both lower and upper principal representations for $\mathbf{c}$.

Proof. The proof is by 'smoothing'. The main point is that the existence of a strictly positive function allows us to limit back and obtain the desired result.

For convenience we shall only consider the case of a lower principal representation and $n = 2m + 1$. Let $\{u_i^\varepsilon\}_{i=1}^n$, $\varepsilon > 0$, be such that $U^\varepsilon = \text{span}\{u_1^\varepsilon, \ldots, u_n^\varepsilon\}$ is a T-system on $[a, b]$, and $\lim_{\varepsilon \to 0+} u_i^\varepsilon = u_i$, $i = 1, \ldots, n$, uniformly on $[a, b]$. U^ε exists from Proposition 6 of Appendix A.

Set $\mathbf{c}^\varepsilon = (c_1^\varepsilon, \ldots, c_n^\varepsilon)$ where

$$c_j^\varepsilon = \int_a^b u_j^\varepsilon d\mu, \qquad j = 1, \ldots, n.$$

From Theorem 14 there exist $a = x_1^\varepsilon < \cdots < x_{m+1}^\varepsilon < b$ and $\lambda_i^\varepsilon > 0$, $i = 1, \ldots, m + 1$, such that

$$c_j^\varepsilon = \sum_{i=1}^{m+1} \lambda_i^\varepsilon u_j^\varepsilon(x_i^\varepsilon), \qquad j = 1, \ldots, n.$$

Since the x_i^ε all lie in $[a, b]$, we can choose a subsequence of ε, tending to zero, on which $x_i^\varepsilon \to x_i$, $i = 1, \ldots, m + 1$, where $a = x_1 \leq \cdots \leq x_{m+1} \leq b$. Since U contains a strictly positive function it is easily proven that $\sum_{i=1}^{m+1} \lambda_i^\varepsilon$ is uniformly bounded for all sufficiently small ε. As such we can also choose a subsequence of ε tending to zero where not only the x_i^ε converge, but also $\lambda_i^\varepsilon \to \lambda_i$, $i = 1, \ldots, m + 1$ with $\lambda_i \geq 0$, all i. Thus

$$c_j = \sum_{i=1}^{m+1} \lambda_i u_j(x_i), \qquad j = 1, \ldots, n.$$

It remains to prove that $\lambda_i > 0$, $i = 1, \ldots, m + 1$, and $a = x_1 < \cdots < x_{m+1} < b$, i.e., $I(x_1, \ldots, x_{m+1}) = n$. If this is not the case there exists (using a 'smoothing' argument and Theorem 4 of Appendix A) a non-negative non-trivial $\tilde{u} \in U$ satisfying

$$\sum_{i=1}^{m+1} \lambda_i \tilde{u}(x_i) = 0.$$

But $\int_a^b \tilde{u} \, d\mu > 0$, a contradiction. $\qquad\qquad\qquad\qquad\qquad\qquad\qquad\square$

Remark. No claim, as in Theorem 14, is being made as to the uniqueness of the principal representations. Uniqueness need not hold. Let μ be Lebesgue measure on $[-2, 2]$, and $U = \text{span}\{1, u_2\}$, where

$$u_2(x) = \begin{cases} x + 1, & -2 \leq x \leq -1 \\ 0, & -1 \leq x \leq 1 \\ x - 1, & 1 \leq x \leq 2. \end{cases}$$

U is a WT-system containing a strictly positive function and yet $4u(x_1)$ is a lower principal representation for U for every $x_1 \in [-1, 1]$.

As a consequence of Proposition 18 and the method of proof of Theorem 15, we have:

Theorem 19. *Let U be an n-dimensional WT-system on $[a,b]$ containing a strictly positive function. Let μ be an 'admissible' measure.*

  a) *Let $\underline{\mu}$ be any lower principal representation for μ and $f \in C^+(U)$. There exists a $u^* \in U$ satisfying $u^* \le f$ and*

$$\int_a^b (f - u^*)d\underline{\mu} = 0.$$

Furthermore, u^* is a best one-sided $L^1(\mu)$ approximant from U to f from below.

  b) *Let $\overline{\mu}$ be any upper principal representation for μ and $f \in C^-(U)$. There exists a $u^* \in U$ satisfying $u^* \le f$ and*

$$\int_a^b (f - u^*)d\overline{\mu} = 0.$$

Furthermore, u^* is a best one-sided $L^1(\mu)$ approximant from U to f from below.

As a consequence of Theorem 19 and (B.3) we also have the extension of the Markov-Krein Inequality in the above setting.

Theorem 20 (Markov-Krein Inequality). *Let U be an n-dimensional WT-system on $[a,b]$ containing a strictly positive function. Let μ be an 'admissible' measure and $f \in C^+(U)$. Then*

$$\int_a^b f \, d\underline{\mu} \le \int_a^b f \, d\mu \le \int_a^b f \, d\overline{\mu}$$

where $\underline{\mu}$ and $\overline{\mu}$ are any lower and upper principal representations, respectively.

We continue to assume that U is an n-dimensional WT-system on $[a,b]$ containing a strictly positive function, and that μ is an 'admissible' measure. Two questions naturally arise. Namely, are the principal representations unique, and how do we construct a best one-sided $L^1(\mu)$ approximant to each $f \in C^{\pm}(U)$? In Part I of this appendix we discussed the analogous questions for the two-sided approximation problem. Different sets of conditions were imposed, and in each case we proved uniqueness. We now establish somewhat parallel results.

Let us recall from Part I that for $a \le t_1 < \cdots < t_n \le b$, we set

$$U[t_1, \ldots, t_n] = \{(f(t_1), \ldots, f(t_n)) : f \in C(U)\}.$$

The following analogue of Proposition 11 is essentially to be found in Micchelli, Pinkus [1977, Corollary 2.2].

Proposition 21. *Let U be an n-dimensional WT-system on $[a,b]$ containing a strictly positive function. Assume that $\dim U[t_1, \ldots, t_n] = n$ for every choice* of $a \le t_1 < \cdots < t_n \le b$. *Then every 'admissible' measure* μ *has unique lower and upper principal representations.*

Proof. For convenience we shall only prove that μ has a unique lower principal representation in the case $n = 2m$. Assume

$$\int_a^b u\,d\mu = \sum_{i=1}^m \lambda_i u(x_i) = \sum_{i=1}^m \alpha_i u(y_i)$$

for all $u \in U$, where $\lambda_i, \alpha_i > 0$, $i = 1, \ldots, m$, and $a < x_1 < \cdots < x_m < b$; $a < y_1 < \cdots < y_m < b$.

From Theorem 20, we have

$$\sum_{i=1}^m \lambda_i f(x_i) = \sum_{i=1}^m \alpha_i f(y_i)$$

for all $f \in \mathcal{C}^+(U)$. Since $\mathcal{C}^-(U) = -\mathcal{C}^+(U)$, equality holds for all $f \in \mathcal{C}(U)$. Because there are at most n distinct points in $\{x_1, \ldots, x_m, y_1, \ldots, y_m\}$, it follows by assumption that if, for example, $y_k \notin \{x_1, \ldots, x_m\}$, we can find an $f \in \mathcal{C}(U)$ which vanishes at all the x_i and y_i, except for y_k. A contradiction ensues. Thus $x_i = y_i$, $i = 1, \ldots, m$. A similar argument implies that $\lambda_i = \alpha_i$, all i. □

Let us now impose a different set of conditions on U. Here we consider only U and f in $C^1[a,b]$. The condition we impose is the demand that U be a unicity space for $C^1_{1+}([a,b], \mu)$.

The interpolation problem intimately connected with one-sided approximation involves interpolating not only function values but also derivative values at interior points. For a given representation

$$\int_a^b u\,d\nu = \sum_{i=1}^r \lambda_i u(x_i)$$

for $u \in U$, we interpret $u(d\nu) = 0$ as the interpolation problem which sets u and its first derivative equal to zero at the points x_i interior to $[a, b]$, while at an endpoint only the value $u(x_i)$ is set equal to zero. Thus, for example, if μ is a lower principal representation for μ, and $u(d\mu) = 0$ has only the zero solution in U, then there exists to each $f \in \mathcal{C}^+(U) \cap C^1[a,b]$ a unique best one-sided $L^1(\mu)$ approximant. It is given by the unique (since the index of the points of the representation is n) $u^* \in U$ which interpolates to f at each point in the lower principal representation, and to the derivative of f at each of these same points in (a, b).

This next proposition, due to Strauss [1982, Theorem 3.2], will be used in proving our last result.

Proposition 22. *Let U be an n-dimensional WT-system in $C^1[a, b]$ containing a strictly positive function, and assume that μ is an 'admissible' measure.*

a) *There do not exist two lower principal representations $\underline{\mu}_1$ and $\underline{\mu}_2$ for which both $u(d\underline{\mu}_1) = 0$ and $u(d\underline{\mu}_2) = 0$ have only the zero solution.*

b) *There do not exist two upper principal representations $\overline{\mu}_1$ and $\overline{\mu}_2$ for which both $u(d\overline{\mu}_1) = 0$ and $u(d\overline{\mu}_2) = 0$ have only the zero solution.*

Proof. For convenience we only prove (a) in the case $n = 2m$. Assume

$$\int_a^b u\,d\underline{\mu}_1 = \sum_{i=1}^m \lambda_i u(x_i)$$

and

$$\int_a^b u\,d\underline{\mu}_2 = \sum_{i=1}^m \alpha_i u(y_i),$$

where $\lambda_i,\,\alpha_i > 0$, $i = 1,\ldots,m$, and $a < x_1 < \cdots < x_m < b$; $a < y_i < \cdots < y_m < b$. Since both $\underline{\mu}_1$ and $\underline{\mu}_2$ are representations for μ, we have

$$\sum_{i=1}^m \lambda_i u(x_i) = \sum_{i=1}^m \alpha_i u(y_i)$$

for all $u \in U$.

Assume, without loss of generality, that $x_i = y_i$, $i = 1,\ldots,k-1$, and $x_k < y_k$. Since $u(d\underline{\mu}_1) = 0$ has only the zero solution in U, there exists a unique $\widetilde{u} \in U$ satisfying $\widetilde{u}(x_i) = 0$, $i = 1,\ldots,m$, and $\widetilde{u}'(x_i) = \delta_{ik}$, $i = 1,\ldots,m$. From continuity considerations there exists for $\varepsilon > 0$, ε sufficiently small (with $x_k + \varepsilon < y_k$), a unique $u^* \in U$ satisfying $u^*(x_i) = 0$, $i = 1,\ldots,m$; $i \neq k$, $u^*(x_k + \varepsilon) = 0$, and $u^{*\prime}(x_i) = 0$, $i = 1,\ldots,m$; $i \neq k$, $u^{*\prime}(x_k + \varepsilon) = 1$. From these same continuity considerations we have $u^*(x_k) < 0$. Since U is a WT-system, it follows from properties of the determinant that $u^*(x) \leq 0$ for all $x \in [a, x_k + \varepsilon]$, and $u^*(x) \geq 0$ for all $x \in [x_k + \varepsilon, b]$. Thus

$$\sum_{i=1}^m \lambda_i u^*(x_i) = \lambda_k u^*(x_k) < 0,$$

while

$$\sum_{i=1}^m \alpha_i u^*(y_i) = \sum_{i=k}^m \alpha_i u^*(y_i) \geq 0.$$

This is a contradiction.   □

We can now state our final result due to Strauss [1982, Lemma 3.1 and Theorem 3.2].

Proposition 23. *Let U be an n-dimensional WT-system in $C^1[a,b]$ containing a strictly positive function, and let μ be an 'admissible' measure. Assume that U is a unicity space for $C_{1+}^1([a,b],\mu)$. Then μ has unique lower and upper principal representations $\underline{\mu}$ and $\overline{\mu}$, respectively. Furthermore, both $u(d\underline{\mu}) = 0$ and $u(d\overline{\mu}) = 0$ have only the zero solution in U.*

Proof. We shall only prove the proposition for the lower principal representation. On the basis of Proposition 22, it suffices to prove that if $\underline{\mu}$ is any lower principal representation for μ, then $u(d\underline{\mu}) = 0$ has only the zero solution in U.

Assume to the contrary that there exists a $u^* \in U$, $u^* \neq 0$, for which $u^*(d\underline{\mu}) = 0$. From Lemma 5.20 there exists an $f \in C^1[a,b]$ satisfying $f \geq |u^*|$ and $f(x) = 0$ for every x in the support of $\underline{\mu}$. It now easily follows (see Theorem 5.2) that αu^* is a best one-sided $L^1(\mu)$ approximant to f from U for every $\alpha \in [-1, 1]$. Thus U is not a unicity space for $C^1_{1+}([a,b],\mu)$. A contradiction.   □

References

Abdelmalek, N. N. [1974]: *On the discrete linear L_1 approximation and L_1 solutions of overdetermined linear equations.* J. Approx. Theory **11**, *38–53*.

Abdelmalek, N. N. [1975]: *An efficient method for the discrete linear L_1 approximation problem.* Math. Comp. **29**, *844–850*.

Abdelmalek, N. N. [1985]: *Chebyshev and L_1 solutions of overdetermined systems of linear equations with bounded variables.* Numer. Funct. Anal. and Optimiz. **8**, *399–418*.

Angelos, J., Schmidt, D. [1983]: *Strong uniqueness in $L^1(X, \Sigma, \mu)$.* In "Approximation Theory IV", eds. C. K. Chui, L. L. Schumaker, J. D. Ward, *297–302*, Academic Press, New York.

Armstrong, R. D., Hultz, J. W. [1977]: *An algorithm for a restricted discrete approximation problem in the L_1 norm.* SIAM J. Numer. Anal. **14**, *555–565*.

Barrodale, I. [1970]: *On computing best L_1 approximations.* In "Approximation Theory", ed. A. Talbot, *205–215*, Academic Press, London.

Barrodale, I., Roberts, F. D. K. [1973]: *An improved algorithm for discrete l_1 linear approximation.* SIAM J. Numer. Anal. **10**, *839–848*.

Barrodale, I., Roberts, F. D. K. [1978]: *An efficient algorithm for discrete l_1 linear approximation with linear constraints.* SIAM J. Numer. Anal. **15**, *603–611*.

Barrodale, I., Young, A. [1966]: *Algorithms for best L_1 and L_∞ linear approximations on a discrete set.* Numer. Math. **8**, *295–306*.

Bartels, R. H., Conn, A. R., Sinclair, J. W. [1978]: *Minimization techniques for piecewise differentiable functions: The l_1 solution to an overdetermined linear system.* SIAM J. Numer. Anal. **15**, *224–241*.

Berdyshev, V. I. [1975]: *Metric projection onto finite-dimensional subspaces of C and L.* Mat. Zametki **18** (1975), *473–488*; see also Math. Notes **18**, *871–879*.

Bernstein, S. [1926]: "Leçons sur les Propriétés Extrémales et la Meilleure Approximation des Fonctions Analytiques d'une Variable Réelle". Gauthier-Villars, Paris; see also Bernstein, S., de la Vallée Poussin, C. (1970): "L'Approximation". Chelsea, New York.

Bojanic, R., DeVore, R. [1966]: *On polynomials of best one-sided approximation.* L'Enseignement Math. **12**, *139–164*.

de Boor, C. [1978]: "A Practical Guide to Splines". Applied Mathematical Sciences 27, Springer-Verlag, New York.

Borsuk, K. [1933]: *Drei Sätze über die n-dimensionale euklidische Sphäre.* Fund. Math. **20**, *177–190.*

Brown, A. L. [1964]: *Best n-dimensional approximation to sets of functions.* Proc. London Math. Soc. **14**, *577–594.*

Carroll, M. P., Braess, D. [1974]: *On uniqueness of L_1-approximation for certain families of spline functions.* J. Approx. Theory **12**, *362–364.*

Cheney, E. W. [1966]: "Introduction to Approximation Theory". McGraw-Hill, New York.

Cheney, E. W., Wulbert, D. E. [1969]: *The existence and unicity of best approximations.* Math. Scand. **24**, *113–140.*

Claerbout, J. F., Muir, F. [1973]: *Robust modeling with erratic data.* Geophysics **38**, *826–844.*

Danzer, L., Grünbaum, B., Klee, V. [1963]: *Helly's theorem and its relatives.* In Proc. Sym. in Pure Math. Vol VII, "Convexity", ed. V. L. Klee, *101–180,* Amer. Math. Soc., Providence.

Deutsch, F. [1983]: *A survey of metric selections.* In "Fixed Points and Nonexpansive Mapping", ed. R. C. Sine, Contemporary Mathematics, **18**, *49–71,* Amer. Math. Soc., Providence.

DeVore, R. [1968]: *One-sided approximation of functions.* J. Approx. Theory **1**, *11–25.*

Dunford, N., Schwartz, J. T. [1958]: "Linear Operators Part I: General Theory". Interscience, New York.

Duris, C. S., Sreedharan, V. P. [1968]: *Chebyshev and l^1-solutions of linear equations using least squares solutions.* SIAM J. Numer. Anal. **5**, *491–505.*

Eggleston, H. G. [1958]: "Convexity". Cambridge Tracts in Math. and Math. Physics, 47, Cambridge University Press, Cambridge.

Galkin, P. V. [1974]: *The uniqueness of the element of best mean approximation to a continuous function using splines with fixed nodes.* Mat. Zametki **15**, *3–14;* see also Math. Notes **15**, *3–8.*

Garkavi, A. [1964]: *On Čebyšev and almost Čebyšev subspaces.* Izv. Akad. Nauk SSSR Ser. Mat. **28**, *799–818;* see also Amer. Math. Soc. Transl. (2) **96** (1970), *153–175.*

Glashoff, K., Schultz, R. [1979]: *Über die genaue Berechnung von besten L^1-Approximierenden.* J. Approx. Theory **25**, *280–293.*

Haar, A. [1918]: *Die Minkowskische Geometrie und die Annäherung an stetige Funktionen.* Math. Ann. **78**, *294–311.*

Havinson, S. Ja. [1957]: *On dimensionality of polyhedra of best approximation in the metric L_1.* Sbornik Trudov MISI **19**, *18–29*.

Havinson, S. Ja. [1958]: *On unicity of functions of best approximation in the metric of the space L^1.* Izv. Akad. Nauk SSSR Ser. Mat. **22**, *243–270*.

Havinson, S. Ja., Romanova, Z. S. [1972]: *Approximation properties of finite-dimensional subspaces in L_1.* Mat. Sb. **89**, *3–15*; see also Math. USSR Sb. **18**, *1–14*.

Helly, E. [1923]: *Über Mengen konvexer Körper mit gemeinschaftlichen Punkten.* Jber. Deutsch. Math. Verein. **32**, *175–176*.

Hettich, R., Zencke, P. [1982]: "Numerische Methoden der Approximation und semi-infiniten Optimierung". Teubner, Stuttgart.

Hobby, C. R., Rice, J. R. [1965]: *A moment problem in L_1 approximation.* Proc. Amer. Math. Soc. **16**, *665–670*.

Hörmander, L. [1983]: "The Analysis of Linear Partial Differential Operators I". Springer-Verlag, Berlin.

Jackson, D. [1921]: *Note on a class of polynomials of approximation.* Trans. Amer. Math. Soc. **22**, *320–326*.

James, R. C. [1947]: *Orthogonality and linear functionals in normed linear spaces.* Trans. Amer. Math. Soc. **61**, *265–292*.

Jones, R. C., Karlovitz, L. A. [1970]: *Equioscillation under nonuniqueness in the approximation of continuous functions.* J. Approx. Theory **3**, *138–145*.

Karlin, S. [1968]: "Total Positivity, Vol. I". Stanford Univ. Press, Stanford.

Karlin, S., Studden, W. J. [1966]: "Tchebycheff Systems: With Applications in Analysis and Statistics". Interscience, New York.

Kiwiel, K. C. [1985]: "Methods of Descent for Nondifferentiable Optimization". Lect. Notes in Math., No. 1133, Springer-Verlag, Berlin.

Köthe, G. [1969]: "Topological Vector Spaces I". Springer-Verlag, Berlin.

Krein, M. G. [1951]: *The ideas of P. L. Čebyšev and A. A. Markov in the theory of limiting values of integrals and their further development.* Uspehi Mat. Nauk **6**, *3–120*; see also Amer. Math. Soc. Transl. (2) **12** (1959), *1–121*.

Krein, M. [1962]: *The L-problem in an abstract linear normed space.* In "Some Questions in the Theory of Moments", by N. I. Ahiezer and M. Krein, *175–204*, Transl. Math. Monographs **2**, Amer. Math. Soc., Providence. Originally published in Russian in 1938.

Krein, M. G., Nudel'man A. A. [1977]: "The Markov Moment Problem and Extremal Problems". Transl. Math. Monographs, **50**, Amer. Math. Soc., Providence.

Kripke, B. R. [1964]: *Best approximation with respect to nearby norms.* Numer. Math. **6**, *103–105*.

Kripke, B. R., Rivlin, T. J. [1965]: *Approximation in the metric of $L^1(X, \mu)$.* Trans. Amer. Math. Soc. **119**, *101–122*.

Kroó, A. [1982]: *Some theorems on unicity of multivariate L_1-approximation.* Acta Math. Acad. Sci. Hungar. **40**, *179–189*.

Kroó, A. [1984]: *Some theorems on best L_1-approximation of continuous functions.* Acta Math. Acad. Sci. Hungar. **44**, *409–417*.

Kroó, A. [1985a]: *On an L_1-approximation problem.* Proc. Amer. Math. Soc. **94**, *406–410*.

Kroó, A. [1985b]: *Some uniqueness problems in best Chebyshev and mean multivariate approximation.* In "Multivariate Approximation Theory III", eds. W. Schempp, K. Zeller, *262–269*, ISNM 75, Birkhäuser Verlag, Basel.

Kroó, A. [1986a]: *Chebyshev rank in L_1-approximation.* Trans. Amer. Math. Soc. **296**, *301–313*.

Kroó, A. [1986b]: *On uniqueness of best L_1-approximation on disjoint intervals.* Math. Z. **191**, *507–512*.

Kroó, A. [1987a]: *Best L_1-approximation with varying weights.* Proc. Amer. Math. Soc. **99**, *66–70*.

Kroó, A. [1987b]: *A general approach to the study of Chebyshev subspaces in L_1-approximation of continuous functions.* J. Approx. Theory **51**, *98–111*.

Kroó, A. [1988]: *On the uniqueness of canonical points in the Hobby-Rice Theorem.* Preprint.

Lazar, A. J., Wulbert, D. E., Morris, P. D. [1969]: *Continuous selections for metric projections.* J. Func. Anal. **3**, *193–216*.

Lewis, J. T. [1970]: *Computation of best one-sided L_1 approximation.* Math. Comp. **24**, *529–536*.

Liapounoff, A. [1940]: *Sur les fonctions-vecteurs complètement additives.* Bull. Acad. Sci. URSS Sér. Math. **4**, *465–478*.

Light, W. A., Cheney, E. W. [1985]: "Approximation Theory in Tensor Product Spaces". Lect. Notes in Math., No. 1169, Springer-Verlag, Berlin.

Light, W. A., Holland, S. M. [1984]: *The L_1-version of the Diliberto-Straus algorithm in $C(T \times S)$.* Proc. Edinburgh Math. Soc. **27**, *31–45*.

Lin, P. K. [1985]: *Remarks on linear selections for the metric projection.* J. Approx. Theory **43**, *64–74*.

Marti, J. T. [1975]: *A method for the numerical computation of best L_1-approximations of continuous functions.* In "Numerische Methoden der Approximationstheorie, Band 2", eds. L. Collatz, G. Meinardus, *79–92*,

ISNM 26, Birkhäuser Verlag, Basel.

Micchelli, C. A. [1977]: *Best L^1 approximation by weak Chebyshev systems and the uniqueness of interpolating perfect splines.* J. Approx. Theory **19**, *1-14.*

Micchelli, C. A., Pinkus, A. [1977]: *Moment theory for weak Chebyshev systems with applications to monosplines, quadrature formulae and best one-sided L^1-approximation by spline functions with fixed knots.* SIAM J. Math. Anal. **8**, *206-230.*

Moroney, R. M. [1961]: *The Haar problem in L_1.* Proc. Amer. Math. Soc. **12**, *793-795.*

Motzkin, T. S., Walsh, J. L. [1956]: *Least* p*th power polynomials on a finite point set.* Trans. Amer. Math. Soc. **83**, *371-396.*

Newman, D. J., Shapiro, H. S. [1963]: *Some theorems on Čebyšev approximation.* Duke Math. J. **30**, *673-681.*

Nürnberger, G. [1985a]: *Unicity in one-sided L_1-approximation and quadrature formulae.* J. Approx. Theory **45**, *271-279.*

Nürnberger, G. [1985b]: *Global unicity in semi-infinite optimization.* Numer. Funct. Anal. and Optimiz. **8**, *173-191.*

Osborne, M. R. [1985]: "Finite Algorithms in Optimization and Data Analysis". J. Wiley and Sons, Chichester.

Papini, P. L. [1978]: *Approximation and strong approximation in normed spaces via tangent functionals.* J. Approx. Theory **22**, *111-118.*

Paszkowski, S. [1957]: *On approximating with nodes.* Rozprawy Mat. **14**, *1-62.*

Phelps, R. R. [1960]: *Uniqueness of Hahn-Banach extensions and unique best approximation.* Trans. Amer. Math. Soc. **95**, *238-255.*

Phelps, R. R. [1966]: *Čebyšev subspaces of finite dimension in L_1.* Proc. Amer. Math. Soc. **17**, *646-652.*

Pinkus, A. [1976a]: *One-sided L^1 approximation by splines with fixed knots.* J. Approx. Theory **18**, *130-135.*

Pinkus, A. [1976b]: *A simple proof of the Hobby-Rice theorem.* Proc. Amer. Math. Soc. **60**, *82-84.*

Pinkus, A. [1986]: *Unicity subspaces in L^1-approximation.* J. Approx. Theory **48**, *226-250.*

Pinkus, A. [1988]: *Continuous selections for the metric projection on C_1.* Constr. Approx. **4**, *85-96.*

Pinkus, A., Strauss, H. [1987]: *One-sided L^1-approximation to differentiable functions.* Approximation Theory Appl. **3**, *81-96.*

Pinkus, A., Strauss, H. [1988]: *Best approximation with coefficient constraints.* IMA J. Numer. Anal. **8**, *1–22.*

Pinkus, A., Totik, V. [1986]: *One-sided L^1-approximation.* Canad. Math. Bull. **29**, *84–90.*

Pinkus, A., Wajnryb, B. [1988]: *Necessary conditions for uniqueness in L^1-approximation.* J. Approx. Theory **53**, *54–66.*

Powell, M. J. D. [1981]: "Approximation Theory and Methods". Cambridge University Press, Cambridge.

Powell, M. J. D., Roberts, F. D. K. [1980]: *A discrete characterization theorem for the discrete L_1 linear approximation problem.* J. Approx. Theory **30**, *173–179.*

Pták, V. [1958]: *On approximation of continuous functions in the metric $\int_a^b |x(t)|dt$.* Czech. Math. J. **8**, *267–273.*

Rice, J. R. [1964]: "The Approximation of Functions, Vol. I, Linear Theory". Addison-Wesley, Reading, Mass.

Rivlin, T. J. [1969]: "An Introduction to the Approximation of Functions". Blaisdell, Waltham, Mass.

Robers, P. D., Ben-Israel, A. [1969]: *An interval programming algorithm for discrete linear L_1 approximation problems.* J. Approx. Theory **2**, *323–336.*

Rozema, E. [1974]: *Almost Chebyshev subspaces of $L^1(\mu; E)$.* Pacific J. Math. **53**, *585–604.*

Rudin, W. [1973]: "Functional Analysis". McGraw-Hill, New York.

Schmidt, D. [1987]: *A theorem on weighted L^1-approximation.* Proc. Amer. Math. Soc. **101**, *81–84.*

Schumaker, L. L. [1981]: "Spline Functions: Basic Theory". John Wiley and Sons, New York.

Shor, N. Z. [1985]: "Minimization Methods for Non-Differentiable Functions". Springer-Verlag, Berlin.

Singer, I. [1956]: *Caractérisation des éléments de meilleure approximation dans un espace de Banach quelconque.* Acta Sci. Math. (Szeged) **17**, *181–189.*

Singer, I. [1960]: *On a theorem of V. Pták concerning best approximation of continuous functions in the metric $\int_a^b |x(t)|dt$.* Czech. Math. J. **10**, *425–431.*

Singer, I. [1970]: "Best Approximation in Normed Linear Spaces by Elements of Linear Subspaces". Springer-Verlag, Berlin.

Sommer, M. [1979]: *L_1-approximation by weak Chebyshev spaces.* In "Approximation in Theorie und Praxis", ed. G. Meinardus, *85–102*, Mannheim,

Bibliographisches Institut.

Sommer, M. [1983a]: *Weak Chebyshev spaces and best L_1-approximation.* J. Approx. Theory **39**, *54–71*.

Sommer, M. [1983b]: *Some results on best L_1-approximation of continuous functions.* Numer. Funct. Anal. and Optimiz. **6**, *253–271*.

Sommer, M. [1985]: *Uniqueness of best L_1-approximations of continuous functions.* In "Delay Equations, Approximation and Application", eds. G. Meinardus, G. Nürnberger, *264–281*, ISNM 74, Birkhäuser Verlag, Basel.

Sommer, M. [1987]: *Examples of unicity subspaces in L_1-approximation.* Numer. Funct. Anal. and Optimiz. **9**, *131–146*.

Sommer, M. [1988]: *Properties of unicity subspaces in L_1-approximation.* J. Approx. Theory **52**, *269–283*.

Sommer, M., Strauss, H. [1977]: *Eigenschaften von schwach tschebyscheffschen Räumen.* J. Approx. Theory **21**, *257–268*.

Sommer, M., Strauss, H. [1981]: *Unicity of best one-sided L_1-approximations for certain classes of spline functions.* Numer. Funct. Anal. and Optimiz. **4**, *413–435*.

Spyropoulos, K., Kiountouzis, E., Young, A. [1973]: *Discrete approximation in the L_1 norm.* Comput. J. **16**, *180–186*.

Stockenberg, B. [1977a]: *On the number of zeros of functions in a weak Tchebyshev-space.* Math. Z. **156**, *49–57*.

Stockenberg, B. [1977b]: *Subspaces of weak and oriented Tchebyshev-spaces.* Manuscripta Math. **20**, *401–407*.

Strauss, H. [1975]: *L_1-Approximation mit Splinefunktionen.* In "Numerische Methoden der Approximationstheorie, Band 2", eds. L. Collatz, G. Meinardus, *151–162*, ISNM 26, Birkhäuser Verlag, Basel.

Strauss, H. [1981]: *Eindeutigkeit in der L_1-Approximation.* Math. Z. **176**, *63–74*.

Strauss, H. [1982]: *Unicity of best one-sided L_1-approximations.* Numer. Math. **40**, *229–243*.

Strauss, H. [1984]: *Best L_1-approximation.* J. Approx. Theory **41**, *297–308*.

Tchakaloff, V. [1957]: *Formules de cubatures mécaniques a coefficients non négatifs.* Bull. Sci. Math. **81**, *123–134*.

Usow, K. H. [1967a]: *On L_1 approximation I: Computation for continuous functions and continuous dependence.* SIAM J. Numer. Anal. **4**, *70–88*.

Usow, K. H. [1967b]: *On L_1 approximation II: Computation for discrete functions and discretization effects.* SIAM J. Numer. Anal. **4**, *233–244*.

Wagner, H. M. [1959]: *Linear programming techniques for regression analysis.* J. Amer. Statist. Assoc. **54**, *206–212.*

Watson, G. A. [1980]: "Approximation Theory and Numerical Methods". J. Wiley and Sons, Chichester.

Watson, G. A. [1981]: *An algorithm for linear L_1 approximation of continuous functions.* IMA J. Numer. Anal. **1**, *157–167.*

Wolfe, J. M. [1976]: *Nonlinear L_1 approximation of smooth functions.* J. Approx. Theory **17**, *166–176.*

Wulbert, D. [1971]: *Uniqueness and differential characterization of approximations from manifolds of functions.* Amer. J. Math. **93**, *350–366.*

Zangwill, W. I. [1969]: "Nonlinear Programming: a unified approach". Prentice-Hall, Englewood Cliffs, N.J.

Zielke, R. [1979]: "Discontinuous Čebyšev systems". Lect. Notes in Math., No. 707, Springer-Verlag, Berlin.

Author Index

Subject Index